AGENTS, GAMES, AND EVOLUTION
Strategies at Work and Play

Steven Orla Kimbrough

The Wharton School
University of Pennsylvania
Philadelphia, USA

CRC Press
Taylor & Francis Group
Boca Raton London New York

CRC Press is an imprint of the
Taylor & Francis Group, an **informa** business

A CHAPMAN & HALL BOOK

CRC Press
Taylor & Francis Group
6000 Broken Sound Parkway NW, Suite 300
Boca Raton, FL 33487-2742

First issued in paperback 2019

ISBN-13: 978-1-4398-3470-1 (hbk)
ISBN-13: 978-0-367-38185-1 (pbk)

Library of Congress Cataloging-in-Publication Data

Kimbrough, Steve.
    Agents, games, and evolution : strategies at work and play / Steven Orla Kimbrough.
      p. cm.
    Includes bibliographical references and index.
    ISBN 978-1-4398-3470-1 (hardcover : alk. paper)
    1. Games of strategy (Mathematics) 2. Cellular automata. I. Title.

QA270.K56 2012
519.3--dc23

2011039832

Visit the Taylor & Francis Web site at
http://www.taylorandfrancis.com

and the CRC Press Web site at
http://www.crcpress.com

Contents

List of Figures

List of Tables

Preface

A game—or as I prefer to say, a *context of strategic interaction* (CSI)—is any situation in which the consequences of one agent's actions depend in part on actions of another agent. Whether a batter gets a hit or not depends on how the batter swings, but also on where the pitcher pitches. How much profit a new product yields depends upon the quality and marketing of the product, as well as on the quality and marketing of its competitors and substitutes. How you are understood depends upon what you say as well as on how your listener interprets what you say. How well your team does depends on what you do and what your teammates do.

CSIs pervade and suffuse our lives, indeed the lives of all organisms. How are we to make sense of, come to grips with, understand these situations?

This book is about the *problems of play* that confront agents who are in, or who would influence, contexts of strategic interaction. I have focused on two groups of related questions. The first group consists of agent-oriented questions. Given a context of strategic interaction, and a collection of strategies, what will happen? How can we evaluate the strategies an agent might use? By what principles can we rationally settle upon a specific strategy for use? How can we discover new strategies, ones we are not aware of? How, in general, can agents learn to play more effectively? Within this group, much of the book's attention is directed at the *strategy selection* problem: Given a context of strategic interaction, how can an agent find a good strategy for itself?

The second group of issues concerns institution-oriented questions. Given a society or system of interacting players—an institution such as a market governed by certain rules and constraints—what will happen? How can we manage its performance? Will it be stable or not? Will it be fair? Efficient? And so on.

Answers to these two groups of questions might be said to constitute a body of *managerial game theory,* accounts of what we understand about how to play in, and how to manage, a strategic context. They constitute accounts, in short, of the problems of play. As such, this is a work in what might be called *post-classical game theory.* Classical game theory is equilibrium game theory. Its program, in short, is to find the equilibria of rigorously specified game representations. This is all well and good, but there are questions of import—particularly the problems of play—that are beyond the ken of equilibrium game theory. It is these questions to which this book is addressed.

My hope is that this book will be interesting for a number of audiences.

I shall certainly use it in the context that occasioned the work, my course "Agents, Games, and Evolution" at the University of Pennsylvania. That course, which is my delight to teach, attracts students principally from the business school, engineering school, biology, psychology, and the Philosophy, Politics, and Economics program at Penn. I believe that any of these departments might offer a course based on the material in this book. I hope as well that both researchers and the reading public will find much of value here. I mean the book to stimulate thought and research.

Strategic interaction is core to us as social beings. There is very much to be learned about strategic interaction by doing the sort of thing done in this book: Put the strategies in play and learn from the outcomes. In the words of Napoléon Bonaparte, *On s'engage et puis on voit.* We commit and then we see what happens.

To these ends, I have minimized what little mathematics is necessary, and located it mostly in appendices, where it need not distract from the flow of discussion. The material is overwhelmingly procedural and computational in nature. Agents are given procedures and placed in strategic contexts; we then observe what happens. All computational materials have been implemented in NetLogo (freely available at `http://ccl.northwestern.edu/netlogo/`), documented, and made available on the Web at the book's Web site: `http://opim.wharton.upenn.edu/~sok/AGEbook/`. As such, these models are in condition to be exercised by the tyro and modified by those with elementary programming skills.

* * *

One hardly knows whether one is a creator or an instrument of expression for any idea, or in what mix. Very many people, too many to list fully, have contributed essentially and extensively to the creation and completion of this work. You know who you are and I thank you most sincerely. I simply must express specifically my heartfelt gratitude to James D. Laing, who got me started on all of this and whose ideas and advice have sustained me throughout; to Frederic H. Murphy, who I am most grateful to have as an active collaborator; and finally to my family and friends who have supported me with unearned levels of tolerance throughout (I'll do my best to make good on it).

Cover illustration: "The Shadow of Society." A small number of unconditional cooperators (in black) playing Prisoner's Dilemma with unconditional defectors (in yellow) prosper and come to dominate the society. See Chapter 4 for a full discussion.

Dedicated to:
James D. Laing
Mentor, colleague, friend

Part I

Starters

Chapter 1

Contexts of Strategic Interaction

Ideas have lives of their own. They arise at surprising times and come from surprising places. Ideas interact with other ideas. They become associated and these associations—or societies—lead to new social structures and to new ideas. These in turn may calve and form new societies. Of course, ideas live and have their being in human minds and cultural artifacts. Without us they would not exist. Nor would most of us exist without the benefit of ideas that have created and sustained our civilization. Nor would there be animals without photosynthesizing plants, or photosynthesizing plants without photosynthesizing bacteria. Interdependence pervades.

The subject of this book is a certain society of ideas. I shall call it the AGE society (Agents, Games, and Evolution), without making any claim that the name ideally describes its subject. What name does? As with any interesting and reasonably complex society, AGE is a product, and continues to be a producer, of history. Its structure, its dynamics, even its constituents are often opaque and puzzling, and everything is in flux. Hence the need for study.

I shall not attempt to define the subject at hand. It likely cannot be done and would not be of much value even if it could. Hyper-precision would only be a distraction at this point. Rather, I will plunge in, immerse us at the center of AGE society, and explore from where we find ourselves. Better to start somewhere reasonable, then ask questions, attend to relevant problems and data, refine and recombine concepts and hypotheses, and build models and conduct experiments. All this is to be undertaken in an iterative, exploring, probing, nondeterministic search for sharper clarity, deeper understanding, and useful results. It shall be our mode throughout. If the process seems to be a sort of groping, adaptive muddling, so be it. The means are informed by the results.

Let us begin, then, by discussing *contexts of strategic interaction* (CSIs), also known as *games*. Here is a—perhaps *the*—main theme in our AGE society of ideas.

1.1   Two Kinds of Decisions

Decision-making individuals, or *agents,* encounter two types of decisions. In *parametric* (or *non-strategic*) decisions, the outcome that an agent receives depends on its choices and states of nature, including states generated by random processes. In *strategic* decisions, the outcome or reward that the agent gets depends in addition on (strategic) decisions made by other agents. The theory of games and this book are about strategic decision making. A game, in this special sense, simply is a context of strategic interaction or CSI. Decision making is *interdependent;* the consequences of a decision by an agent for the agent depend in part on decisions made by other agents.

This characterization of the parametric-strategic distinction is perhaps not entirely adequate. If the second and only other agent has decisions that matter to the first agent, but the second agent's decisions are known to the first agent before it decides, then we might best classify the decision context as parametric. And there are other issues. Even so, our characterization provides a sense good enough for present purposes of the difference between parametric decisions and decisions in games. The key thing in a game is that there are at least two players (agents) the values of whose decisions[1] depend on the decisions of the other player(s) and these latter decisions are not known to the deciding players. Examples, rather than minutely hewed definitions, will serve us best.

Here are some decision situations naturally modeled as parametric:

- A car buyer considers whether to purchase a hybrid automobile. On the negative side, the car is more expensive and less roomy than the non-hybrid under consideration. On the positive side, better mileage will reduce operating expenses and there is a moral benefit from having a smaller carbon footprint.

  What if what other people think of the buyer's decision matters to the buyer? If the buyer can predict their behavior, then it may be best to factor the consequences into the payoffs for the buyer. The decision would remain parametric. On the other hand, if the buyer and her neighbor are purchasing cars this afternoon, are in ignorance of each other's choices, and both gain or lose value, depending on the other's choice, then we have a strategic context.

- A patient is considering whether to have surgery or not. The surgery may or may not be successful, the condition may stabilize or may worsen rapidly without surgery. There are chances all around and the patient's values to take into account. Even though the actions of others matter—the surgeon may or may not do a good job—this is normally best thought

---

[1]Or *choices*. I shall use the two terms interchangeably.

of as a parametric decision. The surgeon has no interest in doing a bad job; the preferences and rewards of that agent are not material for the decision to hand.

Here are some other decision situations naturally modeled as strategic:

- A pitcher and a batter face each other in a baseball game. The count is 3 balls and 2 strikes. The pitcher (and catcher) must decide which pitch to throw (fastball? curveball?) and where to throw it (inside? outside? up? down?). The batter has to decide which pitch he thinks the pitcher will throw and prepare himself for that pitch. Unless the batter is "sitting on" a pitch, he has little chance of hitting it well, but can hope to foul it off if he guesses wrong.

- A deer knows several places that have been safe and where the eating may be good. Other deer have similar knowledge. In addition, transiting between eating places is risky. How long should the deer stay in its present patch, which contains other deer and is being depleted?

These should suffice for establishing a clear enough distinction between parametric and strategic contexts. We shall see many examples of games, and hence strategic situations, in what follows.

1.2   Categorizing Games

It is useful to distinguish various types of games. The generic 2×2 game, and specifications thereof, will serve present purposes. See Figure 1.1 for the general, canonical *strategic form game representation* for two players each having two strategies. In a 2×2 game, there are two players, called Row and

	C_1	C_2
R_1	c_1 r_1	c_2 r_2
R_2	c_3 r_3	c_4 r_4

FIGURE 1.1: Canonical matrix representation for the 2×2 game in strategic form.

Column. The game is played once, and is said to be a *one-shot* game. Each player has two strategies from which it can chose. Row's strategies are labeled

R_1 and R_2 and Column's C_1 and C_2 in the figure. The payoffs are given in the cells of the figure. If Row plays R_1 and Column plays C_1, then Row's payoff is r_1 and Column's is c_1, and similarly for the other cells. Unless stated otherwise, it is assumed that the game is one-shot, that Row and Column do not communicate with each other, cannot form binding agreements,[2] and in fact are not known to each other. Play is anonymous. Also, we assume that the payoffs are specified as numbers and that uniformly more is better.[3]

The game form on display in Figure 1.1 is called the *strategic form* of the game.[4] There are other ways of representing games (such as *extensive form*; see §A.2.5). When we need them, we shall introduce them.

There are different types of games and it is worth distinguishing them. First are *games of pure conflict*. In these games, the interests of the players (agents) are entirely opposed: one's gain is another's loss. One way (there are others) for a game in strategic form to be a game of pure conflict is for the payoffs in each cell to add to a common number. Figure 1.2 from [253] has an example.

	C_1	C_2
R_1	3 2	1 4
R_2	4 1	2 3

FIGURE 1.2: Game #11-ORDINAL, a constant sum game in strategic form.

Second, at the other extreme, in some games the interests of the players are entirely aligned. Such games are said to be *no-conflict* games [253, page 22] or even *inessential* games. Figure 1.3 (from [253]) shows an example.

Observe that both players prefer that the strategy play (R_1, C_1) obtain. If they both know the game and are properly rational, surely that's what will happen. Perhaps surprisingly, we will be very interested in such games.

In our third category of game, the interests of the players are largely in agreement, but there is more than one mutually agreeable outcome. These are called *coordination games*. Figure 1.4 shows an example.

---

[2]Our games, unless otherwise noted, are called *non-cooperative* games, in distinction to *cooperative* games. See §A.2.2, page 428.

[3]Some results in classical game theory assume only an ordering on the payoffs, other results assume the payoffs are given as utility values. This subtlety need not detain us at present. Unless otherwise noted, the payoffs for the games we consider will, we should assume, be in natural goods, such as money, apples, years out of jail, number of offspring, and so on. The natural goods are measured on a ratio scale and have a natural zero. See §A.2.8, page 431.

[4]It is also called the *normal form* of the game, but this expression has fallen out of favor.

	C_1		C_2	
R_1		4		3
	4		3	
R_2		2		1
	2		1	

FIGURE 1.3: Game #1, an inessential game in strategic form.

	C_1		C_2	
R_1		4		1
	4		1	
R_2		1		4
	1		4	

FIGURE 1.4: A 2×2 coordination game in strategic form.

Finally, our fourth category, which will attract most of our attention, is the category of *mixed-motive* games. In these games, the players' interests are in partial conflict. Figure 1.5 shows an example.

	C_1		C_2	
R_1		1		0
	4		0	
R_2		0		4
	0		1	

FIGURE 1.5: A Battle of the Sexes game in strategic form.

The Stag Hunt game is an important mixed-motive game and one we will investigate at length. Figure 1.6 (from [253]) shows an example.

	C_1	C_2
R_1	4 / 4	3 / 1
R_2	1 / 3	2 / 2

FIGURE 1.6: Stag Hunt, aka: game #61-ORDINAL.

1.3   Game Theory and the Theory of Games

The *theory of games,* or better the *theory of strategic interaction* (TSI), seeks to understand games. It aims to explain what happens, predict what will happen, and support intervention in contexts of strategic interaction. It is, in short, the science of strategic interaction.

Classical, or *orthodox,* game theory (exemplified by standard texts, such as [31, 114, 184, 185, 203]) is part of the theory of games (TSI). Very roughly, but sufficient for present purposes, orthodox game theory is about investigating the mathematical properties of models of games, under the assumption of ideal (or at least very strong) rationality on the part of the players. This program of research has been, and remains, remarkably productive, and it has generated indispensable insights into strategic interaction, insights that are invaluable for all other research programs in the theory of games.

Game theory, as presently and foreseeably constituted, is not a complete theory of games. Its scope is much more limited. This is generally recognized among game theorists themselves. See the passage from Binmore's text *Fun and Games* [31, pages 50–1] quoted in Figure 1.7, page 9. Binmore is commendably frank and forthcoming, saying in effect that if you want to understand how to play a real-world game, do not look to game theory, instead "consult a psychologist."[5] Why? Because orthodox game theory is about "rational" play under very strong, usually unrealistic, assumptions of what constitutes rationality. When these assumptions do not hold, game theory is (mostly) silent.

These facts have led some game theorists to highly critical positions on the field. Gintis is especially blunt in print (but others say worse in conversation).

> Ironically, game theory is often hoisted on its own pétard: many of its most fundamental predictions—predictions that would have

---

[5]Reinhard Selten is quoted in [128] as saying "game theory is for proving theorems, not for playing games." This is a widely shared view.

What is important here is that game theory does not pretend to tell you how to make judgments about the shortcomings of an opponent. In making such judgments, you would be better advised to consult a psychologist than a game theorist. Game theory is about what players will do when it is understood that both are rational in some sense. Sometimes, ...this means that an orthodox game-theoretic analysis is not necessarily a very helpful guide on how to play against real people. ...

Does this mean that game theory is useless? Obviously I do not think so or I would not devote my time to it. It is however true that, unless there are good reasons for supposing that the people involved will behave rationally, game theory cannot realistically be used in a naive way to make predictions about what real people will do. As a consequence, a player would often be unwise to use the strategy that a game theory book may label as "optimal" because this will usually only be optimal if *everyone* plans to play optimally. Of course, there are circumstances in which it *is* reasonable to work on the hypothesis that people will behave in a reasonably rational manner. Economics is somewhat shakily founded on the assumption that this will typically be the case in commercial and business transactions. However, it would be skating on very thin ice to use game theory for predictive purposes if none of the following criteria were satisfied:

- The game is simple.
- The players have played the game many times before[a] and hence have had much opportunity for trial-and-error learning.
- The incentives for playing well are adequate.

---

[a]Against different opponents each time. If you are to play a particular game against the same opponent many times, one must model the repeated situation as a single "super-game."

FIGURE 1.7: Ask a psychologist.

been too vague to test with any confidence in the pre-game-theoretic era—are *decisively and repeatedly disconfirmed,* in laboratory settings, with substantial agreement among experimenters, regardless of their theoretical priors. [124, page xxiv] (emphasis in original)

This is due, in large part, to hyper-strong assumptions about the rationality of players. In consequence,

It is better to drop the term "rational" altogether, ...

In the same vein, we do not follow classical game theory in asking how agents "learn" to play optimal strategies, because the cognitive processes involved in "learning" are probably, under most conditions, much less important than the forms of imitation underlying the replicator dynamic... and cultural transmission. .... In short, evolutionary game theory replaces the idea that games have "solutions" that agents "learn," with the idea that games are embedded in natural and social processes that produce agents who play effectively.

Dispensing with the rationality postulate does not imply that people are *irrational* (whatever that means). The point is that the concept of "rationality" does not help us understand the world. [124, pages xxv–xxvi]

The position I take in this book is that rationality is a concept well worth saving, even if we need to develop new varieties of it. That, however, is a long story, one that will unfold as the book does. For now a working distinction will be enough to proceed. Details and elaboration will follow.

Orthodox game theory employs, as I said, a very strong notion of rationality. Players are assumed to choose so as to maximize the possible benefits to them. It is crucial to understand that by "maximize" game theory means "actually achieve the maximum" rather than, say, "aim for, or try to, achieve the maximum." To be rational is to maximize in this strong sense. Classical game theory asks, What will happen in games in which all of the players are rational in this way? That is an important question and game theory has delivered admirably in addressing it.

This book investigates a different question: What will happen in games in which the players lack the ideal rationality of game theory, but are in possession of plausible adaptation or learning regimes which are invoked during strategic interaction and are used to inform future actions?

To sloganize: game theory is about strategic situations in which the players *achieve maximization* for themselves, while agent-based game theory is about players who *seek improvement* for themselves. To use terminology introduced by Herbert Simon, the classical theory is concerned with "substantive rationality." The classical theory is not concerned with *how* agents achieve maximization, just with the consequences of doing so. We are here, with agent-based or procedural game theory, concerned with "procedural rationality" (Simon's contrastive term). Moreover, we do not assume that the procedures of our agents succeed in maximizing anything. The following comments from Herbert Simon are apt and will be useful to us in the sequel.

> In its treatment of rationality, neoclassical economics [as well as game theory] differs from the other social sciences in three main respects: (a) in its silence about the content of goals and values; (b) in its postulating global consistency of behavior; and (c) in its postulating "one world"—that behavior is objectively rational in relation to its total environment, including both present and future environment as the actor moves through time.

> In contrast, the other social sciences, in their treatment of rationality, (a) seek to determine empirically the nature and origins of values and their changes with time and experience; (b) seek to determine the processes, individual and social, whereby selected aspects of reality are noticed and postulated as the "givens" (factual bases) for reasoning about action; (c) seek to determine the computational strategies that are used in reasoning, so that very limited information-processing capabilities can cope with complex realities; and (d) seek to describe and explain the ways in which nonrational processes (e.g., motivations, emotions, and sensory stimuli) influence the focus of attention and the definition of the situation that set the factual givens for the rational processes.

> These important differences in the conceptualization of rationality rest on an even more fundamental distinction: in economics, rationality is viewed in terms of the choices it produces; in the other social sciences, it is viewed in terms of the processes it employs (Simon 1976[291]/1982[292]). The rationality of economics is substantive rationality, while the rationality of psychology is procedural rationality. [293, page S210]

> The rational person of neoclassical economics always reaches the decision that is objectively, or substantively, best in terms of the given utility function. The rational person of cognitive psychology goes about making his or her decisions in a way that is procedurally reasonable in the light of the available knowledge and means of computation. [293, page S211]

We are only at the beginning. What follows in this book is a detailed elaboration of these points.

1.4   Why Study Games?

Why are we interested in contexts of strategic interaction (CSIs)? Games are interesting because CSIs—war, diplomacy, poker, business strategy, coalition formation, bargaining, and so on—are each interesting in themselves and are each examples from a larger pool of important phenomena' meriting, even demanding, attention. Besides interdependent, interactive choice—the characterizing feature of games—we see the interplay of reasoning, calculation and reckoning, deception, skill, bluffing, power, adaptation, flexibility, cooperation, learning, arbitrage, coordination, norms, communication, markets, social organization, and much else that is pervasive in, and fundamental to our understanding of, the social order. (All belonging to the AGE society of ideas.) Games are interesting because they are vortices of many interesting phenomena. Social phenomena manifest themselves—play themselves out—in strategic interactions.

What would we like to know about games? As in any field of science we seek to describe, explain, predict, and intervene [138]. We wish to describe and classify naturally occurring games ("games in the wild") systematically. Chapter 2 hints at a much-needed natural history of strategic interaction. We wish to explain and predict game outcomes. This is often called *solving the game* in the classical literature. We also wish to understand—explain and predict—how it is outcomes are reached. How do agents of various sorts (experienced humans, naïve humans, monkeys, rats, lichens, bacteria, insects, organizations, artificial agents) find and implement their strategies of play? How does play unfold over time (and over space when geography is relevant)? Finally, we seek understanding of games in order to intervene in the world. We might hope to improve our own play in strategic contexts, or to design better social institutions (such as markets for electrical power that resist manipulation—"gaming"—as in the Enron affair [302] or in the 2008 financial collapse, or markets and institutions that will support sustainable exploitation of *common-pool resources* such as fisheries, or even the earth's atmosphere), or to field artificial agents that labor on our behalves (perhaps for negotiation or purchasing over the Internet). The scope of potential investigation is both magnificent and beyond our means. We should be content with modest progress, while keeping ourselves reminded of the larger issues. That, at least, describes my aims in this book.

1.5   Methods of Study

Games in the wild, as they naturally occur, are the primary phenomena that motivate study of contexts of strategic interaction. The games we make up or develop as abstract models are ultimately interesting to the extent they contribute to our understanding of games in the wild. How, in particular then, can we study strategic interaction? Various ways are open to us:[6]

1. *a priori.* CSIs or games in the wild may be abstracted and reduced to formal models, then studied mathematically, typically upon assumption of axioms of rationality. From this perspective, the theory of games is a branch of mathematics. Much of classical game theory proceeds in this mode. Standard textbooks and reference works include [31, 114, 184, 203, 287].

2. *in vivo.* Games, or strategic situations, may be studied as they (more or less) naturally occur "in the field" (hence the applicable term, field study). This is a historical—"natural history"—mode of investigation, but of course the history may be contemporary and the means of study may use techniques from anthropology, sociology, or journalism. Pioneers of this approach include Thomas Schelling (e.g., [277, 278, 279, 280]), Jon Elster (e.g., [89, 90]), and Elinor Ostrom (e.g., [240, 241]).

3. *in vitro.* We can study games by doing experiments with real ("wet") agents, including humans (e.g., [54, 94, 162]), monkeys (e.g., [97]), even blue jays (e.g., [300]). And why not lichens and bacteria? The literature uses such names as *behavioral game theory* and *experimental economics* to refer to these kinds of investigations.

4. *in silico.* There is much to be learned about games by representing agents as procedures, or decision algorithms, that choose their plays, and by studying the behavior of the resulting system. In some ways, this method of simulation or experimental mathematics is a variant of the *a priori* method. Let us call it *procedural game theory,* adverting to Simon's distinction ([293], see above) between substantive and procedural rationality.[7] By allowing ourselves to use computational methods (instead of purely analytic mathematics) we may greatly extend the range, scope, and realism of models addressed, and concomitantly reduce the stringency of the assumptions required.

We shall draw upon each of these methods. Our main focus methodologically, however, will be on procedural studies of artificial agents, using agent-based models (ABMs). Such studies may be, and have been, conducted from a variety of perspectives. Agents may be modeled as naked strategies (what I call *strategy-centric* agents), possibly reactive or adaptive strategies, that

---

[6]I am grateful for the discussion in [233].

[7]Calling this *procedural studies of games* or *procedural analysis of strategic interaction* would be more accurate and properly modest. Also, the term algorithmic, instead of procedural, would be appropriate. However, although usage is hardly uniform or well established, *algorithmic game theory* is being used to describe a program related to, but somewhat different, than that undertaken here, e.g., [235].

play in tournaments (e.g., Axelrod's original and seminal study [15]) or that play in a populated ecology which evolves under the replicator dynamic (e.g., [15, 124, 294]) or that play in a differentiated geography (spatial games, network games; e.g., [93, 135]). Alternatively, agents may be modeled as beings that *have* and may change strategies. Such *identity-centric* agents, as I shall call them, may undergo learning and in consequence change their policies of play.

Again, we shall draw upon these and related studies but focus our efforts especially on these four main conditions.

1. *Finite, non-ideal* contexts of strategic interaction. Agents are finite beings. Their rationality, their abilities to reckon and foresee, are limited. The algorithms with which we model these agents must be computable and computable without exorbitant use of resources. Play unfolds in a finite population, for a finite time, and in a finite space. A major theme will be to compare and contrast results under finite, non-ideal and infinite, ideal regimes of play. Classical game theory employs what philosophers call an *externalist* theory of rationality. Here we are asking different questions and shall be focusing on *internalist* notions of rationality. The upshot of this point will emerge as we proceed.[8]

2. *Identity-centric* more so than strategy-centric agents. Humans, and indeed monkeys, blue jays, and even bacteria in contexts of strategic interaction may be said to *have* strategies (rather than to *be* strategies), and to be capable of changing them in response to experience. These players, and most of the agents we shall consider, may meaningfully be said to have identities distinct from the strategies they employ at any given time. They are more than naked strategies. In particular, they are

3. *Exploring, probing* agents, not merely reactive agents. Humans, monkeys, blue jays, bacteria, and most of our agents face the exploration–exploitation dilemma or tradeoff, addressed throughout the machine learning literature.

Finally, the strategic contexts we will focus on will be

4. *Chronic* and *social* more so than acute and singular, *ongoing* and *widespread* more so than unique. The games may be *repeated* (played many times) or *iterated* (played many times by the same players with each other) or be like other games that will be played, rather than being unique, non-repeatable events. See §A.2.17, page 435.

A word of elaboration and justification for this last aspect of our focus. Contexts of decision, or choice, may be distinguished into *strategic* (or game-theoretic, the principal subject of this book) and *parametric* (not strategic,

---

[8]See [183] for an introduction to the internalism–externalism distinction.

the principal subject of the field of decision analysis).[9] Further, contexts of decision or choice may be distinguished into *acute* ("one-shot" or once-only) and *chronic*. Herrnstein's name for chronic choices [150]—*distributed*—is an apt description, and I shall use it. Chronic decisions or choices are distributed, usually in time. We may think of habits as chronic decisions that are, or become, more or less settled. Table 1.1 summarizes this framework. This book is

	Acute (one-shot)	Chronic (distributed)
Parametric	decision analysis	decision analysis
Strategic	classical game theory	procedural game theory

TABLE 1.1: Framework categorizing decision/choice contexts.

an essay on procedural game theory; it often uses agent-based models, ABMs, to explore the behavior of procedures in contexts of strategic interaction.

Further, as I have noted, decision contexts may be distinguished into *individual* and *social*, although this distinction is perhaps more applicable to strategic than to parametric contexts. In any event, there will be much emphasis in what is to follow on *social* aspects of distributed strategic choice.

To conclude this chapter, I note that often we *are* inclined to think of games in terms of acute, dramatic points of decision. This is captured in the penultimate stanza of "Casey at the Bat" (Ernest L. Thayer, alias Phin, page 4 of the San Francisco *Daily Examiner,* June 3, 1888).

> The sneer has fled from Casey's lip, the teeth are clenched in hate.
> He pounds, with cruel violence, his bat upon the plate.
> And now the pitcher holds the ball, and now he lets it go,
> and now the air is shattered by the force of Casey's blow.

Of course the final stanza is

> Oh, somewhere in this favored land the sun is shining bright.
> The band is playing somewhere, and somewhere hearts are light.
> And, somewhere men are laughing, and little children shout,
> but there is no joy in Mudville–mighty Casey has struck out.

But first, many games, many contexts of strategic interaction, *are* distributed or chronic, or approximately so. Agents do business with a particular merchant, doctor, lawyer, restauranteur repeatedly.[10] Agents have friends, partners, lovers, spouses, colleagues they encounter more than once. Agents have

---

[9]See Paul Bloom's *Descartes' Baby* [34] for an accessible presentation of evidence that human mentality innately is organized in recognition of the distinction between parametric and strategic decisions.

[10]When play is repeated between or among the same agents, the same counter-players, I shall say that it is, in addition to being repeated, *iterated*. See §A.2.17.

competitors in the market for more than a day. Agents are embedded in societies. Very often indeed, strategic contexts cannot be separated from the future or the past.

And second, is Casey's situation really unique, even for Casey? True enough, Casey is in a zero-sum game in the sense that only one team can win. It is also true that in any given at-bat the pitcher in baseball has the advantage; anyone can strike out. Most likely, however, there will be another game tomorrow or the next day. Casey's interest lies in maximizing the expectation of his future contributions to the team. Getting angry, pounding the bat, focusing exclusively on this game and this moment is, perhaps, not the wisest of moves on Casey's part. Better to take the long view. Better to have the pitcher strike you out than for you to strike yourself out. Perhaps the long view can inform the acute. Perhaps, at least sometimes, learned policies of play in the chronic case should drive or at least inform play in the acute case. What follows has among its aims the investigation of such conjectures and their ramifications.

1.6   For Exploration

1. This is the generic payoff matrix for a *symmetric* 2×2 game:

A, A	B, C
C, B	D, D

Let the possible payoffs be 1, 2, 3, 4, with higher numbers being preferred to lower numbers. Then there are $4! = 24$ possible arrangements, but half of these are eliminated by NW-SE symmetry. This leaves 12 possible 2×2 games, assuming payoffs are ranked and without ties.

(a)

$4, 4$	$3, 2$
$2, 3$	$1, 1$

(b)

$4, 4$	$2, 3$
$3, 2$	$1, 1$

(c)

4, 4	2, 1
1, 2	3, 3

(d)

4, 4	1, 2
2, 1	3, 3

(e)

4, 4	3, 1
1, 3	2, 2

(f)

4, 4	1, 3
3, 1	2, 2

(g)

3, 3	4, 2
2, 4	1, 1

(h)

3, 3	2, 4
4, 2	1, 1

(i)

3, 3	4, 1
1, 4	2, 2

(j)

3, 3	1, 4
4, 1	2, 2

(k)

2, 2	3, 4
4, 3	1, 1

(1)

2, 2	4, 3
3, 4	1, 1

For each of these games, considered as a one-shot game played by anonymous players who cannot communicate with each other,

  (a) If you were to play this game under the conditions specified above, how would you play? Justify and explain your answer.

  (b) Identify all of the equilibrium outcomes.

  (c) An outcome, *O*, is *Pareto optimal* if for all other outcomes at least one player does worse than it does in *O*. Identify all of the *Pareto optimal* outcomes. See §A.2.9, page 432.

2. Put in a nutshell, the program of classical game theory is to represent or model games in the wild and then to find or characterize the equilibrium (or equilibria) for the game representation. Players in a game, so conceived, have *strategies* they may play, where a strategy is a complete set of instructions for playing the game. See §A.2.3, page 428. For a 2×2 game in strategic form, a strategy is simply a row, for the Row player, or a column, for the Column player. A player plays by picking a strategy. The outcome is an *equilibrium* if, given the strategy choices of all the players, no player by itself has a strategy that would give it a better reward than the reward from the actual set of strategy choices (i.e., actual outcome). See §A.2.11, page 432.

The core prediction of classical game theory is that ideally rational players will reach an equilibrium when they play. The theory, however, does not provide any general procedure by which players will pick strategies that lead to equilibrium outcomes. Nor, in the case when there is more than one equilibrium, does the basic theory predict which equilibrium will result. Consider again the 2×2 games of question (1), above. For which games do you think it unproblematic that the players will pick strategies that produce a single equilibrium outcome? For which games do you think it is problematic? Explain your reasoning.

1.7   Concluding Notes

By way of background on standard game theory, Kreps's *Game Theory and Economic Modeling* [185] is an excellent choice for an introduction to conventional game theory. *Games and Decisions* by Luce and Raiffa [203], now more than 50 years old, remains in print and well worth reading.

As emphasized in the body of this chapter, we shall be exploring strategic interaction procedurally. By what procedures can agents find good strategies of play? What happens when specified strategies play one another? How can strategies be learned? How should systems of interacting agents be designed to achieve stated goals? And so on. See my essay [169] for thoughts on how procedures can be used to explain. As such, they are are often competitors with equations and other forms of mathematical modeling. The theory of evolution is at bottom expressed as a class of procedures, which may itself explain some of the resistance to it. People think of scientific theories as presenting themselves in the manner of Newton's laws.

Related ideas have been developed by Ian Bogost in the context of video games [36]. He argues that they exemplify the leading edge of rhetorical argument and expression, through the unconventional use of displaying procedures.

Chapter 2

Games in the Wild and the Problems of Play

Games, interactive decision making, or more descriptively *contexts of strategic interaction* (CSIs), are everywhere.[1] They pervade social situations and occur quite naturally (or appear "in the wild" as geneticists say of certain alleles). Two people play backgammon. They are in a game, or context of strategic interaction (CSI), because the reward (winning or losing) for each player depends at least in part on decisions made by the other player. One player cannot make a series of decisions that results in winning or losing, *independently of what the other player does.* The other player has to make decisions, too, and they matter. The context is interactive—two or more players are involved—and it is strategic because both players have interests, which they take into account in making their decisions.

Backgammon is representative of many games in that it is purely competitive.[2] One player's win is the other player's loss. The interests of the players are, we may assume, entirely opposed. In other CSIs (or games) the players' interests are entirely coincident. These are what we call *games of pure coordination.* Two people are conversing by telephone when the connection is suddenly dropped. How should they attempt to resume the conversation? If both call back simultaneously both will get a busy signal or perhaps voice mail. They share a joint interest in mutually divining a decision that results in prompt and unfrustrated resumption of their conversation. Here we may assume the interests of the two agents are identical. Neither really cares who makes the new call, so long as it results in immediate resumption.

Lying between games of pure competition (e.g., backgammon) and games of pure coordination (e.g., resuming a broken phone call) are *mixed motive*

---

[1]The term *game* is perhaps an unfortunate one for a number of reasons. It suggests a certain frivolity, also that only contexts of pure competition are of interest. More importantly, we need a distinction between a situation involving strategic interaction and a model of such a situation. *Game* gets used for both. When necessary to differentiate, I'll use game$_S$ for the situation, not always well defined with all vagueness left out, and game$_M$ for a model, presumably specified with great precision, of a game$_S$. Or, CSI for game$_S$ and game for game$_M$.

[2]At least approximately or often. Consider playing with a tyro and playing to lose for purposes of instruction. An important point lurks: given a game in the wild, game$_S$, or even a representation of it, game$_M$, it remains for the players to interpret and act. The representation cannot, in general, determine how a player will frame and act upon the situation as it understands it.

games (or CSIs). A small group negotiates where to have dinner. No two people have identical preferences, but everyone agrees that failing to come to a congenial decision quickly is the worst outcome. Remarkably subtle moves will typically attend this familiar situation. Bluff, bluster, threat, compromise, accommodation, probing, retreating, appeal to norms, humor, and much else are routinely employed with facile skill by everyone who participates in such groups.

How are we to understand games? In particular, how are we to predict and explain both behavior and outcomes in games? This is a large and important question. I remind the reader that our mode here is to make some progress through an "iterative, probing, nondeterministic search for sharper clarity and deeper understanding." To this end, a rough characterization of our topic will be helpful:

> Games, or CSIs, essentially involve at least two *agents* (or players) who make *choices* and receive *rewards* (or payoffs).[3] The reward to an individual agent is based in part on its choices *and the choices made by the other agent(s)*, as well as the underlying structure of the situation.

Now consider a few representative, idiosyncratically chosen examples of contexts of strategic interaction.

2.1 War

Much more than a pure, brutal contest of strength, war has been recognized from the earliest writings as a field of interactive decision making. Deception especially has been and remains a primary theme; it is inherently a strategic concept. Think of the Trojan horse incident told in the *Iliad* and the story of the Cyclops in the *Odyssey*. Think of the elaborate obfuscations undertaken by the Allies in World War II concerning the time and place of D-Day. Sun Tzu, in *The Art of War* (http://www.chinapage.com/sunzi-e.html), the oldest known military treatise, wrote this:

>   18. All warfare is based on deception.
>
>   19. Hence, when able to attack, we must seem unable; when using our forces, we must seem inactive; when we are near, we must

---

[3]It will be useful to distinguish *rewards,* which are received after each move in a strategic situation, and *returns,* which are the net of the rewards obtained in a multi-move strategic context. Unless otherwise noted, what I say about rewards applies to returns and vice versa. Also, recall that unless otherwise noted the rewards or returns of our games are denominated in natural dimensions; §A.2.8, page 431.

make the enemy believe we are far away; when far away, we must make him believe we are near.

20. Hold out baits to entice the enemy. Feign disorder, and crush him.

21. If he is secure at all points, be prepared for him. If he is in superior strength, evade him.

22. If your opponent is of choleric temper, seek to irritate him. Pretend to be weak, that he may grow arrogant.

23. If he is taking his ease, give him no rest. If his forces are united, separate them.

24. Attack him where he is unprepared, appear where you are not expected.

25. These military devices, leading to victory, must not be divulged beforehand.

Other themes abound, but deception and surprise remain keystones to military strategy. Other works on the short list of classics in military strategy include: *On War,* by Karl von Clausewitz, *The Prince,* by Niccolò Machiavelli, and *A Book of Five Rings,* by Miyamoto Musashi.[4] Liddell Hart, e.g., [200] is an especially persuasive spokesman for the importance of military deception and surprise. Thomas Schelling is uniformly insightful and a joy to read, e.g., [277, 278, 279, 280]. The *Memoirs* of Ulysses S. Grant are chock full of material to stimulate reflection on war and on interactive decision making in general. Here is my favorite passage. Grant is describing his first field command in the American Civil War.

> My sensations as we approached what I supposed might be a 'field of battle' were anything but agreeable. I had been in all the engagements in Mexico that it was possible for one person to be in; but not in command. If someone else had been colonel and I had been lieutenant-colonel I do not think I would have felt any trepidation. . . . As we approached the brow of the hill from which it was expected we would see the enemy. . . my heart kept getting higher and higher until it felt as though it was in my throat. I would have given anything to have been back in Illinois, but I had not the moral courage to halt and consider what to do; I kept right on. When we reached a point from which the valley below was in full view I halted. The place where the Confederates had been encamped was still there but the troops were gone. My heart resumed its place. It occurred to me at once that [Colonel Thomas] Harris had been as much afraid of me as I had been of him. This was a view of the question I had never taken before; but it was

---

[4]See http://www.gametheory.net/books/.

one I never forgot afterwards. From that event to the close of the war, I never experienced trepidation upon confronting an enemy, though I always felt more or less anxiety. I never forgot that he had as much reason to fear my forces as I had his. The lesson was valuable.

2.2   Trading and Investing

"Buy low, sell high" is great advice if (and only if) you know what to do. As the song says about that special form of trade and investment called love, "Nice work if you can get it, And you can get it if you try."

Examples of buying low or selling high? This is from *The Reader's Digest*, June 2003, pages 76–7:

> Customers at The Home Depot who overestimate how much paint they need return the unopened cans, which are stocked in the "Oops Paint" section. The "remnant" paint—perfect for bathrooms and other small projects—sells for $5 a gallon and $1 a quart (regular gallon prices are $21 to $25). "And it's not all chartreuse," says The Home Depot spokesperson Mandy Holton. "There are usually a lot of great neutrals." Best time to buy: Sundays and Mondays, because folks return unwanted paint over the weekend.

More generally, traders and investors are in the business of finding assets that are either under-valued or over-valued in the market. In other words, they seek opportunities for *risky arbitrage*. Risky because—unlike the paint at The Home Depot—the values of the assets in question are typically not known with much certainty. Arbitrage because the traders are looking to buy assets that are under-priced (and then resell them at their proper prices) or looking to sell ("unload") assets that are over-priced. In any event, the trick is to have and exploit knowledge that is superior to what is represented in the market. The nature of this knowledge and the means of getting it vary greatly. An investor in equities may look deeply and carefully at the fundamentals of the companies. Which are and which are not well managed, well positioned, in possession of new products and alliances? An investor may look at the "technical" data, the trends and other movements in prices. In the extreme, so-called day traders do this in real time, attempting to out-guess the market, that is out-do the other traders in discerning what is over-valued or under-valued. Note that investing on analysis of fundamentals would seem to have less strategic content than investing on technical grounds.

On-line, Internet-based examples are readily available for those who wish to trade or just to study and learn. Tradesports was a real-time, on-line trading market that affords an excellent case for study. Academic analogs—but with real money if you want—are available for elections at Iowa Electronic Markets (IEM; `http://www.biz.uiowa.edu/iem/`; `http://www.biz.uiowa.edu/iem/markets/`) and the University of British Columbia Election Stock Market (http://esm.ubc.ca/). A *BusinessWeek* article, "The 'Election Futures' Market: More Accurate than Polls?"[5] presents the case in a popular format that these markets predict election outcomes better than opinion polls. The Bush administration even toyed with creating a similar market for the purpose of gauging intelligence in the Middle East (see "Betting on Terror: What Markets Can Reveal" by Floyd Norris in *The New York Times,* August 3, 2003). The idea was dropped after being exposed to public ridicule. Is it ridiculous? Consider: What would it take to "game" (distort for ulterior purposes) these markets? When would anyone want to do this? What might be done to prevent manipulation? Does it make sense to have an SEC for markets in international affairs?

There are always the public equity markets. Consider this comment on the bond market by a Salomon trader in the 1980s.

> The men on the trading floor [Salomon's bond trading area] may not have been to school, but they have Ph.D's in man's ignorance. In any market, as in any poker game, there is a fool. The astute investor Warren Buffett is fond of saying that any player unaware of the fool in the market probably *is* the fool in the market. In 1980, when the bond market emerged from a long dormancy, many investors and even Wall Street banks did not have a clue who was the fool in the new game. Salomon bond traders knew about fools because that was their job. Knowing about markets is knowing about other people's weaknesses. And a fool, they would say, was a person who was willing to sell a bond for less or buy a bond for more than it was worth. A bond was worth only as much as the person who valued it properly was willing to pay. And Salomon, to complete the circle, was the firm that valued the bonds properly.
>
> —*Liar's Poker,* Michael Lewis [197, page 35]

2.3 Athletic Contests

There are sports, called games in common parlance, that have little or no strategic content. They amount more or less to contests of skill. Among them are golf, bowling, darts, skiing, track and field events, and bobsledding. Still other competitive games, such as billiards, have strategic content only with

---

[5] 1996; `http://www.businessweek.com/1996/46/b3501116.htm`.

fairly advanced play. These are not, for the most part, of interest as CSIs, contexts of strategic interaction, and will not concern us further.

Many other athletic contests most unambiguously count as CSIs. Baseball has given us a wonderful strategic slogan, entirely appropriate for war and other games: "Hit 'em where they ain't." Wee Willie Keeler hit .432 in 1897. Asked how a man of his diminutive size could put together such an average, Keeler responded: "Simple. I keep my eyes clear and I hit 'em where they ain't."[6] Deception—or the fake-out—plays as prominent a role in these athletic contests as it does in warfare. Think of the pitcher-batter duel in baseball, the fake-out moves in basketball, or the mixing of plays in American football.

Management of sports teams is as much a matter of strategic interaction as the play itself. In *Moneyball: The Art of Winning an Unfair Game,* Michael Lewis describes how the Oakland A's baseball team, with consistently small amounts spent on player salaries, is consistently able to contend in major league baseball and reach the playoffs [198]. In two words: risky arbitrage. The A's, and in particular their general manager Billy Bean, have identified predictive measures superior to those used by other teams, for example using on-base percentage instead of batting average to evaluate the worth of a batter. Better measures of value allow the A's to "buy" (hire) under-priced players. They may not have the best team in baseball, but they have one of the best. Their efficiency in the sense of what it costs them to win a game is tops and they operate at a profit in a media market dominated by the San Francisco Giants baseball team.

2.4 Gambling

Many forms of gambling do not involve strategy or even much skill. Examples include playing slot machines and playing roulette. Not so with poker at the professional level. Poker is prototypical. It is to competitive games of strategy what robins are to birds: a standard, familiar, readily available example, displaying in typical form many of the characterizing features of the subject. Everyone over time gets roughly the same quality of hands, yet there is an enormous difference among players in their success rates. Everyone can count cards and figure the odds. What actually matters is bluffing, reading your opponents (discerning their "tells," behavior such as slamming down chips that indicates what is in their hands), and preventing your opponents from reading you. The following fine passage from a master poker player is well worth quoting at length:

> Let's take a quick glimpse at the high-stakes poker world, an enterprise that yields several of my friends over a million dollars a

---

[6] From http://www.baseballtips.com/slang.html.

year! At this level, too, luck is a factor on any given day, week, or month, but what's different is that if you play better poker than your opponents do, pretty consistently, you'll find that over almost any *two*-month period your winnings have exceeded your losses. Furthermore, if you play better poker than your opponents over a *six*-month period, your results will have moved very solidly in the winning direction. Making a few well-timed bluffs each day will add up to a lot of money each year!

In fact, if an inexperienced poker player were to sit down for a few hours with a group of world-class poker players, he would have virtually no chance to win over even an eight-hour period. This very fact is why five or six top pros might be willing to sit down in the same game with this fellow and each other: the money that even one amateur is likely to contribute makes it work their while to do battle with so many respected opponents.

This is why so many of the top poker players today drive fine cars and live in palatial homes [the author of this passage lives with his family in Palo Alto]. Right now, as you're reading this book, there is a \$600–\$1,200-limit poker game at the Bellagio Casino in Las Vegas and a \$400–\$800-limit poker game at the Commerce Casino in Los Angeles. There is ...

If that's not enough action for you, four nights a week in Los Angeles, there is a \$2,000–\$4,000-limit Seven-Card Stud game a Larry Flynt's Hustler Club Casino, with Larry himself often playing. In the \$400–\$800-limit poker game it's easy to take a \$25,000 swing in one hour. In the \$2,000–\$4,000-limit game, where movie stars, former governors, and billionaires play, it's not uncommon for someone to win or lose \$250,000 in one night. In these "nosebleed" poker games (the term refers to the altitude of the stakes), strategy, discipline, calculation of the odds, and practiced observation contribute to a game that involves much more skill. Better play wins more hands in the long run.

—*Play Poker Like the Pros* by Phil Hellmuth, Jr., 2003 [146, pages 4–5]

The society of poker players has given us an important concept—the *tell*—not only for poker but for CSIs generally. Dictionaries have generally not picked up on this sense, nor is it represented in WordNet, http://wordnet.princeton. edu/. The Wikipedia predictably has it (http://en.wikipedia.org/wiki/ Tell_%28poker%29, accessed 2011-1-5): "In poker, a tell is a detectable change in a player's behavior that gives clues to that player's hand." Also see the archeological sense of the word, which is perhaps related to the strategic sense in poker and elsewhere.

The strategic sense of *tell* is lucidly on display in the following passage about the great baseball player and base runner, Rickey Henderson.

But Kennedy knew how devastating stealing could be: he had been with the San Francisco Giants in the 1989 World Series, when Henderson and the A's swept the Giants in four games and Henderson set a post-season record, with eleven stolen bases.

Henderson agreed to give a demonstration, and there was a buzz as Goodman, Johnson, and the other players gathered around first base. Henderson stepped off the bag, spread his legs, and bent forward, wiggling his fingers. "The most important thing to being a good base stealer is you got to be fearless," he said. "You know they're all coming for you; everyone in the stadium knows they're coming for you. And you got to say to yourself, 'I don't give a dang. I'm gone.'" He said that every pitcher has the equivalent of a poker player's "tell," something that tips the runner off when he's going to throw home. Before a runner gets on base, he needs to identify that tell, so he can take advantage of it. "Sometimes a pitcher lifts a heel, or wiggles a shoulder, or cocks an elbow, or lifts his cap," Henderson said, indicating each giveaway with a crisp gesture.

Once you were on base, Henderson said, the next step was taking a lead. Most players, he explained, mistakenly assume that you need a big lead. "That's one of Rickey's theories: Rickey takes only three steps from the bag," he said. "If you're taking a big lead, you're going to be all tense out there. Then everyone knows you're going. Just like you read the pitcher, the pitcher and catcher have read you."

He spread his legs again and pretended to stare at the pitcher. "O.K., you've taken your lead; now you're ready to find that one part of the pitcher's body that you already know tells you he's throwing home. The second you see the sign, then *boom*, you're gone." [130, page 58]

Hellmuth's book has a great deal of information on Texas Hold 'em, which is generally the most popular form of poker in tournaments and is the variety of poker played at the World Series of Poker each year at Binion's Horseshoe Hotel & Casino in Las Vegas.[7] *Positively Fifth Street* [221] by Jim McManus, a published poet, novelist, and professor, describes the 2001 World Series of Poker and the world around it. Strategic insight abounds in both works.

---

[7]See   http://conjelco.com/wsop.html   and   http://www.binions.com/home.asp. Wikipedia has an introduction to Texas Hold 'em: http://www.wikipedia.org/wiki/ Poker/Texas_holdem. At the World Series of Poker, No Limit Texas Hold 'em is the game. "No Limit" means that the largest bet permitted is the size of the current wealth outside the pot of the poorest player still in the hand. Once a player has bet all of his or her chips, the player is said to be "all in," since the player's wealth is all in the pot. Once a player is all in for a particular hand, other players may call but may not raise.

2.5   Business Strategy

The gambit, a term from chess, is a favorable trade, which the opponent may or may not realize is happening. The player offering the gambit offers a comparatively small loss in exchange for a larger gain in position or other form of resource. Here is something very like a gambit played big time in business.

> Analysts called it "Marlboro Friday"—Philip Morris announced on April 2, 1993 that it would reduce the U.S. price of its premium brand of cigarettes by 20%. The tobacco manufacturer also said it would increase the budget for its domestic advertising by a substantial amount. R.J. Reynolds, Philip Morris's biggest competitor, responded by matching the price cut on its own premium brands (Camel and Winston among them) and by pouring more money into its own domestic advertising.
>
> The pricing war that ensued cost both companies tens of millions of dollars. But was the domestic market share the real reason Philip Morris lowered the price of Marlboro cigarettes? Consider that just as R.J. Reynolds had depleted its cash resources trying to keep up with its chief opponent, Philip Morris was expanding aggressively into the Eastern European market, investing $800 million in Russia and other regions that were formerly part of the Soviet Union. R.J. Reynolds was in no position to fight back, having spent so much money to maintain its market share in the United States, and Philip Morris won the battle for Eastern European market share, hands down.
>
> –"Global Gamesmanship" by Ian C. MacMillan et al., *Harvard Business Review*, May 2003 [205]

Sometimes you can even make a profit on a gambit:

> One day earlier in his career [Robert] Dall was in the market to buy (borrow) fifty million dollars. He checked around and found the money market was 4 to 4.25 percent, which meant he could buy (borrow) at 4.25 percent or sell (lend) at 4 percent. When he actually tried to buy fifty million dollars at 4.25 percent, however, the market moved to 4.25 to 4.5 percent. The sellers were scared off by a large buyer. Dall bid 4.5. The market moved again, to 4.5 to 4.75 percent. He raised his bid several more times with the same result, then went to Bill Simon's office to tell him he couldn't buy money. All the sellers were running like chickens.
>
> "Then you be the seller," said Simon.
>
> So Dall became the seller, although he actually needed to buy.

He sold fifty million dollars at 5.5 percent. He sold another fifty
million dollars at 5.5 percent. Then, as Simon had guessed, the
market collapsed. Everyone wanted to sell. There were no buyers.
"Buy them back now," said Simon when the market reached 4
percent. So Dall not only got his fifty million dollars at 4 percent
but took a profit on the money he had sold at higher rates. *That*
was how a Salomon bond trader thought: He forgot whatever it
was that he wanted to do for a minute and put his finger on the
pulse of the market. If the market felt fidgety, if people were scared
or desperate, he herded them like sheep into a corner, then made
them pay for their uncertainty. He sat on the market until it puked
gold coins. *Then* he worried about what he wanted to do.

—*Liar's Poker,* Michael Lewis [197, page 88]

2.6   Negotiation

Negotiation exemplifies strategic interaction *par excellence*. After all, there
is no point in negotiating if your counter-party's actions don't matter to you.
Familiar as negotiation is to everyone, it is useful to be reminded that often
negotiation is not explicit, at least not at first. Here is a description of this
sort of encounter "in the wild."

To begin to negotiate the environment does not, of course, mean
that you enter the negotiation with a clear-cut goal in mind. A
clear-cut goal is not needed even in purely human negotiations.
Suppose you pass a stall in a market every week and notice an
antique ornament for sale. At first it seems ugly, but as it grows
familiar, you catch yourself wondering how it would look on your
shelf. One day it rains while you are crossing the market and you
take shelter in the stall. The ornament is still there; for something
to do you ask its price. Even when a low price is mentioned you
automatically snort in contempt, for you have no intention of buying... or have you? During the week that follows you decide that
the price really was low and think of a friend who has a birthday
soon and might like it. Next week you stop and begin to bargain.

When did the negotiation begin? When you started to bargain?
Or earlier, when you asked the price? Or earlier still, when you
first noticed the ornament among an anonymous heap of others?
Pointless to say, as pointless as to say where mind began.

–*Language and Species* by Derek Bickerton [30, pages 234–5]

2.7   Coordination, Symbiosis, Mutualism, Cooperation

Contexts of strategic interaction are not all adversarial in the sense that one agent's gain is another's loss (so-called *constant-sum* or equivalently *zero-sum* games). In *coordination games* all players gain if they can arrive at a common outcome and lose if they fail. Think of the game of finding someone you have separated from during a shopping trip. You both wish to meet up again, but did not plan for the separation and have no easy means of communication. Schelling's early treatment of such games is masterful and well worth reading today [279].

Biologists have named and studied several kinds of interactive decision making that—in terms of game theory lingo—is not constant-sum. Symbiosis and mutualism are two of the most important for our purposes, and they appear in nature in many forms. See [211] for a recent treatment of this theme by Margulis and Sagan.

Lichens—those familiar greenish splotches on trees and rocks—present a most striking example of symbiosis, since they appear as single organisms, yet in fact are constituted as symbiotic associations between entirely different sorts of microorganisms, fungi and green algae. A form of *emergence*—a topic we will return to at length—occurs with lichens. Surprisingly, what appears to be, and in many ways is, a single individual is actually composed of, arises through the interactions of, individuals from two distinct biological kingdoms.

Next, cooperation is—in its prototypical sense—a human social phenomenon, one that has been much noticed and remarked upon by social scientists, including game theorists. Cooperation, or roughly non-greedy behavior, has been called "the cement of society" [90] (by analogy with causation, which Hume called "the cement of the universe"). Without it, in the pungent phrasing of Thomas Hobbes, there would be

> no place for industry, because the fruit thereof is uncertain; and consequently no culture of the earth; no navigation, nor use of the commodities that may be imported by Sea; no commodious Building; no Instruments of moving and removing such things as require much force; no Knowledge of the face of the Earth; no account of Time; no Arts; no Letters; and which is worst of all, continuall feare, and danger of violent death; And the life of man, solitary, poore, nasty, brutish, and short. (Hobbes, *Leviathon*)

Without cooperation we are lost (or so it may seem). How, then, does it arise and how might it be sustained? Hobbes thought that realistically it was necessary to turn power over to a sovereign (king or powerful government)—a leviathon—who would enforce cooperation on society. Others have thought that perhaps cooperation could emerge and be sustained naturally, without a central authority, much as, say, lichens emerge and are sustained naturally. Is this possible? If so, what is required of the games and the players?

2.8 Conversation

When we speak we have in mind how others will react to what we say and what we do not say. In this regard, a representative news story—"Official's comments set off euro's surge. U.S. Treasury's Snow said a weaker dollar would help U.S. exports. The dollar fell against the euro." by David McHugh—appeared in *The Philadelphia Inquirer* on May 13, 2003. The first sentence says it all: "The U.S. dollar fell to another four-year low against the euro yesterday, inching closer to its all-time low, after U.S. Treasury Secretary John Snow said a weaker dollar would help U.S. exports." Secretary Snow never said he favored letting the dollar fall, but what he did say, as he no doubt understood, led the markets to infer that he favored a decline in the dollar. This form of strategic interaction is rife in linguistic communication and even has a special name: conversational implicature. Examples abound. A sign at Big Sur Lodge, Pfeiffer State Park, near a food counter:

Stressed

Spelled

Backwards

Is

Desserts

Translation: Buy a dessert from us; it'll make you feel good. Or the concluding line in Hitchcock's movie, "Frenzy": "Mr. Rusk, you're not wearing your tie." Translation: You're the necktie murderer and I'm placing you under arrest. Or the use of irony, as in "Rick, Major Strasse is one of the reasons the German Reich enjoys the reputation it has today," from the movie "Casablanca." Translations: (to Strasse) The Reich is an impressive accomplishment and you are a big part of it; (to Rick) Watch out, this guy Strasse is a very bad man. Or the dialog-less eating scene in the movie "Tom Jones" with Albert Finney. Translation: This is just foreplay foreplay; the best is yet to come. See Paul Grice's "Logic and Conversation" [134] for the original treatment, still worth reading.

2.9   Games against Yourself

The long and justly celebrated story from the *Odyssey* of Ulysses and the Sirens continues to enchant and inform us. (Jon Elster has even written an entire book relating the story to modern social science [89].) From the perspective of strategic interactions, the story may be interpreted as a game played by Ulysses at one time against Ulysses at another time. At t_0, before approaching within earshot of the Sirens, Ulysses foresees that Ulysses at t_1, within earshot, will have preferences and inclinations quite at variance from Ulysses at t_0 and from Ulysses at t_2, post the encounter with the Sirens (if he should live that long). So Ulysses at t_0 cleverly prevents Ulysses at t_1 from acting as Ulysses at t_1 would prefer. He hears the Sirens and lives to tell the tale.

Robert Louis Stevenson's familiar story, *Dr. Jekyll and Mr. Hyde,* carries a similar theme. Thomas Schelling tells a fable about a man who is struggling to quit smoking. A friend who smokes arrives at his house, converses, and leaves without incident. The friend, however, forgets his jacket and our protagonist notices the jacket contains a package of cigarettes. Not having an immediate compulsion to smoke and knowing the friend will return tomorrow, he puts the jacket away. Later, upon reflection, he recovers the jacket, removes the cigarettes, and destroys them.

2.10   Confidence Games

The con man (or woman) first gets your trust, your confidence, and then abuses it for profit. "Take the money and run" is the operating creed. Confidence rackets are celebrated in literature, theater, and film. Examples include Herman Melville's novel *The Confidence Man,* Thomas Mann's novel *Confessions of Felix Krull, Confidence Man: The Early Years,* Sinclair Lewis's novel *Elmer Gantry* (movie with Burt Lancaster and Jean Simmons), Jim Thompson's novella *The Grifters* (movie with Anjelica Huston, John Cusack, and Annette Bening), Guy Owen's short story "The Flim-Flam Man" (now published as *The Ballad of the Flim-Flam Man,* Coastal Carolina Press, May 2000; movie with George C. Scott and Sue Lyon), N. Richard Nash's play *The Rainmaker* (movie with Burt Lancaster and Katherine Hepburn), David Mamet's movie *House of Games* (with Lindsay Crouse and Joe Mantegna), and Meredith Wilson's Broadway musical *The Music Man* (movie with Robert Preston and Shirley Jones). This is from a Penn Web site, August 2003:[8]

```
6:30 pm - 8:30 pm     Confidence Games at the GSC

The GSC shows films about con artists:
Catch Me If You Can on 7/31;
The Thomas Crown Affair on 8/7;
The Spanish Prisoner on 8/14; and
The Grifters on 8/21.

Location:  Graduate Student Center, 3615 Locust Walk
Category:  Film
More info:
http://www.upenn.edu/gsc/programs/film.htm#con
```

Con games lie at the core of much detective fiction and fact, as well as recently popular email scams. There is a confidence business, indeed an industry, with its own lessons and skills. (This takes us beyond the scope of the book. Those wishing to go further might consult such works as *How to Become a Professional Con Artist*, by Dennis M. Marlock [214].)

2.11   Statesmanship

Ending this list on a less cynical note, George Washington is understood to have been a politically ambitious man throughout his life. He actively, deliberately sought and schemed for the power, influence, and adulation he ultimately received. Washington notoriously wore his military uniform during the deliberations on the Declaration of Independence, just to remind the other delegates of his availability for command. In pursuing his ambitions Washington consistently and consciously followed a strategy of seeking rewards by actually deserving to get them. Resigning from the army at the end of the Revolution, an unexampled act, was a move calculated to make him fit for political leadership in a democracy. Declining to run for a third term as president was a move calculated to secure the success of the new country and of Washington's legacy.

Napoleon on his deathbed and imprisoned lamented, "They expected me to be another Washington."

---

[8]See http://www.gametheory.net/ for yet another list of game-related movies, as well as lots of useful material on game theory.

2.12   Problems of Play

This has been a whimsical—that is, somewhat haphazard, idiosyncratic, and highly incomplete—tour of games in the wild. There needs to be a systematic inventory and assessment—a natural history—of them, but that is a huge task. In the meantime, we can use what we have to identify, in at least rough and tentative form, the principal aspects of, the main issues that arise in strategic interaction. I call these the problems of play. Approximately, they are constituted by answering the question: What, if done better (or worse) in a context of strategic interaction would result in a better (or worse) outcome for a given stakeholder in the game?

It is useful to divide the problems of play into two groups of stakeholders: the problems individual agents face in dealing with a context of strategic interaction, and the problems a society faces in managing a strategic situation in its midst. Here is a starter list.

1. Agent-view problems

   Given a well-defined game, the *strategy selection* problem for a player is the problem of choosing which strategy to play. This is a vexed problem because the number of strategies may be biologically large,[9] and the value of any strategy depends upon the strategy choices of the other players, yet the agent's strategy must be chosen before these other strategies are fully revealed. Classical game theory has surprisingly little to say about this, predicting only that the players will choose their individual strategies so that the resulting set of chosen strategies (one for each player) will be in equilibrium. That is, given how the other players have chosen their strategies, no single player has a better strategy than the one that player actually chose. How this will come about when there is more than one equilibrium, or a very large number of equilibria, is not explained. In any event, real players in the wild need to decide how they are to play. What can be said by way of advising them?

   If the game is not well defined, players face the problems of *determining the scope* of the game and of *determining the possible moves and strategies* in the game. Can we find situations in the wild when framing and conceptualizing the game is been an important problem, upon which the outcome depends?

   Thinking ahead is nearly always a problem in real-world games. Failure to do so well has a proverbial name: the law of unintended consequences. What can be done to help agents foresee the consequences of events in games? How can it be done effectively?

   Agents in real-world games will often seek and be materially assisted by

---

[9]Biologically large is larger by far than astronomically large.

intelligence, that is by information that is helpful in decision making. What can and do they do to gain advantage? To counter the intelligence-seeking activities of the counter-players?

2. Social-view problems

   The constitution of a context of strategic interaction is often influence-able by the larger society, and often the larger society will have an interest in how the game is played. In *social dilemmas* [70, 71], which we shall discuss at length, individual agents have incentive to behave in ways that are ultimately destructive for the larger society and indeed for the agents themselves. This arises in the well-known *tragedy of the commons* [141] in which, for example, it is in the interest of each household to pasture another cow on the municipal commons, but when enough households follow through the commons is destroyed for all.

   Contemporary examples include dealing with the ozone hole in the earth's atmosphere, with fisheries, and with climate change. The general problem, called the *institutional design* problem is one of designing social arrangements (called *institutions*) that will effectively produce desired or at least acceptable outcomes. One form of the problem is called the *problem of cooperation.* Can it arise and be maintained naturally? If so, how and under what circumstances? If it cannot arise and be sustained without intervention, which interventions are effective?

These are all large and important problems. They will accompany us throughout what follows.

2.13   For Exploration

1. Who was it who first recognized that lichens were symbiotic associations of microorganisms, rather than a single kind of organism?

2. A measure of astronomically large is the number of atomic particles in the universe, which is estimated at about 10^{80}. A smallish protein in a living organism is 1000 amino acids long. There are 20 naturally occurring amino acids. Proteins are linear combinations of amino acids. How many possible proteins are there that are 1000 amino acids long and consist of only the 20 naturally occurring amino acids? How does this biological number compare with our astronomical number?

3. Reconsider the example games in the wild, discussed above, in light of the problems of play discussed in §2.12. Given a game (context of strategic interaction), how do the problems of play manifest themselves in it? For example, in the base stealing game of baseball, the Ricky

Henderson story teaches us that intelligence is a, if not the, main factor. The moves and strategies available are not much of a challenge. What made Ricky Henderson a great base stealer was his ability to read the pitcher, to know the tells. That more than speed or anything else was instrumental to his success.

Analyze one or more games in the wild with respect to the problems of play? What matters? What does not? Are there important factors not covered by the lists in §2.12?

4. Here briefly retold is a well-known strategic joke, widely circulating with great approval in business schools. Dave and Fred are having a pleasant stroll in the woods when they come across a very cute bear cub. Fascinated, they quietly approach and observe the cub. After a moment the angry mother bear appears. For a short time, everyone freezes. Then Dave takes off running. Fred yells, "Don't run. It's foolish. You can't outrun a bear." To which Dave, still running, replies "Yes, but I can outrun you."

   Got any others?

5. See Figures 2.1–2.3. How is the task described strategic? The words were randomly drawn from Ogden's Basic English, `http://ogden.basic-english.org/words.html`. Do you think this is a good source? Can you think of other promising sources that would facilitate high-quality innovative findings? For what, if anything, might this kind of process be useful for?

   For comparison purposes you may want to look at Richard Dawkins's book *The Blind Watchmaker* [73]. There, in chapter three, he presented his Biomorph program, one of the earliest and most influential examples of using human judgment to direct an evolutionary process. There since has been steady development in what is called IEC (interactive evolutionary computation), which has been applied for architectural design and creating art. See [305] for a comprehensive review, up to 2001. PicBreeder, a more recent program `http://picbreeder.org/`, is also fun.

Instructions
Please read carefully

This is a two-part exercise. In Part I, working alone, you will be presented with two tables of words, similar to the example tables appearing below. As in the examples below, one of the tables will consist entirely of English adjectives, the other of nouns.

Adjectives:

married	happy	boiling	tall	dear	bent	violent
male	feeble	round	early	grey	regular	brown
fat	private	foolish	wide	wrong	cheap	bad
open	last	stiff	physical	bright	wise	hollow
cruel	solid	different	straight	able	first	possible
mixed	same	great	like	right	dependent	fertile
tight	future	simple	hard	red	slow	public

Nouns:

expansion	finger	guide	agreement	side	map
sense	position	quality	record	instrument	jewel
spring	power	seat	earth	market	skin
machine	approval	poison	island	window	run
view	meeting	branch	watch	help	transport
adjustment	bath	plough	meat	iron	tray
flower	attempt	journey	property	shock	pain
servant	time	support	leaf	plow	laugh
advertisement					

Your task is to draw upon the words in the tables (and only the words in the tables) to form two- and/or three-word phrases that are interesting. You may propose

1. an adjective-noun pair with words drawn from the two tables, such as **married expansion**, or

2. a noun-noun pair, with the first noun serving as an adjective, such as **market advertisement**, or

3. an adjective-noun-noun triple, such as **different plough bath**, or

4. a noun-noun-noun triple, such as **meat watch island**.

Instructions continued on the next page.

FIGURE 2.1: Instructions.

Instructions Continued

Please read carefully

In Part II of this exercise you and the other individuals participating in Part I will work as a group, facilitated by a coordinator. You working individually, as well as everyone else, may submit up to five phrases, along with explanatory comments, for Part II.

The coordinator will convene the group as a panel, charged with evaluating and ranking all submitted entries on the basis of being

1. novel

2. useful

3. reducible to practice (implementable in the real world)

You may, and probably should, attach a short note to each of your submitted phrases. The note can serve to communicate the nature of your idea and intent to the panel that will judge the submissions.

If you have any questions, now is the time to ask them.

Your name (print): ⸺⸺⸺⸺⸺⸺⸺⸺⸺⸺⸺

Your email address: ⸺⸺⸺⸺⸺⸺⸺⸺⸺⸺⸺

FIGURE 2.2: Further instructions.

Task 1

Please read carefully

Drawing on and only on the following two tables to identify up to five phrases for submission to the panel.

Adjectives:

bitter	sad	narrow	regular	military	automatic
chemical	shut	married	good	solid	hollow
probable	cruel	clear	simple	quick	wet
flat	fixed	political	wrong	different	loose
present	cold	smooth	dependent	delicate	frequent
separate	second	long	able	stiff	safe
general	sweet	parallel	full	opposite	responsible
material	healthy	awake	like	right	high
beautiful					

Nouns:

gun	snake	wave	soup	step	punishment	oil
learning	baby	animal	lift	square	comparison	quality
increase	mind	rat	song	needle	prose	color
tongue	brake	hand	attempt	seat	brother	chin
night	danger	low	design	shade	meeting	wax
journey	ear	sister	garden	stem	throat	roof
picture	top	cake	poison	request	wind	hate

Phrase 1: _____

Phrase 2: _____

Phrase 3: _____

Phrase 4: _____

Phrase 5: _____

FIGURE 2.3: Task 1.

2.14   Concluding Notes

The definitive natural history of games, of games in the wild, is yet to be written. Until—and even after—it is, Dixit and Nalebuff's very readable *Thinking Strategically* [78] and their updated version, *The Art of Strategy* [79], will be useful to anyone with a general interest in strategic interaction. Because games can make good stories, a number of popular books, aimed at the general reader, have appeared, offering informed and realistic accounts of particular sorts of strategic contexts. Brandenburger and Nalebuff's *Co-opetition* [41] addresses business strategy. Michael Lewis's *Liar's Poker* [197] is an often hilarious memoir of a bond trader. William Poundstone's *Fortune's Formula* [248] is a fascinating and insightful take on investing and its relation to gambling and the world of gambling. Lewis's *Moneyball: The Art of Winning an Unfair Game* [198] applies the investor's concept of arbitrage to the game of baseball. James McManus's *Positively Fifth Street: Murderers, Cheetahs, and Binion's World Series of Poker* [221] tells a lurid, amusing, and ultimately insightful story about the Texas Hold 'Em variety of poker and its milieu. Poundstone's *Prisoner's Dilemma: John von Neumann, Game Theory and the Puzzle of the Bomb* [247] is an intellectual history for the general reader. It describes the people and the ideas (game theory, computation, and nuclear weapons) present and in play at the creation of game theory. (See also http://en.wikipedia.org/wiki/John_von_Neumann.) Jared Diamond's *Collapse: How Societies Choose to Fail or Succeed* [77] pushes the envelope of historical method in examining entire societies and their strategic choices. Robert M. Sapolsky is also a biologist whose writings often shed light on contexts of strategic interaction, for example "A Natural History of Peace" [275] and *Monkeyluv: And Other Essays on Our Lives as Animals* [274]. Robert H. Frank regularly produces readable and insightful books on strategic interaction from an economic perspective, for example *Passions within Reason: The Strategic Role of Emotions* [106], *The Winner-Take-All Society: Why the Few at the Top Get So Much More Than the Rest of Us* [107], *Luxury Fever: Why Money Fails to Satisfy in an Era of Excess* [108], and *What Price the Moral High Ground? Ethical Dilemmas in Competitive Environments* [109].

Histories are often rich in strategic detail. Allison's *Essence of Decision: Explaining the Cuban Missile Crisis* [7] is a notable example. Ernest May's *Strange Victory: Hitler's Conquest of France* [215] is revisionist, controversial and fascinating in its interpretations. Especially interesting in our context is the German use of war games to plan the invasion of France. Margaret MacMillan's *Paris 1919* [206] tells the story of the Peace Conference at the conclusion of World War I and makes fascinating reading for anyone interested in strategic interaction.

The management (business) literature is suffused with works on strategy. Michael E. Porter's 1980 book, *Competitive Strategy: Techniques for Analyz-*

ing Industries and Competitors, [246] is recognized as a classic of the genre. *The Strategy and Tactics of Pricing: A Guide to Growing More Profitably,* by the consultants Nagle and Hogan [231], is chock full of strategically interesting discussion on a topic that is strangely under-represented in the academic literature. General-audience management periodicals, such as the *Harvard Business Review,* the *Sloan Management Review,* and the *California Management Review,* very often publish readable, accessible articles pertaining to management and strategy. Two recent examples are "If Brands Are Built over Years, Why Are They Managed Over Quarters?" [201] and "Managing Our Way to Economic Decline" [143].

Mitchell Zuckoff's article, "The Perfect Mark: How a Massachusetts Psychotherapist Fell for a Nigerian E-mail Scam" [337] and David Lewis's "The Ballad of Big Mike" [199] are two eminently readable accounts of successful fraud by e-mail spam and of strategic aspects of American football, respectively. In addition, Lewis's story is a touching and uplifting account of the rise of an improbable player.

Michael Pollan's book *The Omnivore's Dilemma* [244], particularly chapter 16, is insightful journalism on fundamentals of food and eating, with much emphasis on the strategic underpinnings. See also his short article, "Unhappy Meals" [245].

David Owen's piece in *The New Yorker* [243], delightful itself, reviews *The Backwash Squeeze & Other Improbable Feats: A Newcomer's Journey into the World of Bridge* by Edward McPherson.

Operation Mincemeat: How a Dead Man and a Bizarre Plan Fooled the Nazis and Assured an Allied Victory, by Ben Macintyre [204], is the story of a master deception from World War II.

Online video games now have to be counted as important examples of games in the wild. See Jane McGonigal's TED talk for openers (http://www.ted.com/talks/lang/eng/jane_mcgonigal_gaming_can_make_a_better_world.html). Follow that up with her book [220], and other books, including [36] and [57] Online games are now a matter of serious business (http://en.wikipedia.org/wiki/Serious_game). *Serious games* are with us, and with us to stay it would seem.

Part II

Mixed Motives

Chapter 3

Playing Prisoner's Dilemma

3.1 Introduction

A *social dilemma* is a context of strategic interaction in which each participating agent has incentive to take a decision called its *defecting choice*. Further, each agent has an option to take a decision called its *cooperating choice*. Also, all things being equal, the defecting choice is more attractive for each player than its cooperating choice. Finally, if all (or even most) agents take their defecting choices, all are worse off than if all (or even most) had made their cooperating choices. See [70, 71].

Social dilemmas are pervasive. It is convenient to litter at will, but most inconvenient if everyone litters. It is convenient to drive aggressively and flout rules of right-of-way, and most inconvenient if everyone behaves similarly. It is cheaper not to tip for routine services provided well and gracefully, and most annoying to suffer poor service because the provider has no expectation of reward for a good job. Increasing military expenditures will make a nation safer, unless of course all (or many) nations follow suit and everyone has the bomb (and biological warfare capability and chemical warfare capability and ...). And so on and on.

Social dilemmas come in many sizes. The number of participating agents may range from two to a few to a few score to more or less the entire human population. That's when the agents are all humans. The numbers are larger when, say, bacteria are involved.

We shall discuss social dilemmas often and in depth. Starting simply, we focus in this chapter on two-player, two-choice social dilemmas.

3.2 Iterated Prisoner's Dilemma

The Prisoner's Dilemma is the prototypical game representation for two-player social dilemmas. We can represent the game in strategic form; see Figure 3.1. Note we require for Prisoner's Dilemma that $T > R > P > S$ and $2R > (T + S)$. Further, [N] labels an outcome of Nash equilibrium play, and

	C	D
C	R [P] R	T [P] S
D	S [P] T	P [N] P

	C	D
C	3 [P] 3	5 [P] 0
D	0 [P] 5	1 [N] 1

FIGURE 3.1: Canonical and default (Axelrod) Prisoner's Dilemma.

[P] labels a Pareto-optimal outcome.(See §A.2.11 and §A.2.9 for information on Nash equilibria and Pareto optimal outcomes.) It is transparent from the representation that each player has a cooperating choice (labeled C) and a defecting choice (labeled D). The D strategy strictly dominates the C strategy: no matter what the counterpart player chooses, the player does better with D than with C. Both players should choose D by this reasoning, yet both would do better if both choose C rather than both choosing D. In addition, we assume that each player has full knowledge of the game.[1]

There are many who describe social dilemmas, and in particular the Prisoner's Dilemma, as paradoxes. Those who do find it paradoxical that rational choice (ideally rational choice, in the strong sense of Rational Choice Theory[2]) should result in a Pareto-inefficient outcome. After all, both players can do better. Is it not paradoxical that ideal rationality should lead to such an inferior outcome? Worse, careful experimental studies (Sally [269] reviews hundreds of experiments, conducted during a 35-year period) robustly find that subjects in fact often choose the cooperating strategy in Prisoner's Dilemma and will on average achieve a higher return than merely P. Sally in his meta-analysis of 130 experiments found "a mean cooperation rate for the entire sample of 47.4%, with a standard deviation of 23.7%" [269]. Are we to believe that irrational behavior produces better results than ideally rational behavior? Does Rational Choice Theory enjoin us to be, in Sen's words [283], "rational fools"? Note that we are discussing Prisoner's Dilemma played as a one-shot game: the game is played once, the players are unknown to each other, and everyone knows this.

Perhaps cooperating players are confused. Perhaps in small-size social dilemmas in real life the players have learned that cooperation pays off because of the prospect of ongoing association between the players. I wish to pass on

---

[1]The players have common knowledge of the game.. See §A.2.14.

[2]I capitalize here, and in similar cases, when the name of the theory or view constitutes a contested description. In the present case, not everyone agrees that Rational Choice Theory really is the only, or even an adequate, theory of rational choice. Saying so doesn't make it so; capitalization will help reduce confusion. See §A.2.13, page 433.

these questions for the present and move on to the often more realistic case in which there is such continuing interaction between the players.

In the terminology of game theory, the Prisoner's Dilemma is now a *stage game* in a *supergame* consisting of sequential plays of the stage game. Further, I want to focus on the case of *iterated play*: sequences of stage games are played with fixed counterparts. That is, Prisoner's Dilemma (in this case) is played repeatedly (multiple times) between the same players. (See §A.2.17, page 435.) As in the one-shot case (above) the players have common knowledge of the game. Now this includes all of the game sequence, all of what has happened up to the present.

Finally, there are two cases of iterated play to consider, pertaining to the stopping condition. In the first case a specific number of rounds of play is announced to the players. We call this the Definitely Iterated Prisoner's Dilemma game. In the second case, the players are not told how long a given session will run, how many rounds of play the session will have will have before it stops. We call this the Indefinitely Iterated Prisoner's Dilemma game. It is usual to terminate the game randomly.[3] The method I use in the simulations described here is to pick an expected number of rounds of play. This determines the value of a variable, called Prob-continue. After each round of play, a random number is drawn. If it is less than Prob-continue, another round is played; otherwise the session terminates.

These two cases—Definitely and Indefinitely Iterated Prisoner's Dilemma— are understood quite differently in classical game theory, which focuses on "solving" a game by finding its equilibria. The Definitely Iterated Prisoner's Dilemma has only one equilibrium: both players always defect.[4]

The argument for universal defection in the Definitely Iterated Prisoner's Dilemma game is based on backward induction. Let us agree that in the one-shot case what is rational is to defect. Suppose now we are about to play the last round in a Definitely Iterated Prisoner's Dilemma game. Surely we should defect, since the game will be over and we cannot possibly suffer any consequences from our uncooperative behavior in this last round. What about the penultimate round? Since we figure that both players will defect on the last round, there can be no untoward consequences for defecting on the penultimate round. So that's what it is rational to do. And so on back to the first round of play. What is rational is for each player to defect at every stage of play. This is an argument we shall have occasion to revisit. For the present let us take it as given, as a finding in classical game theory.

On the other hand, the Indefinitely Iterated Prisoner's Dilemma game has, by the so-called Folk Theorem(s) of game theory, an exceedingly large number of equilibria. Because there is no definite end of play known ahead of choosing strategies, the backward induction argument cannot apply. We can work with the strategy GRIMTRIGGER in order to get a sense of why

---

[3]With a geometric distribution, since it is memoryless.

[4]Technically, only one subgame perfect equilibrium, a detail that need not concern us for present purposes. See any standard text on game theory, e.g., [31, 114].

indefinitely (or infinitely) iterated Prisoner's Dilemma games can have very large numbers of equilibria. Under GRIMTRIGGER, the agent cooperates on the first round of play and continues to cooperate so long as the counterpart player cooperates; otherwise it defects and continues to defect until the game terminates, regardless of what the counterpart does. Assuming a reasonable discount rate for future returns, it is easy to see that the best response to GRIMTRIGGER is to always cooperate.[5] Many strategies will achieve this, including TIT FOR TAT and GRIMTRIGGER itself.

Consider now a variation on GRIMTRIGGER: play GRIMTRIGGER except on round 17, in which you defect, after which you return to GRIMTRIGGER. Call this GRIMTRIGGER-17. What is the best response to GRIMTRIGGER-17? Well, GRIMTRIGGER-17 will do. So the strategy pair (GRIMTRIGGER-17, GRIMTRIGGER-17) is an equilibrium of the Indefinitely Iterated Prisoner's Dilemma game. If a player chooses the GRIMTRIGGER-17 strategy, its counterpart can do no better than choosing it as well.[6] No doubt the reader can imagine many other equilibrium strategy pairs for Indefinitely Iterated Prisoner's Dilemma.[7]

The fact that many of the strategy pairs are incredible, even silly, does not prevent them from being equilibria. We have a problem, then, of deciding how we are to select a strategy of play from among the many that are available. We have this problem whether or not we confine ourselves to the assumptions of classical game theory.

3.3   Pragmatic Strategic Rationality

The Strategy Selection Framework, Figure 3.2, abstracts and generalizes what we do—explicitly or implicitly—when we go about choosing a strategy for behaving in a context of strategic interaction. It will be useful to compare classical game theory (and Rational Choice Theory) with what I'll call *Pragmatic Strategic Rationality* in the context of this framework.

First, classical game theory. There, we begin with a formalized game model of a strategic situation. We should think of the model as including both the formal game structure—given say as a game in strategic form or a game in extensive form, or in other ways—and the associated assumptions. These assumptions normally include that payoffs are given in utilities for the individual players, that every player is ideally rational (as described by Rational Choice

---

[5]This neglects discounting of future returns, or rather assumes that the discount rate is large enough.

[6]Again, neglecting discount rate considerations.

[7]Again, see any standard textbook on game theory for discussion of the Folk Theorem, which applies to any indefinitely iterated game, not just the Prisoner's Dilemma.

1. Given a CSI (context of strategic interaction), specify G, a representation of it.

   Think: G is the game.

2. Specify a collection of strategies, $\mathcal{U}$, as a strategy *consideration set* for G.

   Think: the strategies that are under consideration; the universe of strategies.

3. From $\mathcal{U}$, specify a collection of one or more strategies, $\mathcal{C}$, as contenders for strategy selection.

   Think: the contenders are strategies under consideration that are chosen for good reasons.

4. Pick one of the contenders and play it.

FIGURE 3.2: SSF: Strategy Selection Framework.

Theory), that the players have common knowledge of the setup, and so on.[8] Together, these constitute the game representation, G.

The second step, specifying $\mathcal{U}$, the consideration set of strategies for G, is nominally trivial in classical game theory. $\mathcal{U}$ contains every possible strategy for G. As such $\mathcal{U}$ need not be finite or even countably infinite.[9]

The third step, specifying $\mathcal{C}$, the contending strategies that rationally commend themselves to us, is also in principle trivial in classical game theory. $\mathcal{C}$ contains every strategy that belongs to some equilibrium of G. A little more specificity may be helpful. Implicit in G is a list of players. The strategies in $\mathcal{U}$ and $\mathcal{C}$ are indexed by player, since not every strategy need be available to every player. This is mere bookkeeping. So we may think of $\mathcal{C}$ as containing tuples (indexed by player) of all the equilibria of G. It may in practice be quite challenging to characterize $\mathcal{C}$, that is all the equilibria for a game. Nevertheless, the underlying concept is entirely straightforward.

Finally, the fourth step—selecting one of the outcomes in $\mathcal{C}$ for actual play—is for the most part *not* addressed in classical game theory. The theory simply predicts that one equilibrium or another will be chosen in any actual play of the game. The theory has little to say about which of the equilibria will be chosen. (But see the 'equilibrium selection' literature, seminally [142], although these results are neither comprehensive nor generally accepted.) Nor does the theory have much to say about how players might coordinate their

---

[8]See standard game theory texts for the details.

[9]That is, it may have at least as many members as there are real numbers.

choices so that the strategies selected by different players actually belong to a common equilibrium.

The story for Pragmatic Strategic Rationality is rather different. First, the range of possible game representations, G, is much broader. Essentially any usable description may be employed, including of course verbal descriptions of real-world strategic contexts. This is not to suggest that rigorously articulated formal models are not valuable. They are and they will normally be preferred because they are more useful for the strategy selection process. Rather, the point is that there are often useful things to be done that can be achieved with less formal models when fully formal ones are unavailable. The range of formal models is simply broader with pragmatic rationality. Payoffs may be given in any quantity of interest, not just utilities. Players may be assumed to have limited computational and epistemic powers. And so on, none of which militates against fully formal game representations.

Second, pragmatic rationality requires only that the contents of the consideration set U be warranted, or justified appropriately. This is a subtle issue, one we will explore and discuss as we proceed. The key notion is that empirical evidence, say regarding what the players are likely to come up with, as well as pragmatic criteria may with justification be used to constitute the consideration set U. Pragmatic criteria include considerations such as cost, computational tractability, and what can be discovered by a favored heuristic, all of which may limit the strategies that can effectively be considered. It is not the case, however, that "anything goes." A consideration set will rarely if ever be pragmatically warranted if there are known or readily discoverable strategies of high warrant that are not in the consideration set. They should be added. This is not the place to articulate a finished theory of pragmatic rationality; I am merely essaying to sketch the notion. Better to proceed by example, keeping in mind the main issues.

What I said about the second step, about the advisability of proceeding case by case before attempting to draw any general picture, applies to the third step as well. We do need to start somewhere, so here are starter conditions to be met for membership in C, the collection of strategies contending for pragmatically rational acceptance. Any member of C should

1. Perform well when played against any other member of C, including itself, and

2. Be robustly resistant to exploitation by strategies in U but not in C.

Provisionally, any strategy in C so constituted is pragmatically rational for use in playing G. In consequence, for step four, picking any of the strategies in C is permissible on grounds of pragmatic rationality.

Again, this is a sketch of a notion. I offer the particular principles only as starting points for discussion and careful elaboration. This will ensue in conjunction with the examples and cases that follow.

Game theoretic rationality and Rational Choice Theory are well established and widely accepted concepts, constituting the incumbent theory. In offering pragmatic strategic rationality as an alternative, I do not mean to suggest that they are false or somehow irrevocably muddled. Criticisms of the incumbent theory certainly abound and have been made by many others. I am inclined to the view that the two theories—the incumbent theory and pragmatic rationality—are best seen as complements, not competitors, and that it is the job of researchers to discover their proper scopes and roles.

What is to come in this book has been undertaken in that spirit. It is the spirit evident in the answer to an interviewer's question once given by Robert Altman, a movie director with very un-Hollywood-like sensibilities. Asked "Why do you hate Hollywood?", Altman replied "I don't hate Hollywood. They make shoes. I make gloves. We're in different businesses."

I don't hate game theory. It seeks equilibria of games. I seek pragmatically warranted heuristics for strategic decision making. We're in different businesses.

Promissory note issued. Capital raised. Now down to business.

3.4   Axelrod's Tournaments

In the late 1970s, Robert Axelrod, a political scientist at the University of Michigan, conducted two computerized tournaments of strategies for playing Iterated Prisoner's Dilemma. He investigated play with a single stage game, having the payoffs shown in Figure 3.1. Participation in each of the tournaments was solicited and open to the general public. See [15, 18].

Participants in the first tournament were given a report describing and analyzing a preliminary tournament, consisting of strategies generated in Axelrod's lab. The report also provided instructions for how to write the computer code necessary for implementing a submitted strategy. Conditions of the tournament—including that there would be exactly 200 rounds of play and that each submission would play every other submission—were announced in the report as well.

There were 14 entries submitted for the first tournament. Adding RANDOM (the strategy of random play, of randomly cooperating or defecting), Axelrod played each of the resulting 105 pairs (=15 × 16/2) in one session of length 200 and reported the results. The winning strategy was TIT FOR TAT. It cooperates on the first round of play and thereafter to plays what the counterpart played on the previous round. TIT FOR TAT was submitted by one participant, Anatol Rapoport.

After analyzing the first tournament, and writing up and publicizing the results, Axelrod held a second tournament. Again, participation was open and was openly solicited. Participants were given a report analyzing the results of the first tournament. The only other difference of any significance with the first tournament was that the length of the second tournament was announced to be

200 rounds in expectation. That is, the second tournament was an Indefinitely Iterated Prisoner's Dilemma, with an expected length of 200 rounds of play. In actually conducting the play, Axelrod played each pair five times with randomly determined session lengths of 63, 77, 151, 156, and 308 rounds of play [15, page 217]. A strategy's score for its play in the pair was the average of its score from the five sessions. Its score for the tournament was the sum of its pairs scores, including play with itself as the counterpart.

There were 62 entries for the second tournament, to which Axelrod added RANDOM as before. Once again, only Anatol Rapoport submitted TIT FOR TAT and TIT FOR TAT won the tournament.

Axelrod, in analyzing the results of these two tournaments [15, 18], notes that TIT FOR TAT is nice (it is never the first to defect), retaliatory (it responds to defection by defecting), and forgiving (it responds to renewed cooperation by cooperating). These are, on the face of it, surely credible as virtues for a strategy in Iterated Prisoner's Dilemma. Not only does TIT FOR TAT have them, but most of the more successful strategies in the second tournament have them as well, at least to some degree.

There is no universally best strategy for IPD, independent of the counterpart's strategy.[10] It is in consequence pleasing and broadly very useful to discover general properties of attractive strategies. Putting this on the table as a live prospect, applying beyond Prisoner's Dilemma to any game of interest, is perhaps the most profound contribution arising from Axelrod's tournaments. We shall continue by exploring this idea.

3.5   Further IPD Tournaments

The NetLogo program 2x2-Tournaments-Det.nlogo is, as usual, available on the book's Web site. It is designed to be used for conducting computerized tournaments for iterated games composed from 2×2 stage games with deterministic ('Det') payoffs. We can use it to conduct studies similar to Axelrod's tournaments. We shall do just that in this section: conduct a small IPD experiment with the same stage game that Axelrod used.

2x2-Tournaments-Det.nlogo names the two players A0 (agent 0, or Row) and A1 (agent 1, or Column). It names the two strategies for the stage game S0 (strategy 0, whether for Row or Column, or Invest) and S1 (strategy 1, or NotInvest). We shall use the mnemonics—Row and Column, Invest and NotInvest—throughout. See Figure 3.3.

---

[10]This is assuming that the discount rate on the stream of returns from play is sufficiently high. See Axelrod's proposition 1 [15, Chapter 1]. The demonstration is straightforward.

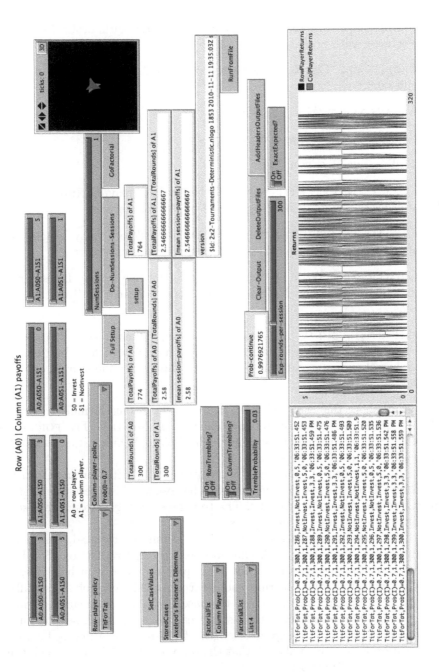

FIGURE 3.3: 2x2-Tournaments-Det.nlogo interface.

The strategies in our consideration set are collected in the program's "List 1." They are:

- PROB(I)=0.7. (In the stage game) play Invest with probability 0.7, NotInvest with probability 0.3.

- PROB(I)=0.2. Play Invest with probability 0.2, NotInvest with probability 0.8.

- ALWAYSINVEST. Always play Invest.

- NEVERINVEST. Always play NotInvest.

- TITFORTAT. TIT FOR TAT.

- FF2TITSFOR1TAT. TIT FOR TAT, but defect (play NotInvest) twice after every play of defect (NotInvest) by the counterpart.

- FF1TITFOR2TATS. TIT FOR TAT, but only defect (play NotInvest) after two consecutive plays of NotInvest by the counterpart.

- GRIMTRIGGER. Begin by playing Invest. Continue to play Invest until the counterpart plays NotInvest, then play NotInvest until the session ends.

- M1FICTITIOUSPLAY. A limited memory approximation of FICTITIOUS PLAY. This strategy responds in each round of play with its best choice against the counterpart's current history of play. See the source code for details.

Table 3.1 shows the results of an IPD tournament among the nine strategies in List 1. Measured by mean return achieved, we see that TIT FOR TAT performed well, but actually only came in third place, behind GRIMTRIGGER and FF2TITSFOR1TAT.

List 2 was formed by collecting the strategies that got a mean return of more than 2.5 on the List 1 tournament. The results of the resulting List 2 tournament are shown in Table 3.2. The top three strategies from the List 1 tournament are now tied for first place.

Both of these tournaments are in fact Definitely Iterated Prisoner's Dilemma games, since they both end after exactly 300 rounds of play. Because none of the strategies condition their play on end-of-run behavior, it should not matter whether the runs end deterministically or probabilistically. Table 3.3 confirms that this is indeed the case. The data summarize a tournament with 100 replications (sessions) for each pairing. Each session ends probablistically and has an expected length of 300 rounds. Allowing for random variation we see, as predicted, that the data in Table 3.3 correspond well with the results in Table 3.1.

How are we to interpret these results? Is there something misleading about Axelrod's tournaments? On the latter question, the answer is no. Axelrod is

Strategy	Min.	1st Qu.	Median	Mean	3rd Qu.	Max.
TITFORTAT	0.0	3.0	3.0	2.574	3.0	5.0
GRIMTRIGGER	0.0	3.0	3.0	2.933	3.0	5.0
P(I)=0.7	0.0	0.0	3.0	2.081	3.0	5.0
P(I)=0.2	0.0	1.0	1.0	2.353	5.0	5.0
ALWAYSINVEST	0.0	3.0	3.0	2.274	3.0	3.0
NEVERINVEST	1.0	1.0	1.0	2.504	5.0	5.0
FF2TITSFOR1TAT	0.0	3.0	3.0	2.649	3.0	5.0
FF1TITFOR2TATS	0.0	3.0	3.0	2.439	3.0	5.0
M1FICTITIOUSPLAY	0.0	1.0	1.0	2.505	5.0	5.0

TABLE 3.1: List 1 tournament results, PD, Axelrod's payoffs, 1 session, fixed termination, 300 rounds exactly.

Strategy	Min.	1st Qu.	Median	Mean	3rd Qu.	Max.
TITFORTAT	0.0	3.0	3.0	2.599	3.0	5.0
GRIMTRIGGER	0.0	3.0	3.0	2.599	3.0	3.0
NEVERINVEST	1.0	1.0	1.0	1.808	1.0	5.0
FF2TITSFOR1TAT	0.0	3.0	3.0	2.599	3.0	5.0
M1FICTITIOUSPLAY	0.0	1.0	1.0	1.803	1.0	5.0

TABLE 3.2: List 2 tournament results; ≥ 2.5 mean on List 1 tournament, PD, Axelrod's payoffs, 1 session, fixed termination, 300 rounds exactly.

quite clear and correct in noting that there is no general optimal strategy for IIPD. Success will always depend on the other strategies in play. Axelrod also has been clear and correct in noting that strategies other than TIT FOR TAT can prevail against the strategies in his tournaments. He even used evolutionary computation to find such strategies [16].

Besides introducing 2x2-Tournaments-Det.nlogo and illustrating how to use it for tournaments, several important points arise from the results being reported in this section.

1. It is indeed not difficult to find consideration sets of strategies for IIPD for which TIT FOR TAT is not the winner.

2. P(I)=0.7, P(I)=0.2, NEVERINVEST, and ALWAYSINVEST are not responsive strategies. Their behavior in a session (a sequence of rounds) is independent of the counterpart's behavior. In Prisoner's Dilemma the best response to them is always to defect (NeverInvest). This goes far to explain the superior performance of GRIMTRIGGER and NEVERINVEST in the List 1 tournaments.

3. The excellent performances by FF2TITSFOR1TAT and M1FICTITIOUSPLAY

Strategy	Min.	1st Qu.	Median	Mean	3rd Qu.	Max.
TITFORTAT	0.0	3.0	3.0	2.598	3.0	5.0
GRIMTRIGGER	0.0	3.0	3.0	2.834	3.0	5.0
P(I)=0.7	0.0	0.0	3.0	2.096	3.0	5.0
P(I)=0.2	0.0	1.0	1.0	2.378	5.0	5.0
ALWAYSINVEST	0.0	3.0	3.0	2.335	3.0	3.0
NEVERINVEST	1.0	1.0	1.0	2.546	5.0	5.0
FF2TITSFOR1TAT	0.0	3.0	3.0	2.65	3.0	5.0
FF1TITFOR2TATS	0.0	3.0	3.0	2.44	3.0	5.0
M1FICTITIOUSPLAY	0.0	1.0	1.0	2.482	5.0	5.0

TABLE 3.3: List 1 tournament results, PD, Axelrod's payoffs, 100 sessions, stochastic termination, 300 rounds expected.

	C_1		C_2	
R_1		3 [N]		1 [P]
	2		4	
R_2		2		4 [P]
	1		3	

FIGURE 3.4: Game #47.

in the List 1 tournaments may be explained similarly. They are better than TIT FOR TAT at exploiting unresponsive strategies.

And there is a larger lesson. In conducting two levels of tournaments—a general tournament, followed by a contest of the winners—we illustrate a quite sensible process for finding good strategies. Given, that is, that the original consideration set was ample and warrantably comprehensive, which it was not in this case. Nonetheless, the subsequent process is, in the small, one that is proper. The first tournament served to weed out generally poor performing strategies and strategies that are exploitable. The second tournament, among the better performers of the first tournament, identified the strong performers from among those that generally perform well and are not (very) exploitable. In short, we see here a small example that illustrates the Strategy Selection Framework in action. In play against a field of our List 1 strategies we have identified three strong performers that are robust across the entire field. These are strategies to be recommended knowing that other players will also be choosing strategies from List 1.

	C_1	C_2
R_1	2 [N] 2	1 [P] 4
R_2	3 1	4 [P] 3

FIGURE 3.5: Game #48.

	C_1	C_2
R_1	3 [N] 2	2 [P] 4
R_2	1 1	4 [P] 3

FIGURE 3.6: Game #57.

3.6 Other 2×2 Social Dilemma Games

The Prisoner's Dilemma is not the only 2×2 game constituting a social dilemma. Figures 3.4, 3.5, and 3.6 (all from from [253]) present games #47, #48, and #57, respectively, with their [N] (Nash equilibrium) and [P] (Pareto efficient) outcomes labeled. Each is a social dilemma.

Using 2x2-Tournaments-Det.nlogo, let us look at a few tournaments with these asymmetric games as stage games with iterated play. In Tables 3.4 and 3.5 we see the result of a list 3 tournament with Game #47. (The strategies in list 3 appear in the Strategy columns of the two tables.) We see that the best 6 strategies for the row player in Table 3.4 are TIT-FORTATCOMPLEMENT, with a mean score of 2.743, NEVERINVEST (2.630), FF2TITSFOR1TATCOMPLEMENT (2.727), FF1TITFOR2TATSCOMPLEMENT (2.702), GRIMTRIGGERCOMPLEMENT (2.776), and M1FICTITIOUSPLAY (2.572). Remarkably, these are also the best six strategies for the column player (see Table 3.5).

By way of explaining the strategies, TIT FOR TAT begins by playing Invest, or R_1 if Row and C_1 if Column. After that, TIT FOR TAT mimics the counter-party. If the counter-party played Invest in the previous round, then TIT FOR TAT plays Invest in the present round, and so on. TIT FOR TAT COMPLEMENT begins by playing Not Invest, or R_2 if Row and C_2 if Column, and subsequently mimics the play of the counter-party, just as in TIT FOR TAT. It is important to appreciate that in these asymmetric games labels like Invest or Defect have at best problematic meaning.

Strategy list 3	Median	Mean	Column Mean
TITFORTAT	2.0	2.356	3.086
TITFORTATCOMPLEMENT	3.0	2.743	3.477
GRIMTRIGGER	2.0	2.324	3.315
GRIMTRIGGERCOMPLEMENT	3.0	2.776	3.15
P(I)=0.7	2.0	2.479	2.387
P(I)=0.2	3.0	2.525	2.972
ALWAYSINVEST	2.0	2.398	2.602
NEVERINVEST	3.0	2.630	3.63
FF2TITSFOR1TAT	2.0	2.360	3.186
FF2TITSFOR1TATCOMPLEMENT	3.0	2.727	3.339
FF1TITFOR2TATS	2.0	2.324	2.729
FF1TITFOR2TATSCOMPLEMENT	3.0	2.702	3.542
M1FICTITIOUSPLAY	2.0	2.572	2.425

TABLE 3.4: Results for Row, playing the List 3 tournament (against Column), Game #47, 1 session, fixed termination, 300 rounds exactly.

Strategy list 3	Median	Mean	Row Mean
TITFORTAT	3.0	3.005	2.256
TITFORTATCOMPLEMENT	4.0	3.401	2.661
GRIMTRIGGER	3.0	2.952	2.829
GRIMTRIGGERCOMPLEMENT	3.0	3.216	2.224
P(I)=0.7	3.0	2.483	2.202
P(I)=0.2	3.0	2.69	3.002
ALWAYSINVEST	3.0	2.798	1.798
NEVERINVEST	4.0	3.216	3.261
FF2TITSFOR1TAT	3.0	3.065	2.448
FF2TITSFOR1TATCOMPLEMENT	3.0	3.323	2.484
FF1TITFOR2TATS	3.0	2.84	1.998
FF1TITFOR2TATSCOMPLEMENT	4.0	3.278	3.002
M1FICTITIOUSPLAY	4.0	3.468	2.535

TABLE 3.5: Results for Column, playing the List 3 tournament (against Row), Game #47, 1 session, fixed termination, 300 rounds exactly.

In round two of the tournament, we form two (coincidentally) identical lists of strategies. List 4 is the best six for the Row player and list 5 is the best six for the Column player. Tables 3.6 and 3.7 report the round two results for Row and Column, using these truncated lists. We see considerable convergence. Row is getting 3 or close to it, Column is getting 4 or close to it.

Strategy list 4	Median	Mean	Column Mean
TITFORTATCOMPLEMENT	3.0	2.999	3.997
GRIMTRIGGERCOMPLEMENT	3.0	2.839	3.828
NEVERINVEST	3.0	2.999	3.999
FF2TITSFOR1TATCOMPLEMENT	3.0	3.0	3.996
FF1TITFOR2TATSCOMPLEMENT	3.0	3.0	4.0
M1FICTITIOUSPLAY	2.0	2.509	2.493

TABLE 3.6: Results for Row, playing the list 4 tournament (against Column list 5), Game #47, 1 session, fixed termination, 300 rounds exactly.

Strategy list 5	Median	Mean	Row Mean
TITFORTATCOMPLEMENT	4.0	3.833	2.835
GRIMTRIGGERCOMPLEMENT	4.0	3.832	2.834
NEVERINVEST	4.0	3.502	3.166
FF2TITSFOR1TATCOMPLEMENT	4.0	3.832	2.834
FF1TITFOR2TATSCOMPLEMENT	4.0	3.666	3.001
M1FICTITIOUSPLAY	4.0	3.648	2.674

TABLE 3.7: Results for Column, playing the list 5 tournament (against Row list 4), Game #47, 1 session, fixed termination, 300 rounds exactly.

Our third round lists are the best from Row in the second round, list 6, and the best from Column, list 7. Tables 3.8 and 3.9 show the results of this third tournament. Note that lists 6 and 7 are now different.

So, this multi-stage tournament has brought us to consideration sets, Cs for Row and Column that have little to distinguish among their members. The process has led to a sustained payoff of 3 for Row and 4 for Column. Should we think of this as a natural outcome of play? Or are there strategies, heretofore not considered, by which Row could induce Column to yield Row more than 3 on average?

Finally, note that the process has led to a kind of optimization: the players have settled on play that extracts the maximum return of 7 (3 + 4) from each round.

Strategy list 6	Median	Mean	Column Mean
TITFORTATCOMPLEMENT	3.0	3.0	4.0
NEVERINVEST	3.0	3.0	4.0
FF2TITSFOR1TATCOMPLEMENT	3.0	3.0	4.0
FF1TITFOR2TATSCOMPLEMENT	3.0	3.0	4.0

TABLE 3.8: Results for Row, playing the list 6 tournament (against Column list 7), Game #47, 1 session, fixed termination, 300 rounds exactly.

Strategy list 7	Median	Mean	Row Mean
TITFORTATCOMPLEMENT	4.0	4.0	3.0
GRIMTRIGGERCOMPLEMENT	4.0	4.0	3.0
FF2TITSFOR1TATCOMPLEMENT	4.0	4.0	3.0

TABLE 3.9: Results for Column, playing the list 7 tournament (against Row list 6), Game #47, 1 session, fixed termination, 300 rounds exactly.

3.7   Discussion

Time to step back and survey the scene. Games—contexts of strategic interaction—present their stakeholders with a number of problems of play. In this chapter we have begun to explore and discuss two major themes—each an important problem of play—that will be with us throughout.

The first of these is social dilemmas and more broadly the problem of cooperation. We may think of cooperation as a social problem. How, especially in the face of social dilemmas, can a society of agents arrange to cooperate? We have focused on the simplest cases, in particular the well-known and ever-present Prisoner's Dilemma game played iteratively, although we saw that other simple games constitute social dilemmas as well. I want to emphasize that the two-player case of social dilemmas is only a small part of this very important subject. Two dimensions of generalization, touched upon lightly in this chapter, will draw more of our attention in the sequel. The first is the number of players. It is one thing to support cooperation between two players interacting with each other over time and quite another when the number of players is large. How does size of the interacting community affect the outcomes of play? How does size affect the prospects for intervening and directing the outcomes of play? And so on.

Our second dimension of generalization is the conditions of play, including the makeup of the players. How should we describe real-world social dilemmas in order to capture the main factors affecting the outcomes? What are these main factors and what are the outcomes? What can agents learn and how best can they learn in social dilemmas and generally regarding cooperation? And so on.

Both dimensions of generalizing from the simple Prisoner's Dilemma to larger issues of cooperation and social dilemmas raise many important questions. These are questions that will concern us throughout what follows.

The second major theme—and problem of play—broached in this chapter is the problem of strategy selection. Given a context of strategic interaction, how should and how will a player play? The problem is severe when the number of possible strategies that belong to some equilibrium is very large. How is a player to choose from among them?

In this chapter I begin to sketch an account of what I am calling Pragmatic Strategic Rationality for addressing problems of strategy selection. My presupposition is that the full resources of reasoning and deliberation should be available for strategy selection. The account starts with the Strategy Selection Framework (SSF), Figure 3.2. We see the warrant, or at least the beginning of the motivation, for this account by reflecting upon what we learn from Axelrod's and other tournaments' strategies for iterative play in simple social dilemmas (mainly Prisoner's Dilemma). The notion I want to advance is that in complex strategic environments (and Iterated Prisoner's Dilemma, simple as it is, is one) selection of a particular strategy is warranted if it scores acceptably or well on certain criteria and does so on balance as well as or better than alternative strategies. Put in terms of the SSF, a strategy becomes a contender C if it scores acceptably well on certain criteria. A strategy in C may be chosen with Pragmatic Strategic Rationality if it compares well with the other members of C. What are those criteria? Here is an initial list.

To be warranted (to be choosable with Pragmatic Strategic Rationality) a strategy must:

- Be accessible to any player that would use it.

You can't play a strategy you don't have. To be playable at all, a strategy must be embodied and available to the agent who would use it. This does not mean that the agent must have the strategy in mind; writing down the strategy on paper or storing it on a computer may perfectly well suffice. Strategies that are not discoverable or cannot be effectively computed, however, fail the accessibility condition for rationality. A game might have but one equilibrium, but if the strategies for that equilibrium are not discoverable or computable (even in a finite, practical sense), then there is no reason to think play will result in the equilibrium outcome.

To be warranted (to be choosable with Pragmatic Strategic Rationality) a strategy should preferably:

1. Do well in competition with alternative strategies.

2. Do well when played against itself.

3. Do well when played against other strategies that do well against themselves.

4. Do well when played with other high-quality strategies.

5. Be robust to starting conditions.

6. Be robust to conditions of play.

7. Not be exploitable: not do very much worse than any strategy it encounters.

8. Avoid self-destructive behavior (a problem for GRIMTRIGGER).

9. Be viable in small numbers.

10. Be an effective signaler.

A few comments now on each of these criteria.

3.8   Strategy Selection Criteria

3.8.1   Do Well Overall

Doing well in a well-designed tournament is sufficient for a strategy to be judged prima facie high quality. In terms of the SSF, this gets a strategy into $\mathcal{C}$. This naturally raises the questions of what constitutes a well-designed tournament and of whether there are non-tournament alternatives that also suffice. These are larger questions that will be addressed as we proceed. Note, however, our use of multi-round tournaments, in contrast with Axelrod's single-round tournaments. For now let us explore criteria for refining $\mathcal{C}$.

3.8.2   Do Well against Itself

If all players in a game face exactly or even roughly the same context of play, then any very attractive strategy is likely to be chosen by more than one player. If so, then it had better do well when played against itself. TIT FOR TAT in Prisoner's Dilemma is of course a prime example of a strategy meeting this criterion. When games are highly asymmetric or in general present different players with very different situations (think: game #57, etc.), then this criterion loses its relevance.

3.8.3   Do Well When Played with Others That Do Well with Themselves

If strategies A and B do well when played against themselves, but A performs poorly when played against B, then warrant for A is considerably weakened. TIT FOR TAT in Prisoner's Dilemma, for example, does do well against ALWAYS COOPERATE and GRIM TRIGGER.

Note that this criterion also loses its relevance if the strategic situation is highly asymmetric.

3.8.4   Do Well against Other High-Quality Strategies

To meet this criterion, a strategy must in addition do well against the other strategies that did well in the tournament (§3.8.1). Notice that this criterion does *not* rely on symmetric games.

3.8.5   Be Robust to Starting Conditions

A strategy may, for example, do well in a given tournament (consisting of the $\mathcal{U}$ strategies) but not do well in a tournament consisting of only some of the strategies from the original tournament. One way to test for this criterion is to repeat the tournament a number of times on random samples from $\mathcal{U}$. This may be called a cross validation test. Repeating this many times affords a variance-based form of sensitivity analysis [270, 271].

3.8.6   Be Robust to Conditions of Play

This is very similar in concept to being robust to starting conditions (§3.8.5), but pertains to examining slightly different forms of the game. For example, we can perturb the game payoffs by randomly altering them, and then repeat the tournaments. Or, me may introduce noise so that the strategy played at any round by a player may be randomly altered. This could be used to model the so-called "trembling hand" of a player who occasionally makes mistakes. Repeating these kinds of tests many times also affords a variance-based form of sensitivity analysis [270, 271].

3.8.7   Not Be Exploitable

If a strategy is exploitable by a high-quality strategy, then it will fail the previous criterion, §3.8.4. The point of this criterion is that a good strategy should do reasonably well against, at least not be exploitable by, strategies in the consideration set $\mathcal{U}$ and not in $\mathcal{C}$. See the Strategy Selection Framework, Figure 3.2.

3.8.8   Avoid Self-Destructive Behavior

A good strategy (in $\mathcal{C}$) should not participate in unilaterally or mutually poor performance against any of the strategies in $\mathcal{U}$ but not in $\mathcal{C}$. Of course, as in the "not be exploitable" criterion, §3.8.7, a strategy in $\mathcal{C}$ that is self-destructive against another strategy in C will fail the "high quality" criterion, §3.8.4.

This may often be a problem for GRIM TRIGGER.

3.8.9 Be Viable in Small Numbers

If a small number of players using a given strategy can be successful, situated among a large number of players using other strategies, then that strategy has at least the beginning of an account for its eventual success. Put conversely, if a strategy depends for its success upon its widespread adoption by others, then that strategy is fragile.

3.8.10 Be an Effective Signaler

By signaling I mean to include both behavior that conveys information (in the interests of the sender) and behavior that deceives counterparts (in the interests of the sender). Let us call these positive and negative signaling. To illustrate, Axelrod touts the simplicity of TIT FOR TAT, but the reason why simplicity is a virtue in the context of Iterated Prisoner's Dilemma is that it is easy for the counterpart to understand the strategy and then to get in line.

A more general point here arises from the fact that in Prisoner's Dilemma if you are playing TIT FOR TAT you might well want to be able to reveal your strategy to your counterpart. You want your strategy to be transparent, which makes cooperation optimal for your counterpart. The problem, of course, is that you cannot always effectively reveal your strategy. How do you (a) communicate what the strategy is and (b) prove that you are committed to it for the duration? This is why the prisoners of Prisoner's Dilemma are in jail.

3.9   Comments on the Strategy Selection Criteria

It cannot be said that these criteria exhaust the properties of interest for strategy selection or that the discussion of those we have is complete.

1. It is difficult (impossible?) to even conceive what it would mean in general to have a demonstration that a complete list of criteria was present. Instead, we simply have to make lists, work with them, and revise them based on experience.

2. The criteria are not formalized and precise. This is intentional. What exactly does it mean "to do well against other strategies," etc.? I have not said. My suggestion is that this of course needs to be made explicit but how that is done must be conditioned on the circumstances present.

For the most part, this should not be terribly problematic. There will be difficult borderline calls, but these cannot be resolved from first principles.

3. How are the criteria to be combined for making an actual selection of a strategy? Formal methods are available, such as multi-attribute utility modeling [167], and these should always be considered. In general, however, we should assume that any valid form or tool of deliberation, formal or not, may be used.

4. How do or how should probabilities figure in to strategy selection? The strategy selection problem would be considerably simplified, or even solved, if we were able to assign probabilities to the universe of possible strategies, $\mathcal{U}$. In nearly all cases, however, the difficulties here are insurmountable.

A distinction is traditionally drawn between decisions under risk (where probabilities of outcomes are available) and decisions under uncertainty (where probabilities of outcomes are not available). In game theoretic analysis we thus normally assume that strategy selection is a decision under uncertainty. If so, then by what principles should we make decisions? Under risk, we can calculate expected values, but under uncertainty this makes no sense. The one-shot Prisoner's Dilemma seems straightforward because defecting is best no matter what; it is said to be dominant. But this is a rare case. If there are no dominant strategies, what can we do? One approach, called MAXIMIN or MINIMAX depending on circumstances, advises us to choose the strategy that has the least worst downside possibility. Or if we wanted to be very optimistic (or risk seeking), we might choose the strategy with the best upside potential, MAXIMAX. Of course there is no guarantee that one's counterparts will make congenial choices.

We should recognize as well that there are intermediate conditions between risk and uncertainty. For example, instead of MAXIMIN a decision maker might take a good look at, say, the strategies in the best quartile on worst possible performance. The criteria above can be considered—viewed and expanded—in light of this remark.

Finally for now, and most fundamentally, from the perspective of pragmatic rationality we can understand and assess strategy selection criteria for their contributions to finding *robust strategies*. Something is robust if it performs well under varying conditions. Wagner's characterization is representative and applies to any system [326, page 1], "A biological system is robust if it continues to function in the face of perturbations," although often we will want to add "well" after "function."

It is important to contrast equilibrium strategies with robust strategies. An equilibrium of a game, or context of strategic interaction, is an assignment of strategies, one for each player, such that no single player would be better off with a different strategy assigned to it. An equilibrium strategy for a player is any strategy that belongs to some equilibrium. If there is more than one equilibrium, it is entirely possible for each player to choose one of its equilibrium strategies and the resulting assignment of strategies *not* to be an equilibrium. If this happens—if each player plays one of its equilibrium strategies but the result is not an equilibrium of the game—we can say there has been a mis-coordination.

Mis-coordination can happen whenever there is more than one equilibrium in a game. Even when the game has one equilibrium—as is the case for Prisoner's Dilemma and generally for social dilemmas—it is typically the case that when the game is indefinitely iterated—our topic in this chapter—the number of equilibria is manifold (see the Folk Theorem, above). Classical game theory has very little to offer about the consequent problems of mis-coordination. If we predict that in the face of very many equilibria play will be at some equilibrium, how is it that the players coordinate on that equilibrium? It's a problem.

Pragmatic rationality as forwarded here takes a very different tack on the strategy selection problems that players actually face. The suggestion is that players do and should seek robust strategies, strategies that do well for their players under varying conditions and assumptions, including various strategies that might be played by counter-parties. The strategy selection criteria are heuristics for finding robust strategies, heuristics that can be tested through tournaments, experiments, and other broadly empirical means.

3.10   For Exploration

1. Axelrod in reviewing the results of his Prisoner's Dilemma tournaments [15] touts four properties of successful strategies. They are: nice (not the first to defect), retaliatory (they punish defection by the counterpart), forgiving (they will resume cooperation if the counterpart does), and simple. I discussed simplicity as justified by signaling considerations, in §3.8.10. What about the other three of Axelrod's criteria? Should they be added to our list? Are they already covered by items on the list? Are they special virtues, applying only to Prisoner's Dilemma?

2. Consider the following statement and comment upon it. Do you agree with it? Why or why not?

   > We do not need all the criteria for Pragmatic Strategic Rationality. If we hold a well-designed tournament, that should be enough. Go with the winner.

3. In §3.8.4, discussing "do well against other high-quality strategies" it

was asserted that doing well in a tournament is sufficient to be judged prima facie high quality.

There is more than one way to run a tournament. Discuss some of the different ways and comment on their values for strategy selection.

4. Again, in §3.8.4, discussing "do well against other high-quality strategies" it was asserted that doing well in a tournament is sufficient to be judged prima facie high quality.

   Are there other conditions that might be used to establish prima facie high quality? If so, what are they?

5. Can you think of other criteria that should be judged important for strategy selection in a complex game? Discuss.

6. In §3.9 we find this passage:

   > The strategy selection problem would be considerably simplified, or even solved, if we were able to assign probabilities to the universe of possible strategies, $\mathcal{U}$. In nearly all cases, however, the difficulties here are insurmountable.

   Discuss this. Do you agree? Why or why not?

7. Discuss single-round versus multi-round tournaments. What, if anything, can we learn from one that we cannot learn from the other?

8. With the Prisoner's Dilemma tournaments in view, we abstracted and generalized to two problems of play. We abstracted tournaments to the Strategy Selection Framework and to the several criteria for strategy selection. We generalized Prisoner's Dilemma to social dilemmas and to the problem of cooperation. These may also be seen as problems of play, that is, as important general issues associated with strategic interaction. Cooperation in turn might be further generalized to problems of managing strategic interactions. What are some more specific forms of the problem of managing strategic interaction, other than cooperation?

9. Where do candidate strategies come from? One possibility is that they can be discovered through a machine learning process. Read and report on Axelrod's venture in this regard with respect to his Prisoner's Dilemma tournaments [16]. What did he find?

10. Using the NetLogo program 2x2-Tournaments-Det.nlogo explore tournaments with games other than Prisoner's Dilemma. Note that 2x2-Tournaments-Det.nlogo outputs its results to a file, which may conveniently be read into R [250] as follows:

```
> results <-
      read.table("2x2-Tournaments-Deterministic-output.txt",
```

```
          header=T,sep=",")
> summary(results)
```

11. Explore the effects of risk attitudes by transforming outcomes of tournament play with appropriate utility functions. One such function is

$$u(x) = \left(\frac{(x-w)}{(b-w)} \right)^r \tag{3.1}$$

where b is the best possible outcome, w the worst, and r the risk attitude. When $r = 1$ the attitude is risk neutral. When $0 < r < 1$ the attitude is risk averse, and when $r > 1$, risk seeking.

12. Writing in a biological context Wagner [326, page 1] identifies "other names" in use for or related to robustness: "buffering, canalization, developmental stability, efficiency, homeorhesis, tolerance, etc." The terms resilient and resilience are similarly appearing in the popular vocabulary—in blogs and Web sites—for something like robustness. Consider these and any other related terms you can find for the context of strategy selection in games. How do the ideas and concepts associated with these terms unpack for purposes of strategy selection? How do they differ in meaning from robustness and should these differences be included in what we should count as pragmatic rationality?

3.11   Concluding Notes

See Sally [269] for a very useful meta-analysis of Prisoner's Dilemma experiments, up to 1995. See `http://plato.stanford.edu/entries/prisoner-dilemma/` for a nice, accessible discussion of Prisoner's Dilemma. Also google "rational cooperation in Prisoner's Dilemma."

There is important and valuable work on the problem now known as *equilibrium selection* that addresses what I called above the mis-coordination problem. See [142, 217] for starters.

Wagner's excellent book *Robustness and Evolvability in Living Systems* [326] has much to say and is quite stimulating on the general subject of system robustness.

Chapter 4

Fanning out: 2×2 Games and Models

4.1 Fanning out

While Prisoner's Dilemma is an important game, there are many other interesting games. In fact that universe is indefinitely large. We cannot hope to explore all of it, so the question is where to begin. A natural starting place is symmetric 2×2 games (in strategic form). 2×2 Prisoner's Dilemma is both a member of that class and a mixed motive game; the latter class is very much in our focus. If we look beyond the class of symmetric 2×2 games we open up a vast universe of games, a consideration set too large to consider comprehensively. So the plan is to focus on the simplest cases and seek insights there. What we learn here will be useful in later forays into the much larger universe of mixed motive games.

Sticking then to 2×2 symmetric games, what are they and which among them are mixed motive? We need still a further simplification: let us assume that the payoffs for a given player are strictly ranked. The best is a 4, the worst is a 1. We can label the strategies as A and B, and we can take the perspective of just one of the players, since the games are symmetric. Taking Row's perspective, there are four possible outcomes: AB (Row plays A, Column plays B), AA, BA, and BB. Given strict rankings, there are then 4! = 24 different possible games. We can reduce this in half by assuming that BA > AB. (These 12 games are symmetrically equivalent to the 12 in which AB > BA.) We are now down to 12 games:

1. BA > AA > BB > AB (Prisoner's Dilemma, game #12 in [253])

2. AA > BA > BB > AB (Stag Hunt, game #61 in [253])

3. BA > AA > AB > BB (Chicken, game #66 in [253])

4. BA > BB > AA > AB (Deadlock, game #9 in [253])

5. BA > AB > AA > BB (game #68 in [253])

6. BA > AB > BB > AA (game #69 in [253])

7. BA > BB > AB > AA (Stable Chicken, game #7 in [253])

8. AA > BB > BA > AB (game #63 in [253])

9. BB > AA > BA > AB (game #60 in [253])

10. AA > BA > AB > BB (game #6 in [253])

11. BB > BA > AB > AA (game #3 in [253])

12. BB > BA > AA > AB (game #5 in [253])

Some of them, as we see, have been given names in the literature. They all, however, appear in Rapoport et al.'s enumeration of 2×2 games [253], where they have unique numbers. I shall often advert to those numbers, which I shall preface with a # sign, as in the parenthetical remarks in the above list.

We can pare this list a bit. Starting at the bottom, item 12 (game #5) goes into strategic form as in Figure 4.1. This is really quite an unproblematic game,

	A	B
A	2 2	3 1
B	1 3	4 [N,P] 4

FIGURE 4.1: Game #5 (item 12).

played once or played many times. Strategy B is not only strictly dominant, but it leads to the unique Pareto optimal outcome. With no real conflict here, we can ignore this game for the present, although we shall return to a similar game in Chapter 7. We can dismiss game item 11 (game #3) on the same grounds; see Figure 4.2.

	A	B
A	1 1	3 2
B	2 3	4 [N,P] 4

FIGURE 4.2: Game #3 (item 11).

Game item 10 (game #6), Figure 4.3, is nearly a Stag Hunt. Because mutual defection is the worst outcome for both players, we can also do without further analysis of this game. There is no plausible temptation for either player to play B.

	A	B
A	4 [N,P] 4	3 2
B	2 3	1 1

FIGURE 4.3: Game #6.

Game #60 (item 9) in Figure 4.4 is also similar to Stag Hunt, but lacks the temptation to defect. We will skip this one now, although note that it has two Nash equilibria in pure strategies, only one of which is Pareto optimal.

	A	B
A	3 [N] 3	2 1
B	1 2	4 [N,P] 4

FIGURE 4.4: Game #60.

Game #63 (item 8) likewise, Figure 4.5, is very much like game #60 (item 9, Figure 4.4).

	A	B
A	4 [N,P] 4	2 1
B	1 2	3 [N] 3

FIGURE 4.5: Game #63.

Notice that in each of the five games, which we just eliminated as (provisionally) unproblematic, there is a unique Pareto optimal outcome which is also a Nash equilibrium.

Now to game items 1–7 and to exploring them with computational models.

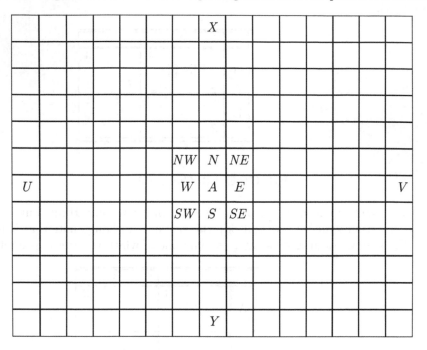

FIGURE 4.6: Gridscape schema.

4.2 A Simple Gridscape Model

In a gridscape model, agents are arrayed on a two-dimensional lattice of *cells*; think: squares on a checkerboard. Each agent, except possibly those on the boundaries has a neighbor to the north, to the south, to the east, and to the west. These four neighbors (plus the center cell) constitute what is called the *von Neumann neighborhood.*[1] Each player (except possibly those on a boundary) has four additional neighbors, to the northeast, southeast, southwest, and northwest. Together the eight neighbors (plus the center cell) constitute the *Moore neighborhood.*[2] Figure 4.6 illustrates. The von Neumann

---

[1]Of range 1, which suffices for our purposes. See http://mathworld.wolfram.com/vonNeumannNeighborhood.html.

[2]Of range 1, which suffices for our purposes. See http://mathworld.wolfram.com/MooreNeighborhood.html.

neighbors of A are N, S, E, and W. The Moore neighborhood of A is NW, N, NE, W, E, SW, S, and SE plus A itself. If the gridscape is not bounded east-west, but wraps around, then U and V are neighbors; and if bounded, not. Similarly, if the gridscape is not bounded north-south, but wraps around, then X and Y are neighbors; and if bounded, not. Unless otherwise indicated, all of our gridscape models will be unbounded in both directions, so the gridscape has the form of a torus (technical term: doughnut).

Symmetric-2x2.nlogo, available on the book's Web site, implements a simple gridscape model in which agents play a 2×2 game repeatedly with their neighbors. NetLogo, and in consequence Symmetric-2x2.nlogo, contains a *world* (NetLogo terminology) which contains a gridscape of *patches* (Net-Logo's term for gridscape cells). In each round of play, the agents (realized as patches) in Symmetric-2x2.nlogo play a one-shot 2×2 game with each of their eight Moore neighbors (not themselves), keeping track of the total number of points they obtain. The one-shot, symmetric, 2×2 game is specified on the user interface. At the end of each round of play, each agent compares its total point value for that round with those of its Moore neighbors. If any of its neighbors have a higher score than itself, the agent switches its strategy to that of the highest-scoring neighbor, with ties broken by chance. This completes one round of play.

Figure 4.8, page 74, shows Symmetric-2x2.nlogo initialized with 18% co-operators (black patches on the word/gridscape) and 82% defectors (white patches). There are 51,653 patches in all and exactly 9,408 cooperators (play-ers strategy 0) and 42,245 defectors (players of strategy 1). Figure 4.7 shows the Prisoner's Dilemma payoffs for the game being played. These payoffs are set in the sliders in the southwest area of Figure 4.8.

		0=C		1=D	
			3		3.2
0=C		[P]		[P]	
	3		0		
			0		1
1=D		[P]		[N]	
	3.2		1		

FIGURE 4.7: Prisoner's Dilemma game of Figure 4.8.

Figure 4.10, page 78, shows the system after the Go button has been clicked and five generations (rounds) of play have transpired. We see that the black patches, the cooperators, have been nearly wiped out. Cooperation appears headed for extinction. This is hardly surprising, since whenever a cooperator plays a defector the defector gets a larger reward. Who would not want to defect under the circumstances? But three clusters remain. They have grown from much smaller clusters, as may be seen by comparing with Figure 4.8.

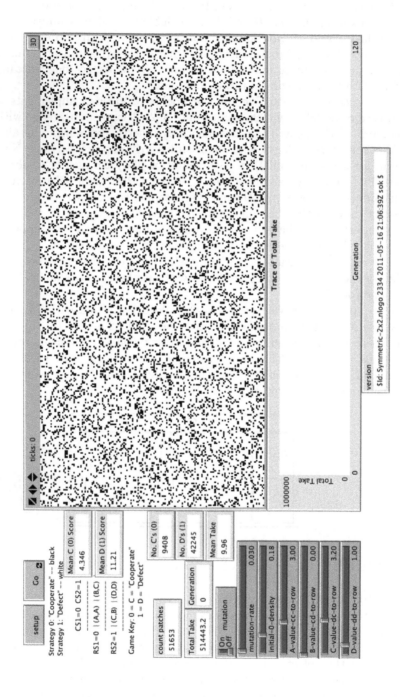

FIGURE 4.8: Symmetric-2x2.nlogo initialized with few cooperators and a PD game of low temptation.

If we resume play the system stabilizes by generation 189, which Figure 4.11, page 79, shows. Cooperators do manage to hang on. Not only do they hang on, they go on to preeminence in the world!

What we see in Figure 4.11 is that the cooperators have recovered and now are permanently ensconced as the majority party. At settlement of this run there are 44,764 cooperators and 6,889 defectors. I emphasize that the strategies here are completely simple: cooperate or defect. No learning is going on; there is no memory of prior play, as in TIT FOR TAT. Cooperators lose every encounter they have with defectors, yet they somehow manage to take over and keep most of the network. How is this possible? Is it inevitable?

The result—conquest by the cooperators—is certainly not inevitable. It depends on the right payoff structure and on the spontaneous appearance of a 3×3 (or larger) rectangle of cooperators. The latter will likely appear in a large network providing the initial probability of cooperation is high enough. If it doesn't, the defectors will eliminate the cooperators. Even assuming the appearance of 3×3 (or larger) rectangles of cooperators, whether cooperators can survive or prosper in the network depends on the game payoffs.

		S_1	S_2
S_1		A $\quad$ A	C $\quad$ B
S_2		B $\quad$ C	D $\quad$ D

FIGURE 4.9: Canonical game matrix for the symmetric 2×2 game in strategic form.

A simple counting argument establishes the basic properties. Note for notational purposes our canonical game matrix for symmetric 2×2 games in strategic form, Figure 4.9. In d dimensions, a hypercube of S_1s can invade an abutting field of S_2s if

$$A[2 \cdot 3^{d-1} - 1] + B3^{d-1} > \max\{D[2 \cdot 3^{d-1} - 1] + C3^{d-1}, D[3^d - 1]\} \quad (4.1)$$

Similarly, a hypercube of S_2s can invade an abutting field of S_1s if

$$D[2 \cdot 3^{d-1} - 1] + C3^{d-1} > \max\{A[2 \cdot 3^{d-1} - 1] + B3^{d-1}, A[3^d - 1]\} \quad (4.2)$$

Mapping these to Prisoner's Dilemma (now 0s for defectors, 1s for cooperators): $A \Rightarrow P, B \Rightarrow T, C \Rightarrow S$, and $D \Rightarrow R$. (Mnemonic: R is the reward for mutual cooperation, T is the temptation to defect, S is the sucker's payoff, and P is the penalty for mutual defection.) Then a cube of the cooperators will invade an abutting field of the defectors if

$$R[2 \cdot 3^{d-1} - 1] + S3^{d-1} > \max\{P[2 \cdot 3^{d-1} - 1] + T3^{d-1}, P[3^d - 1]\} \quad (4.3)$$

Setting $S = 0$ and noting that $P[2 \cdot 3^{d-1} - 1] + T3^{d-1} > P[3^d - 1]$, this simplifies to

$$R[2 \cdot 3^{d-1} - 1] > P[2 \cdot 3^{d-1} - 1] + T3^{d-1} \tag{4.4}$$

or

$$R > P + \frac{T3^{d-1}}{[2 \cdot 3^{d-1} - 1]} \tag{4.5}$$

In 2-d:

$$5R > 5P + 3T \tag{4.6}$$

In our example we have $R = 3$, $T = 3.2$ and $P = 1$. With $5R = 15 > 5P + 3T = 14.6$, we predict correctly that 2-d cubes (3×3 or larger rectangles) of cooperators will expand into a field of defectors. There is more to be said, but this captures the essential point: with the right payoffs for Prisoner's Dilemma (or any other game), cooperators, S_1 players, will expand at the expense of defectors, provided there are initial formations of appropriate hypercubes.

Further points arising:

1. Remarkably, starting with only 18% of the patches, the cooperators have moved to a stable situation in which they constitute more than 86% of the population.

2. Emergence is often non-obvious and may be in defiance of treasured principles. Figures 4.8–4.11 embody a kind of emergence or *self-organization* of a system. From a random scattering of cooperators and defectors a distinctive pattern emerges of areas of cooperation expanding at the expense of defectors, confining them to a very limited existence. That the cooperators—losers in every cooperator-defector encounter—should be doing the pushing must count as surprising, even though it can be predicted from mathematical analysis.

3. What we may call the *shadow of society* appears and affects the outcome. The policies played by one's neighbors may be, and usually are, influenced by policies played by players who are not one's neighbors.

   When agents play games iteratively it is well known that the *shadow of the future* (discount rate; Axelrod's nice expression [15]) may greatly affect play. Similarly, the gridscape society has greatly affected the play we have just examined and can do so in general. It is important to see that these are two distinct factors. The agents in Symmetric-2x2.nlogo have no memory and no capacity to respond to experience. They do or die and that's all; the future means nothing to them.

   The system behavior we have seen is due to the structure of the games played, the strategies played by the agents, *and* the imposed gridscape structure with its attendant rules. The latter brings to the table inherently social factors. A *society*, at least as I shall use the term technically, is also a context of strategic interaction: agents (players) interact with one another and receive returns as a function of their decisions as well as the decisions (the strategies played) by their counter-players. Societies, however, have an additional, special feature: the games played by an agent in a society are affected—via their payoff structures or via the strategies employed by the agent's counter-players—by *games the agent does not play*, i.e., by games played by other agents in the society. Games require at least two players, societies at least three.[3] The pattern of who plays whom—the social network—constitutes (or at least is a major feature of) the *social structure* of a society.

4. It is *not* generally the case for Prisoner's Dilemma on the gridscape that the cooperators will conquer the world. Special conditions are necessary. These conditions, which apply quite generally, are described in more detail by the *invasion inequalities* in §B.5.

4.3   A Richer Gridscape Model

Like Symmetric-2x2.nlogo, M1-Symmetric-2x2-wID.nlogo, available on the book's Web site, implements a gridscape model for symmetric 2×2 games. Unlike Symmetric-2x2.nlogo, in which there are only two strategies possible and in each epoch only one round of play is conducted per pair of counter-players, in M1-Symmetric-2x2-wID.nlogo the players (again implemented as patches in a NetLogo world) have available to them the eight strategies with one period of memory, and the players iterate with each other 10 times in a single epoch (that is they play 10 rounds with each neighbor and then update their strategies).

Figure 4.13 shows M1-Symmetric-2x2-wID.nlogo after being initialized (by clicking the setup button) with game #61, shown in Figure 4.12. In the northeast quadrant of Figure 4.13 we see eight sliders.The eight sliders correspond to the eight memory-1 strategies and are coded as follows. Slider zero-weight has a code of 000 (binary 0), meaning that it cooperates on the first round of play and always thereafter. Patches with this strategy are colored black. Slider one-weight has a code of 001 (binary 1), meaning that it cooperates on the first round of play, and after that if the counter-player played 0 in the previous round (cooperated) then it plays 0 in this round and if the counter-player played 1 in the previous round (defected) then it plays 1 this time. So the one-weight strategy is TIT FOR TAT. One more example. The five-weight strategy has a code of 101 (binary 5). It defects on the first round of play, then

---

[3]I am reminded of a recorded interview with Bertrand Russell I once heard. The interviewer began by discussing Russell's family background (privileged). "It would not be accurate to describe the Russells as well connected. Instead, the Russells were the people to whom the well connected were connected."

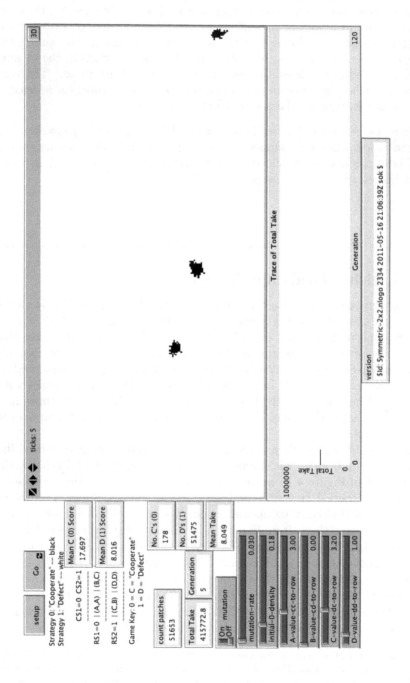

FIGURE 4.10: Symmetric-2x2.nlogo initialized with few cooperators and a PD game after five rounds of play.

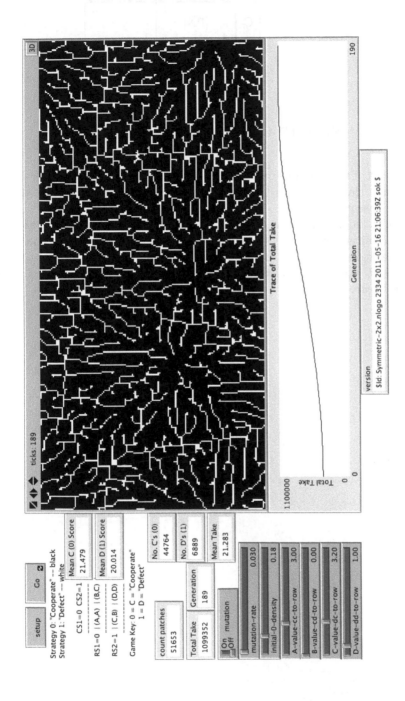

FIGURE 4.11: Symmetric-2x2.nlogo initialized with few cooperators and a PD game after 189 rounds of play.

	C	D
C	[PN] 4 4	3 1
D	1 3	[N] 2 2

FIGURE 4.12: Game #61 (Stag Hunt).

like TIT FOR TAT it mimics play by the counter-party. If the counter-party defected last time, it defects this time; if the counter-party cooperated last time, it cooperates this time.

All eight sliders are set at a value of 10.0. The proportion of a strategy's presence at initialization is its set weight divided by the sum of the set weights of all eight strategies. Thus, if as we have here all strategies have the same initial weight, then at initialization their frequencies are equal. That is what we see in Figure 4.13.

FIGURE 4.13: M1-Symmetric-2x2-wID.logo initialized for game #61 (Stag Hunt) with a uniform distribution of all eight memory-1 strategies.

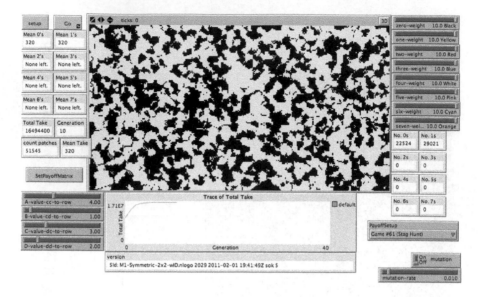

FIGURE 4.14: M1-Symmetric-2x2-wID.logo initialized for game #61 (Stag Hunt) with a uniform distribution of all eight memory-1 strategies, stable after 10 generations.

Figure 4.14 shows the state of the system after 10 generations. The zero-weight and one-weight strategies have conquered the (NetLogo) world. Stag hunters prevail! The shadow of society remains very much at work.

4.4   An Evolutionary Model

Biological evolution provides inspiration for a kind of tournament to compare strategies. For a symmetric game in strategic form, we could create a pool with copies of various strategies. Initially, if there are n strategies under consideration, we might form a population consisting of say 100 copies of each strategy, totaling *PopulationSize* $= 100n$ strategies in all. We might then draw (with replacement) two strategies at random, S_i and S_j, play them against each other, and add the payoffs they receive to their respective score accumulators, $A(S_i)$ and $A(S_j)$. If we continue this process for a large number of samples, say *SampleSize* $= 5,000$, we could hope to get a reasonably good estimate of how well the strategies would perform with infinite sampling. In any event, after *SampleSize* rounds of play between randomly drawn strategies, we can estimate the "fitness" of a strategy, $W(S_i)$, in the population as the ratio of its accumulated score to the total for all strategies:

$$W(S_i) = \frac{A(S_i)}{\sum_i A(S_i)} \tag{4.7}$$

We then calculate W for each strategy and form the population for the next generation by creating (in expectation) $W(S_i) \times PopulationSize$ copies of strategy S_i, for all i. We may then run the process for many generations and observe what happens.

Note that if $W(S_i)$ for a strategy S_i is larger than the strategy's frequency in the population, the frequency of the strategy will (in expectation) increase for the next generation; and similarly if $W(S_i)$ is smaller, the frequency of S_i will decrease. Evolution rewards performance. This process may be called a *discrete replicator dynamics* after the continuous version, the replicator dynamics [153, 238, 307, 330]. Axelrod used a version of the discrete replicator dynamics in [15], calling it an ecological simulation.

Figure 4.16 presents pseudocode for this discrete replicator dynamics process and ReplicatorDynamicsWithBalking.nlogo implements it. Figure 4.17, page 85, shows a run of Game #66 (Chicken), as shown in Figure 4.15.

		A		B	
			3	[N,P]	4
A	3			2	
			2		1
B	[N,P]				
	4			1	

FIGURE 4.15: Game #66 (Chicken).

Chicken is a difficult game under the circumstances. In their interests the players should coordinate on (A,A) or take turns playing B. In the run under display, the system was initialized with 90% weight-0 strategies (000), which always play A, and 10% weight-7 strategies (111), which always play B. After 24 generations the system has reached a stable mixture of 50% each. If all agents play (000) they average a take per agent of 3 per round of play. At 50:50 (000:111) the average drops to about 2.5. Under the circumstances the evolutionary dynamic was reduced social welfare.

1. Preparatory steps:

    (a) Choose the game, G, with its payoffs, and the available strategies, $\mathcal{S}$.

    (b) Set *PopulationSize*, the (fixed) population size to be used in each generation.

    (c) Set $W(S_i)$, the initial frequency of each strategy $S_i \in \mathcal{S}$.

    (d) Set *SampleSize*, the number of pairwise games to be played in each generation.

    (e) Create the initial population of *PopulationSize* strategies drawn from $W(S_i)$.

    (f) Create and set to 0 reward accumulators, $A(S_i)$, for each strategy $S_i \in \mathcal{S}$.

2. Process generations until done:

    (a) Do *SampleSize* times:

        i. Draw uniformly at random and with replacement two members of the population, noting their types (which strategies they are).

        ii. Play the two members against each other in the game G; allocate the outcome rewards to their respective accumulators, $A(S_i)$.

    (b) Determine $W(S_i)$, the frequency distribution of strategies for the next generation, as

    $$W(S_i) = \frac{A(S_i)}{\sum_i A(S_i)} \qquad (4.8)$$

    (c) Create the next generation population of *PopulationSize* strategies according to $W(S_i)$.

    (d) Set the reward accumulators, $A(S_i)$, for each strategy $S_i \in \mathcal{S}$ to 0.

FIGURE 4.16: Pseudocode for a discrete replicator dynamics with a finite population.

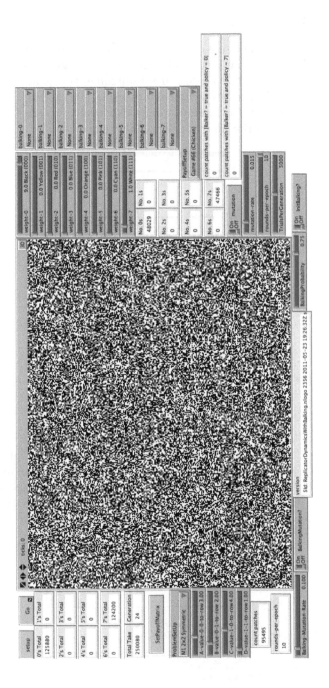

FIGURE 4.17: ReplicatorDynamicsWithBalking.nlogo, playing game #66, initialized 90% As and 10% Bs, after 24 generations.

4.5   Game #9 (Deadlock)

I do not know where the name comes from or why it was given. Superficially similar to Chicken (game #66), this game is easy. With a single Nash equilibrium, coinciding with the single Pareto outcome, this game should be easy. And it is. Initialized with 90% weight-0 strategies (000) and 10% weight-7 strategies (111), as in Figure 4.17, the system quickly evolves to 100% weight-7 strategies (111). Now evolution maximizes social welfare.

		A		B	
			2		4
A					
	2		1		
			1		3
B				[N,P]	
	4		3		

FIGURE 4.18: Game #9 (Deadlock).

4.6   Game #68

Game #68 is Chicken-like in its apparent difficulty. Playing it with M1-Symmetric-2x2-wID.nlogo produces poor per-patch results. See Figure 4.20. After 51 generations the system is reasonably stable. The mean take of 168.72 amounts to just over 1 point per round of play per agent. Each agent accumulates points in an epoch from playing its eight neighbors twice each and doing this 10 times. This amounts to 160 one-shot games played per epoch.

		A		B	
			2		4
A				[N,P]	
	2		3		
			3		1
B		[N,P]			
	4		1		

FIGURE 4.19: Game #68.

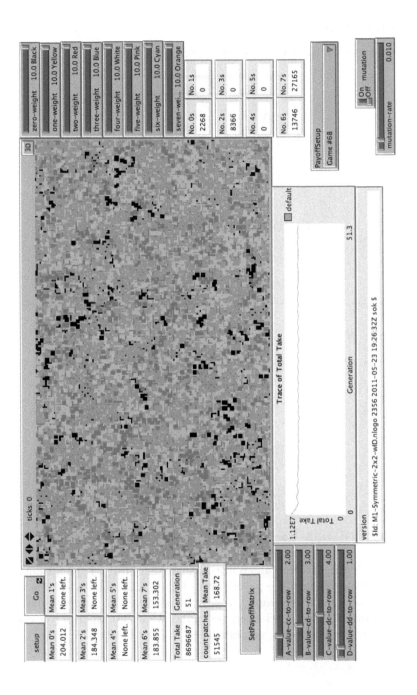

FIGURE 4.20: M1-Symmetric-2x2-wID.nlogo, playing game #68.

4.7 Game #69

There is not much to distinguish here from game #68.

	A	B
A	1 1	4 [N,P] 3
B	3 [N,P] 4	2 2

FIGURE 4.21: Game #69.

4.8 Game #7

This one is also known as Stable Chicken. Figure 4.23 shows the user

	A	B
A	1 1	4 [P] 2
B	2 [P] 4	3 [N,P] 3

FIGURE 4.22: Game item 7 (game #7): Stable Chicken.

interface from ReplicatorDynamicsWithBalking.nlogo after 24 generations of play of Stable Chicken with the payoffs from Figure 4.22. The system is well on its way to dominance by strategy 7 = 111, meaning the strategy plays B on the first round of play and B after, no matter what the other player does. Recall that these results are for games with 10 rounds of play.

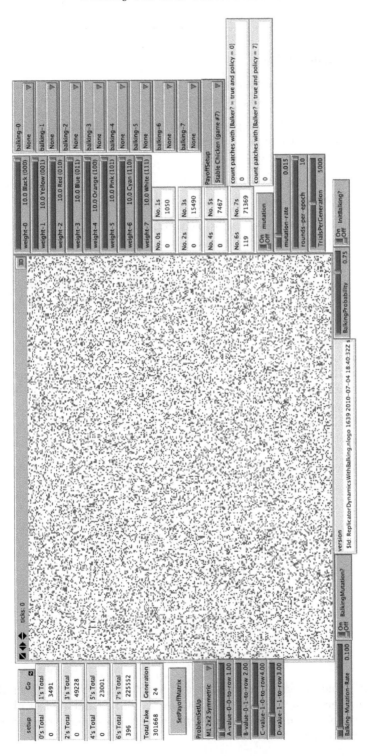

FIGURE 4.23: Game #7 (Stable Chicken) after 24 generations playing all M1 strategies under replicator dynamics.

4.9   For Exploration

We have now introduced and seen in action four NetLogo programs that are useful for exploring strategies in games:

- 2x2-Tournaments-Det.nlogo (Chapter 3), for exploring iterated asymmetric (or symmetric) 2×2 games and conducting tournaments among user-programmed strategies.

- ReplicatorDynamicsWithBalking.nlogo, for exploring evolutionary competitions among simple strategies in iterated, symmetric 2×2 games.

- Symmetric-2x2.nlogo, a two-dimensional spatial model for exploring simple interactions among agents arrayed on a gridscape and playing symmetric 2×2 games with their neighbors.

- M1-Symmetric-2x2-wID.nlogo, a generalization of Symmetric-2x2.nlogo that affords play with strategies having a memory of one round of previous play.

In addition, we have seen how the invasion inequalities for symmetric spatial games can be used to provide insight into social behavior.

All of the 12 symmetric 2×2 games we have discussed in this chapter can profitably be explored further with these tools. Pay particular attention to changes in payoffs. Remember that the games are specified only up to the rankings of the outcomes. As we saw in our discussion of Prisoner's Dilemma, and is apparent from the invasion inequalities, the payoffs can matter greatly.

1. Consider the asymmetric 2×2 game known as Bully [247] (game #10 in [253]).

	A	B
A	[P]　　2 3	[N,P]　　4 2
B	1 4	3 1

FIGURE 4.24: Bully.

If the game is played iteratively and indefinitely, what do you think will happen? Why? How would you use 2x2-Tournaments-Det.nlogo to investigate the game? How would you modify the other three programs to accommodate the asymmetry?

2. Consider Battle of the Sexes. If the game is played iteratively and in-definitely, what do you think will happen? Why? How would you use 2x2-Tournaments-Det.nlogo to investigate the game? How would you modify the other three programs to accommodate the asymmetry?

3. Reconsider games #9, Figure 4.4, and #63, Figure 4.5. If the game is played iteratively and indefinitely, what do you think will happen? Examine them with all four NetLogo programs.

4. Does Stable Chicken merit its name? Why?

4.10   Concluding Notes

See [253] for enumeration and exploration of all 78 2×2 games with strictly ranked payoffs, as well as reports on behavior by subjects in the laboratory in playing them. How do these results compare with those from our compu-tational models?

Chapter 5

Stag Hunt

5.1 Introduction

Can trust arise spontaneously—by an invisible hand as it were—among strategically interacting individuals? If so, under what conditions will it arise? When will it be stable and when will it not be stable? What interventions might be effective in promoting or undermining stability? If trust is established, under what conditions will it be destroyed? What are the roles of social structure, game structure, and cognition in establishing or disestablishing trust? These questions belong to a much longer list of important and challenging issues that the problem of trust presents. Answering them fully and adequately constitutes a research program to challenge a community over a period of many years.

I aim in this chapter to contribute in two ways to that program. In the end, although progress will be made, more work will have been added. First, I shall present findings, focusing on the Stag Hunt game, that bear more or less directly on at least some of these questions. I shall focus on the Stag Hunt game for several reasons. The game does capture well and succinctly certain aspects of the problem, the dilemma, of trust. It has the happy virtue of not being the (overworked but still worthwhile) game of Prisoner's Dilemma. Also, it has fruitfully received new attention of late (e.g., [297]), so that what I will add here will, I hope, enrich a topic that is very much in play.

The second way in which I aim to contribute to the trust research program is more indirect. Trust is a problem or a puzzle, even paradox, in part because there seems to be more of it naturally occurring than can be explained by received theory (classical game theory). I shall say only a little by way of documenting this claim because space is limited and I take it that the claim is widely accepted. (Chapter 3, "Mutual Aid," of [294] is a discussion of how and why nature is *not* "red in tooth and claw." Also behavioral game theory, reviewed in [54], amply documents the imperfect fit between theory and observation in this domain. Those with a taste for blunt talk might consult [124].) Instead, I hope to say something about how the puzzle of trust might be investigated. I will submit that the puzzle of trust arises, at least in part, because of a presupposed account of agent rationality. This account goes by different names, among them *expected utility theory* and *Rational Choice The-*

ory. I want to propose, in outline form, a very different approach to conceiving of rationality. By way of articulating this general approach, which I shall call a *theory of exploring rationality*, I shall present a model of agent behavior which, in the context of a Stag Hunt game (as well as other games), explains and predicts the presence of trust.

5.2   Stag Hunt and a Framing of the Program

The Stag Hunt game (also known as the Assurance game [118]) gets its name from a passage in Jean Jacques Rousseau's *A Discourse on the Origin of Inequality*, originally published in 1755.

> Was a deer to be taken? Every one saw that to succeed he must faithfully stand to his post; but suppose a hare to have slipped by within reach of any one of them, it is not to be doubted but he pursued it without scruple, and when he had seized his prey never reproached himself with having made his companions miss theirs. [266, Second Part]

Here is a representative summary of the Stag Hunt game.

> The French philosopher, Jean Jacques Rousseau, presented the following situation. Two hunters can either jointly hunt a stag (an adult deer and rather large meal) or individually hunt a rabbit (tasty, but substantially less filling). Hunting stags is quite challenging and requires mutual cooperation. If either hunts a stag alone, the chance of success is minimal. Hunting stags is most beneficial for society but requires a lot of trust among its members. [119]

This account may be abstracted to a game in strategic form. Figure 5.1 on the top presents the Stag Hunt game with payoffs that are representative in the literature.[1] Let us call this our *reference game*. On the bottom of figure 5.1 we find the Stag Hunt game presented in a generic form. Authors differ in minor ways. Often, but not always, the game is assumed to be symmetric, in which case $R = R'$, $T = T'$, $P = P'$, and $S = S'$. I will assume symmetry. It is essential that $R > T > P \geq S$.[2]

Thus formalized, the Stag Hunt game offers its players a difficult dilemma, in spite of the fact that their interests coincide. Each does best if both hunt

---

[1]For example, although it is not called the Stag Hunt, the game with these payoffs is discussed at length in [253], where it is simply referred to as game #61.

[2]Usually, and here, $P > S$. Some authors allow $T \geq P$ with $P > S$. None of this matters a great deal for the matters to hand.

	Hunt stag (S)	Chase hare (H)
Hunt stag (S)	4 4	3 1
Chase hare (H)	1 3	2 2

	Hunt stag (S)	Chase hare (H)
Hunt stag (S)	R' R	T' S
Chase hare (H)	S' T	P' P

FIGURE 5.1: Stag Hunt (aka: Assurance game).

stag (S,S). Assuming, however, that the game is played once and that the players lack any means of coming to or enforcing a bargain,[3] each player will find it tempting to "play it safe" and hunt hare. If both do so, (H,H), the players get 2 each in our reference game, instead of 4 each by playing (S,S). Both of these outcomes—(S,S) and (H,H)—are Nash equilibria. Only (S,S), however, is Pareto optimal; in fact it is Hicks optimal.[4] There is a third Nash equilibrium for Stag Hunt: each player hunts stag with probability

$$\frac{P-S}{((R+P)-(T+S))}. \tag{5.1}$$

For our reference game, this amounts to a probability of $\frac{1}{2}$ for hunting stag (and $\frac{1}{2}$ for hunting hare). At the mixed equilibrium each player can expect a return of $2\frac{1}{2}$. Notice that if, for example, the row player hunts hare with probability 1, and the column player plays the mixed strategy, then the row player's expected return is $2\frac{1}{2}$, but the column player's expected return is $1\frac{1}{2}$. Uniquely, the safe thing to do is to hunt hare, since it guarantees at least 2. Hunting hare is thus said to be *risk dominant* and according to many game theorists (H,H) would be the predicted equilibrium outcome.[5]

---

[3]In the jargon of game theory, this is a *noncooperative* game.

[4]The concept is named after John Hicks. It is a measure of efficiency. An outcome of a game is Hicks optimal if there is no other outcome that results in greater total payoffs for the players. Thus, a Hicks optimal outcome is always the point at which total payoffs across all players is maximized. A Hicks optimal outcome is always Pareto optimal. See §A.2.10, page 432.

[5]I'm using the term risk dominant in a general way, since adverting to its precise meaning would divert us. See [142] for the precise meaning.

We can use the Stag Hunt game as a model for investigation of trust. A player hunting stag trusts the counter-player to do likewise. Conversely, a player hunting hare lacks trust in the counter-player. Deciding not to risk the worst outcome (S) is to decide not to trust the other player. Conversely, if trust exists then risk can be taken. There is, of course, very much more to the subject of trust than can be captured in the Stag Hunt game. Still, something is captured. Let us see what we can learn about it.

Before going further it is worth asking whether Rousseau has anything else to say on the matter to hand. Typically in the game theory literature nothing else in this *Discourse* or even in other of Rousseau's writings is quoted. As is well known, Rousseau wrote in praise of the state of nature, holding that people were free of war and other ills of society, and on the whole were happier. That needn't concern us here. What is worth noting is that Rousseau proceeds by conjecturing (his word, above) a series of steps through which mankind moved from a state of nature to the present state of society. Rousseau is vague on what drives the process. The view he seems to hold is that once the equilibrium state of nature was broken, one thing led to another until the present. Problems arose and were solved, one after the other, carrying humanity to its modern condition. He laments the outcome, but sees the process as more or less inevitable. With this context in mind, the passage immediately before the oft-quoted origin of the Stag Hunt game puts a new light on Rousseau's meaning. He is describing a stage in the passage from the state of nature to civil society.

> Such was the manner in which men might have insensibly acquired some gross idea of their mutual engagements and the advantage of fulfilling them, but this only as far as their present and sensible interest required; for as to foresight they were utter strangers to it, and far from troubling their heads about a distant futurity, they scarce thought of the day following. Was a deer to be taken? ... [266, Second Part]

Rousseau is describing behavior of people not far removed from the state of nature. Language, for example, comes much later in his account. These people end up hunting hare because "as to foresight they were utter strangers to it, and far from troubling their heads about a distant futurity, they scarce thought of the day following." If we *define* the game to be one-shot, then there is no future to worry about. Rousseau says: if there is no future or if the players cannot recognize a future, then stag will roam unmolested. Rousseau is also presuming that in the later development of civil society the future matters, agents can recognize this, and much coordination and hunting of stag occurs. Rousseau is *not* agreeing with contemporary game theorists in positing hunting of hare as the most rational thing to do in the Stag Hunt game.

More generally, trust happens. Our question is to understand how and why. We assume that there is a future and we model this (initially) by iterating play of a basic game, called the *stage game*, here the Stag Hunt. Given iterated play of a stage game, there are two kinds of conditions of play that call out for investigation. The first is condition is the *social aspect* of play. We investigate a simple model of this in §5.3. The second condition might be called the *cognitive aspect* of play. How do learning and memory affect game results? We discuss this second aspect in §5.4.

5.3   The Gridscape: A Simple Society

We shall work with a very simple model of social aspects of strategic interaction, called the *gridscape*. The gridscape is a regular lattice—think of a checkerboard—which we will assume is two dimensional and wraps around on itself (is technically speaking a torus). Agents or players occupy cells on the gridscape and each has eight neighbors. Figure 5.2(a) illustrates this. Cell (3,2) has neighbors (2,1), (2,2), (2,3), (3,1), (3,3), (4,1), (4,2), (4,3).[6] Every cell has eight neighbors. Thus, the neighbors of (1,1) are (6,6), (6,1), (6,2), (1,6), (1,2), (2,6), (2,1), and (2,2). With the gridscape as a basis it is now possible to undertake a variety of experiments. We'll confine ourselves to a simple one.

Protocol 1 (Basic Gridscape Stag Hunt Protocol) *Each cell in the gridscape is initialized by placing on it either a hare hunter, H, or a stag hunter, S. H and S are the consideration set of policies of play for the experiment. Players using the H policy always hunt hare; and players using the S policy always hunt stag. Initialization is random in the sense that every cell has the same probability of being initialized with an H (or an S). After initialization, play proceeds by discrete generations. Initialization concludes generation 0. At the start of each subsequent generation each player (at a cell) plays each of its eight neighbors using its policy in use from the consideration set. The player records the total return it gets from playing the eight neighbors. After all players have played their neighbors, each player updates its policy in use. A player changes its policy in use if and only if one of its neighbors is using the counter-policy and has achieved a strictly higher total return in the current generation than has the player or any of its neighbors with the same policy in use achieved. (This is called the* IMITATE THE BEST NEIGHBOR *policy.) Policy updating completes the generation. Play continues until a maximum number of generations is completed.*

The table in Figure 5.2(b) shows representative data when the gridscape is seeded randomly (50:50) with stag hunters (Ss) and hare hunters (Hs) and

---

[6]This is called the *Moore neighborhood*. The *von Neumann neighborhood*, consisting of the four neighbors directly above, below, to the left, and to the right, is also widely studied. The results we report here are not sensitive to which of the two neighborhood definitions is in force.

	1	2	3	4	5	6
1						
2						
3						
4						
5						
6						

(a) Labeled 6×6 gridscape.

Generation	Ss	Hs	Total Points
0	7167	7233	344960.0
1	8715	5685	451224.0
2	12288	2112	599424.0
3	13891	509	668312.0
4	14306	94	686720.0
5	14392	8	690712.0
6	14400	0	691200.0

(b) Representative run on a 120×120 gridscape. $R = 3$, $T = 2$, $P = 1$, $S = 0$.

	1	2	3	4	5	6
1	S	S	H	S	S	S
2	S	S	H	S	S	S
3	S	S	H	S	S	S
4	S	S	H	S	S	S
5	S	S	H	S	S	S
6	S	S	H	S	S	S

(c) Stable 6×6 gridscape, for $R = 3$, $T = 2.5$, $P = 1$, $S = 0$.

FIGURE 5.2: Labeled 6×6 gridscape.

	$1 - \varepsilon$: TFT	ε: ALLD	Total
$1 - \varepsilon$: TFT	$(1 - \varepsilon)(l_e R)$ $(1 - \varepsilon)40$	$\varepsilon((l_e - 1)P + S)$ 9ε	$40 - 31\varepsilon$
ε: ALLD	$(1 - \varepsilon)(T + (l_e - 1)P)$ $12(1 - \varepsilon)$	$\varepsilon(l_e P)$ 10ε	$12 - 2\varepsilon$

TABLE 5.1: State 1: Payoffs to Row in Stag Hunt example when the system is in state 1 and $l_e = 10$.

protocol 1 is executed. Here, after five generations the Hs go extinct. The stag hunters conquer the board. This is the usual case, but it is not inevitable. To see why, consider the examples in Figure 5.3, which shows a 3×3 block of stag hungers (Ss). The remaining cells are blank and for the purposes of the discussion may be filled in as needed.

The first thing to notice is that in Figure 5.3(a) at (3,3) we have an S that is completely surrounded by Ss. This cell obtains a total reward of $8 \times R = 8 \times 4 = 32$ in our reference example. It is impossible to do equally well or better, given the setup. In consequence, given the protocol, once a 3×3 block of Ss is created none of its members will ever change to H. This is true for all versions of the Stag Hunt game (under protocol 1). We say that a 3×3 block of Ss cannot be invaded. More generally, it is easy to see that no rectangular block of Ss larger than 3×3 can be invaded either. (Note further that blocks of hare hunters are not so advantaged. An internal hare hunter gets $8T < 8R$ in all Stag Hunt games.)

Can a block of stag hunters grow? Assume that in Figure 5.3(a) the blank cells are all hare hunters. In general, the stag hunter at (2,3) will get a return of $5R + 3S$ which is $5 \times 4 + 3 \times 1 = 23$ in our reference game. The hare hunter at (1,3) will get $5P + 3T$ in general and $5 \times 2 + 3 \times 3 = 19$ in the reference game. And in general, so long as $5R + 3S > 5P + 3T$ a hare hunter in this position will convert to stag hunting. Note that not all Stag Hunt games will support this conversion. For example, R = 101, T = 100, P = 99, and S = 0 will not. Figures 5.3(b) and (c) show the next two generations and the pattern is clear: the stag hunters will drive the hare hunters to extinction.

Is conquest by stag hunters inevitable if a 3×3 block is created and the game rewards are sufficient for it to grow in a field consisting entirely of hare hunters? Equivalently, if $5(R - P) > 3(T - S)$ and a 3×3 block (or larger) of stag hunters forms, is it inevitable that hare hunters are driven to extinction? No, it is not. For example, the configuration in figure 5.2(c) is stable for the given payoff values.

There are many nice questions to ask and many interesting variations on the basic gridscape model for the Stag Hunt game. For present purposes, however, the following points are most on topic.

1. The gridscape Stag Hunt results described above are robust. What happens—whether Hs come to dominate or not, whether a stable mix-

	1	2	3	4	5	6	7	8
1								
2		S	S	S				
3		S	S	S				
4		S	S	S				
5								
6								
7								

(a) Generation x.

	1	2	3	4	5	6	7	8
1			S					
2		S	S	S				
3	S	S	S	S	S			
4		S	S	S				
5			S					
6								
7								

(b) Generation x+1.

	1	2	3	4	5	6	7	8
1		S	S	S				
2	S	S	S	S	S			
3	S	S	S	S	S			
4	S	S	S	S	S			
5		S	S	S				
6								
7								

(b) Generation x+2.

FIGURE 5.3: Growth of a block of stag hunters when R = 4, T = 3, P = 2, and S = 1.

ture results and so on—depends on the game payoffs (What is it worth if both players hunt stag? etc.) and the initial configuration. For a broad range of cases, however, hunting stag will eventually dominate the society. Trust—in the form of hunting stag predominately—can arise spontaneously among strategically interacting individuals. In fact, this is far from an implausible outcome.

2. Trust in Stag Hunt on the gridscape is also robust in a second sense: once it is established in large part, it is not easily dislodged. Mutations occurring at a small rate in a field of stag hunters will create mostly isolated hare hunters who will convert to S in the next generation. If stag hunters do well without noise they will do reasonably well with it.

3. The gridscape model under protocol 1 evidences a clear social effect. What we may call the *shadow of society* appears and affects the outcome. The policies played by one's neighbors may be, and usually are, influenced by policies played by players who are not one's neighbors. Recalling Figure 5.3(a), what happens to a hare hunter at (1,3) depends very much on the fact that the neighboring stag hunters are themselves adjacent to other stag hunters. Thus, while the hare hunter at (1,3) beats the stag hunter at (2,3), in the sense that it gets more points in the one-on-one play, the stag hunter at (2,3) in aggregate does better and it is the hare hunter who is converted.

4. The Prisoner's Dilemma game arguably presents a trust dilemma in more extreme form than does the Stag Hunt. Can cooperators largely take over the gridscape? Yes, under certain, more restricted conditions. In Prisoner's Dilemma, we require $T > R > P > S$ and $2R > T + S$. Further, we relabel the policies. Hunt Stag becomes Cooperate and Hunt Hare becomes Defect. On the gridscape, the 3×3 (and larger) block is key in the analysis. If, for example, we set $T = R + 1$, $P = 1$ and $S = 0$ in Prisoner's Dilemma, then so long as $R > 4$, a 3×3 (and larger) block of cooperators will be able to expand in a field of defectors. Defectors may not be eliminated, but they may become very much minority constituents of the gridscape. Note further that if Prisoner's Dilemma games are repeated (either infinitely with discounting or finitely), and the TIT FOR TAT policy replaces ALWAYS COOPERATE, then the payoff structure will, under broad conditions, become a Stag Hunt (cf. [54, chapter 7], [297, chapter 1]).

5. The agents on the gridscape have an update policy, IMITATE THE BEST NEIGHBOR, which they use to choose policies for play from their consideration sets. Under this update policy, from protocol 1, agents change their policies of play if, after a round of play, one of their neighbors has used the alternative policy and gotten more points than either the player or one of its neighbors playing with the player's policy of play. This is a

reasonable update policy, but there are reasonable alternatives. Is there a sense in which it is optimal? Is some other update policy optimal?

These are all interesting questions, well worth investigation. A larger issue raised is this. Any reasonable update policy, including ours, may be interpreted as taking a stand on the uncertainty faced by the agent.[7] Agents in games may be interpreted as seeking maximum return. That indeed is a presumption underlying the strategic framework. It is not a presumption that the agents are conscious or have intentions.

Regarding the stand on uncertainty, agents following our update policy are engaging in risky behavior whenever they opt for hunting stag. Yet when all agents so behave collective stag hunting robustly follows. Note the "Total Points" column in Figure 5.2(b). In the first generation the agents collectively garnered 344960 points from the gridscape. If the agents had not updated their policies of play this is where it would stay. As we see, with the update policy used (IMITATE THE BEST NEIGHBOR) the agents collectively more than doubled their take from the gridscape. IMITATE THE BEST NEIGHBOR has this to commend itself: it does well when playing against itself. In Prisoner's Dilemma the same can be said for TIT FOR TAT and other robust strategies, as we saw in Chapter 3. Notice as well that NEVER UPDATE, ALWAYS HUNT STAG does well against itself in Stag Hunt and ALWAYS COOPERATE does well against itself in Prisoner's Dilemma. Further, IMITATE THE BEST NEIGHBOR does well against NEVER UPDATE, ALWAYS HUNT STAG, as TIT FOR TAT does well against ALWAYS COOPERATE. Before pursuing these comments further, indeed as a means of doing so, let us turn to a more sophisticated model of learning by agents in games.

	ε: TFT	$1 - \varepsilon$: ALLD	Total
$1 - \varepsilon$: TFT	$\varepsilon(l_e R)$ $\varepsilon 40$	$(1 - \varepsilon)((l_e - 1)P + S)$ $9(1 - \varepsilon)$	$9 + 31\varepsilon$
ε: ALLD	$\varepsilon(T + (l_e - 1)P)$ 12ε	$(1 - \varepsilon)(l_e P)$ $10(1 - \varepsilon)$	$10 + 2\varepsilon$

TABLE 5.2: State 2: Payoffs to Row in Stag Hunt example when the system is in state 2 and $l_e = 10$.

---

[7]I am using *uncertainty* here in its technical sense, which contrasts with risk [203]. In a decision under risk we have an objectively supported probability distribution (or density) on the outcomes. Not so in a decision under uncertainty.

	$1 - \varepsilon$: TFT	ε: ALLD	Total
ε: TFT	$(1 - \varepsilon)(l_e R)$ $(1 - \varepsilon)40$	$\varepsilon((l_e - 1)P + S)$ 9ε	$40 - 31\varepsilon$
$1 - \varepsilon$: ALLD	$(1 - \varepsilon)(T + (l_e - 1)P)$ $12(1 - \varepsilon)$	$\varepsilon(l_e P)$ 10ε	$12 - 2\varepsilon$

TABLE 5.3: State 3: Payoffs to Row in Stag Hunt example when the system is in state 3 and $l_e = 10$.

	ε: TFT	$1 - \varepsilon$: ALLD	Total
ε: TFT	$\varepsilon(l_e R)$ $\varepsilon 40$	$(1 - \varepsilon)((l_e - 1)P + S)$ $9(1 - \varepsilon)$	$9 + 31\varepsilon$
$1 - \varepsilon$: ALLD	$\varepsilon(T + (l_e - 1)P)$ 12ε	$(1 - \varepsilon)(l_e P)$ $10(1 - \varepsilon)$	$10 + 2\varepsilon$

TABLE 5.4: State 4: Payoffs to Row in Stag Hunt example when the system is in state 4 and $l_e = 10$.

5.4   A Model for Exploring Rationality

We turn now to a more sophisticated model for learning by two agents engaged in repeated play of a stage game.[8] The model—MLPS: Markov Learning in Policy Space—is highly stylized and has quite unrealistic assumptions. It is, however, valid as an approximation of realistic conditions; I ask for the reader's indulgence.

The key idea is that agents have a *consideration set of policies for play*, S. The *supergame* consists of an indefinitely long sequence of *games*, each of which is a finite sequence of *rounds of play* of a stage game (e.g., Stag Hunt). Agents draw elements from their Ss and use them as *focal policies* for a period of time, or number of rounds of play, called a *game*. Each game is divided into n_e epochs of length l_e. Thus, the number of rounds of play in a game is $n_e l_e$. During an epoch an agent plays its current focal policy with probability $(1 - \varepsilon)$, and other policies from its consideration set the rest of the time, with probability ε.

At the end of each game, g_{t-1}, a player, p, picks a focal policy, f_p^t, from its consideration set, S, for play in game g_t. The players use the *fitness-proportional* choice rule. Let $\widehat{V}(p, i, j, k)$ be the average value per round of play returned to player p for policy i, when p has focal policy j and $-p$ (the counter-player) has focal policy k. (Similarly, $V(p, i, j, k)$ is the value realized

---

[8]More extensive treatment of this model may be found in [171].

in a particular round of play.) Then

$$\Pr(f_p^{t+1} = i | f_p^t = j, f_{-p}^t = k) = \widehat{V}(p, i, j, k) / \sum_i \widehat{V}(p, i, j, k) \qquad (5.2)$$

That is, the probability that a player chooses a policy for focus in the next game is the proportion of value it returned per round of play, compared to all the player's policies, during the previous game.

There is nothing egregiously unrealistic about these assumptions. The MLPS model strengthens them for the sake of mathematical tractability. Specifically, it is assumed that a mechanism is in place so that the two players are exactly coordinated. Each has its games begin and end at the same time (round of play in the sequence). Further, each supergame is neatly divided into epochs and the random choices are arranged so that each player's $\widehat{V}$ values exactly realize their expected values. The upshot of this is that the $\widehat{V}$ values seen by the players are constant, as are the underlying expected values. The resulting system is a stationary Markov process with states $\mathcal{S}_p \times \mathcal{S}_{-p}$ and the equilibrium distribution of states can be analytically determined.

To illustrate, assume the stage game is Stag Hunt with $R = 4, T = 3, P = 1$, and $S = 0$. Assume that each player has a consideration set of two policies of play: (1) TIT FOR TAT (TFT) in which the player begins (in the epoch) by hunting stag and subsequently mimics the behavior of the counter-player on the previous round of play, and (2) ALWAYS DEFECT (ALLD) in which the player always hunts hare. This system has four possible states: (1) both players in the game have TFT as their focal policy, (TFT, TFT), (2) player 1 (Row) has TFT as its focal policy and player 2 (Column) has ALLD as its focal policy, (TFT, ALLD), (3) (ALLD, TFT), and (4) (ALLD, ALLD). With $l_e = 10$ we get the payoffs for the various states as shown in Tables 5.1–5.4.

Letting $\varepsilon = 0.1$, routine calculation leads to the transition matrix indicated in table 5.5.

	s(1)=(1,1)	s(2)=(1,2)	s(3)=(2,1)	s(4)=(2,2)
s(1)	$0.7577 \cdot 0.7577$ $= 0.5741$	$0.7577 \cdot 0.2423$ $= 0.1836$	$0.2423 \cdot 0.7577$ $= 0.1836$	$0.2423 \cdot 0.2423$ $= 0.0587$
s(2)	$0.5426 \cdot 0.7577$ $= 0.4111$	$0.5426 \cdot 0.2423$ $= 0.1315$	$0.4574 \cdot 0.7577$ $= 0.3466$	$0.4574 \cdot 0.2423$ $= 0.1108$
s(3)	$0.7577 \cdot 0.5426$ $= 0.4111$	$0.7577 \cdot 0.4574$ $= 0.3466$	$0.2423 \cdot 0.5426$ $= 0.1315$	$0.2423 \cdot 0.4574$ $= 0.1108$
s(4)	$0.5426 \cdot 0.5426$ $= 0.2944$	$0.5426 \cdot 0.4574$ $= 0.2482$	$0.4574 \cdot 0.5426$ $= 0.2482$	$0.4574 \cdot 0.4574$ $= 0.2092$

TABLE 5.5: Stag Hunt transition matrix data assuming fitness proportional policy selection by both players, based on previous Tables 5.1–5.4. Example for $\varepsilon = 0.1 = \varepsilon_1 = \varepsilon_2$.

At convergence of the Markov process:

Pr(s(1))	Pr(s(2))	Pr(s(3))	Pr(s(4))
0.4779	0.2134	0.2134	0.0953

So 90%+ of the time at least one agent is playing TFT. Note the expected take for Row per epoch by state:

1. $(1 - \varepsilon)(40 - 31\varepsilon) + \varepsilon(12 - 2\varepsilon) = 34.39$

2. $(1 - \varepsilon)(9 + 31\varepsilon) + \varepsilon(10 + 2\varepsilon) = 11.91$

3. $\varepsilon(40 - 31\varepsilon) + (1 - \varepsilon)(12 - 2\varepsilon) = 14.31$

4. $\varepsilon(9 + 31\varepsilon) + (1 - \varepsilon)(10 + 2\varepsilon) = 10.39$

Further, in expectation, Row (and Column) gets $(0.4779\ 0.2134\ 0.2134\ 0.0953) \cdot (34.39\ 11.91\ 14.31\ 10.39)' = 23.02$ (per epoch of length $l_e = 10$, or 2.302 per round of play), much better than the 10.39 both would get if they played ALLD with ε-greedy exploration. Note that even the latter is larger than the return, 10 per epoch or 1 per round, of settling on the risk-dominant outcome of mutually hunting hare. There is a third, mixed, equilibrium of the one-shot Stag Hunt game. For this example it occurs at $((\frac{1}{2}S, \frac{1}{2}H), (\frac{1}{2}S, \frac{1}{2}H))$. At this equilibrium each player can expect a return of 2 from a round of play. Players playing under the MLPS regime learn that trust pays. A few points briefly before we turn to the larger lessons to be extracted from these examples.

1. Markov models converge rapidly and are quite robust. The results on display here hold up well across different parameter values (e.g., for ε). Further, relaxation of the mechanism of play so that agents get imperfect, but broadly accurate, estimates of the expected values of the V quantities will not produce grossly different results. We get a nonstationary Markov process, but in expectation it behaves as seen here.

2. The MLPS model also has attractive behavior for different kinds of games. Players in Prisoner's Dilemma games will learn a degree of cooperation and do much better than constant mutual defection. In games of pure conflict (constant sum games) the outcomes are close to those predicted by classical game theory. And in coordination games players go far by way of learning to coordinate. See [171] for details.

3. If we retain the core ideas of the MLPS model, but entirely relax the synchronization conditions imposed by the game mechanism, simulation studies produce results that qualitatively track the analytic results: the players learn to trust and more generally the players learn to approach Pareto optimal outcomes of the stage game [176].

5.5   Discussion

Neither the gridscape model nor the MLPS model with protocol 1 nor the two together are in any way definitive on the emergence of trust in iterated play of Stag Hunt games. They do tell us something: that trust can arise spontaneously among strategically interacting agents, that this can happen under a broad range of conditions, that it can be stable, and so on. The models and their discussion here leave many questions to be investigated and they raise for consideration many new questions. Much remains to be done, which I think is a positive result of presenting these models. I want now to make some remarks in outline by way of abstracting the results so far, with the aim of usefully framing the subject for further investigation.

LPS models: learning in policy space. Both the gridscape model and the MLPS model with protocol 1 are instances of a more general type of model, which I call an LPS (learning in policy space) model. In an LPS model an agent has a consideration set of policies or actions it can take, S, and a learning or update, L/U, policy it employs in selecting which policies to play, or actions to take, at a given time. In the gridscape model, $S = \{H, S\}$ for every player. In the MLPS model with protocol 1, $S = \{\text{TFT}, \text{ALLD}\}$ for both players. In the gridscape model the L/U policy employed by all players was IMITATE THE BEST NEIGHBOR. In the MLPS model, the players used the fitness-proportional update rule, in the context of the mechanism described in the previous section.

LPS models categorize strategies. In classical game theory the players are conceived as having *strategies,* complete instructions for play, which they can be thought of as choosing before the (super)game starts. The possible strategy choices constitute what we call the consideration set, S. Because strategies are picked *ex ante* there is no learning, although the strategies can be conditioned on play and can mimic any learning process. The agents employ what we might call the *null learning/update rule,* $L/U_\emptyset$. In an LPS model with a non-null L/U policy, the consideration set of policies of play does not include all possible strategies in the game. Policies in S are tried sequentially and played for a limited amount of time, then evaluated and put into competition with other members of S. The L/U policy constitutes the rules for comparison, competition, and choice. The total number of possible strategies is not affected by imposition of the LPS framework, but the strategies are implicitly categorized and the agents choose among them during the course of play (instead of *ex ante*). The consideration set of *strategies* used by an agent is implicit in its consideration set of policies, its L/U policy, the structure of the game, and the play by the counter-players. Thus, LPS models subsume standard game-theoretic models. A *proper* LPS model, however, has a non-null L/U policy. Normally, when I speak of an LPS model I shall be referring to a proper LPS model.

Folk Theorem undercuts. According to the Folk Theorem,[9] nearly any set of outcomes in an indefinitely repeated game can be supported by some Nash equilibrium. In consequence, the Nash equilibrium becomes essentially worthless as a predictive or even explanatory tool, in these contexts. The problems of trust arise against this backdrop and against the following point.

Refinements unsatisfying. Refinements to the classical theory, aimed at selecting a subset of the Nash equilibria in predicting outcomes, have been less than fully satisfying. This is a large subject and it takes us well beyond the scope of the present chapter. However, the favored refinement for Stag Hunt would be universal hunting of hare, because it is the risk dominant equilibrium. (For a general discussion see [319, 320].) Agents playing this way might well be viewed as "rational fools" [283] by LPS agents.

LPS agents may be rational. At least naïvely, the L/U regimes employed by our gridscape and MLPS agents are sensible, and may be judged rational, or at least not irrational. Exploring the environment, as our LPS agents do, probing it with play of different policies, informed by recent experience, is on the face of it entirely reasonable. Why not try learning by experience if it is not obvious what to do in the absence of experience? I shall now try to articulate a sense in which LPS agents may be judged rational, even though they violate the rationality assumptions of classical game theory and rational choice theory.

Contexts of maximum taking (MT). Given a set of outcomes whose values are known, perhaps under risk (i.e., up to a probability distribution), given a consistent, well-formed preference structure valuing the outcomes, and given a set of actions leading (either with certainty or with risk) to the outcomes, rational choice theory (or utility theory) instructs us to choose an action that results in our taking the maximum expected value on the outcomes. Presented with valued choices under certainty or risk, we are counseled to take the maximum value in expectation. Although the theory is foundational for classical game theory and economics, it has also been widely challenged both from a normative perspective and for its empirical adequacy.[10]

Contexts of maximum seeking (MS). In an MS context an agent can discriminate among outcomes based on their values to the agent, but the connection between the agent's possible actions and the resulting outcomes is uncertain in the technical sense: the agent does not have an objectively well-grounded probability distribution for associating outcomes with actions. In seeking the maximum return for its actions, the agent has little alternative

---

[9] A genuine theorem, described in standard texts, e.g., [31].

[10] Good, wide-ranging discussion can be found in [112, 131]. A classic paper [186] develops a model in which for the finitely iterated Prisoner's Dilemma game it is sometimes rational for a player to cooperate, *provided the player believes the counter-player is irrational.* Since both players would benefit by mutual cooperation it seems a stretch to call all attempts to find it irrational.

but to explore, to try different actions and to attempt to learn how best to take them.[11]

Exploring rationality is appropriate for MS contexts. The claim I wish to put on the table is that in MS as distinct from MT contexts, rationality is best thought of as an appropriate learning process. An agent is rational in an MS context to the extent that it engages effectively in learning to obtain a good return. In doing so, it will be inevitable that that agent engages in some form of trial-and-error process of exploring its environment. Rationality of this kind may be called an *exploring rationality* to distinguish it from what is often called *ideal rationality,* the kind described by rational choice theory and which is, I submit, typically not appropriate in MS contexts. See [170] for further discussion of the concept of an exploring rationality.

Evaluate exploring rationalities analytically by performance. LPS models with their articulated L/U regimes afford an excellent framework for evaluating forms of exploring rationality. Such evaluation will turn largely on performance under a given L/U regime. For starters and for now informally, an L/U regime may be assessed with regard to whether it is generally a strong performer. Rational admissibility is a useful concept in this regard.

General Definition 1 (Rational Admissibility) *A learning (update) regime for policies of play in an indefinitely repeated game is* <u>*rationally admissible*</u> *if*

1. *It performs well if played against itself (more generally: it performs well if universally adopted).*

2. *It performs well if played against other learning regimes that perform well when played against themselves (more generally: the other learning regimes perform well if universally adopted).*

3. *It is not vulnerable to catastrophic exploitation.*

To illustrate, in the gridscape model IMITATE THE BEST NEIGHBOR performs well against itself in that when everyone uses it, as we have seen, trust breaks out and stag hunting prevails robustly. The null L/U policy of ALWAYS HUNT STAG also does well against itself, and both IMITATE THE BEST NEIGHBOR and ALWAYS HUNT STAG will do well against each other. ALWAYS HUNT STAG, however, is catastrophically vulnerable to ALWAYS HUNT HARE. IMITATE THE BEST NEIGHBOR on the other hand will do better, although how much better depends on the payoff structure of the stage game. Some stag hunters may open themselves up to exploitation because they have one neighbor who hunts stag and is surrounded by stag hunters. In sum, with reference to the set of these three L/U policies, IMITATE THE BEST NEIGHBOR

---

[11]Classical game theory seeks to finesse this situation by assuming classical rationality and common knowledge. The present essay may be seen as an exploration of principled alternatives to making these very strong assumptions.

is uniquely rationally admissible (robustly, across a wide range of stag game payoff structures). A similar point holds for the MLPS model discussed above.

Two additional comments. First, "not vulnerable to catastrophic exploitation" is admittedly vague. It is not to my purpose to provide a formal specification here. I believe that more than one may be possible and in any event the topic is a large one. The motivating intuition is that a learning regime is vulnerable to exploitation if it learns to forego improving moves for which the counter-players have no effective means of denial. Thus, an agent that has learned to hunt stag in the face of the counter-player hunting hare is being exploited because it is foregoing the option of hunting hare, the benefits of which cannot be denied by the counter-player. Similarly, agents cooperating in Prisoner's Dilemma are not being exploited. Even though each is foregoing the temptation to defect, the benefits of defecting can easily be denied by the counter-player following suit and also defecting. Second, the similarity between the definition, albeit informal, of rational admissibility and the concept of an ESS (evolutionarily stable strategy, [217]) is intended. In a nutshell, a main message of this chapter is that for repeated games it is learning regimes and consideration sets of policies, rather than strategies alone, that are key to explanation. (And dare one suggest that rational play in one-shot games may sometimes draw on experience in repeated games?)

Evaluate exploring rationalities empirically, for descriptive adequacy. As noted, it is well established that rational choice theory (ideal rationality) is not descriptively accurate at the individual level. In light of the results and observations given here, one has to ask to what degree subjects at variance from the received theory are perceiving and responding to contexts of maximum seeking (MS), rather than the postulated MT contexts. In any event, it is worth noting that foraging by animals—for food, for mates, for shelter or other resources—is a ubiquitous natural form of behavior in an MS context [125, 301], for which models under the LPS framework would seem a good fit. Experimental investigation is only beginning. I think it shows much promise.

* * *

In conclusion, the problems and paradoxes of trust are vitally important on their own. Trust is the "cement of society."[12] Understanding it is crucial to maintenance and design of any social order, including and especially the new social orders engendered by modern communications technologies, globalization, global warming, and all that comes with them. I have tried to contribute in a small way to understanding how and when trust can emerge or be destroyed. The gridscape and MLPS models are helpful, but they can be only a small part of the story and even so their depths have barely been plumbed. But it's a start; it's something. The more significant point, I think, is that the problems of trust lead us, via these very different models, to a common pattern that abstracts them: LPS, learning in policy space, and contexts of maximum seeking (MS), as distinguished from contexts in which maximum

---

[12]Elster's term [90], after Hume who called causation the cement of the universe.

taking (MT) is appropriate. The fact, demonstrated here and elsewhere, that agents adopting this stance generate more trust and improve their take from the environment, is encouraging. So is the observation that such behavior is analogous to, if not related to or even a kind of, foraging behavior.

5.6   For Exploration

1. Compare the definition of Rationality Admissibility, General Definition 1, on page 108, with the strategy selection criteria for Pragmatic Strategic Rationality, §3.8 (starting on page 62). Does one list need to be lengthened? Shortened?

2. Use 2x2-Tournaments-Det.nlogo to investigate strategies for playing Iterated Stag Hunt.

3. Use ReplicatorDynamicsWithBalking.nlogo to investigate Stag Hunt played in an evolutionary context.

4. Use Symmetric-2x2.nlogo to investigate Stag Hunt.

5. Use M1-Symmetric-2x2-wID.nlogo to investigate Stag Hunt.

6. The Stag Hunt has been investigated behaviorally, in the laboratory, cf., [23, 54]. Summarize the empirical findings.

5.7   Concluding Notes

Brain Skyrms has written an entire book on the Stag Hunt game and the issues associated with it [297]. Highly recommended.

This chapter was originally published as [172]. Permission to republish this material is gratefully acknowledged, to Springer. Many thanks to Alex Chavez and James D. Laing for comments on an earlier version.

Chapter 6

Pareto versus Nash

6.1 Background

Contexts of strategic interaction (CSIs) appear in nearly every social situation. They are characterized by interdependent decision making: two or more agents have choices to make and the rewards an individual receives in consequence of its choices depend, at least in part, on the choices made by other agents. Such contexts, when abstracted and formalized in certain ways, are the subject of game theory, which seeks to "solve"—predict and explain the outcomes of—games (i.e., of CSIs abstracted and formalized in certain stylized fashions).

Any solution theory for CSIs (or games) must make and rely upon two kinds of assumptions:

1. *SR (Strategic Regime) assumptions.* There are assumptions about the representation and structure of the CSI (or game), including the rules of play and the payoffs to the players. Typically, these assumptions are expressed as games in strategic form, games in extensive form, characteristic function games, spatial games, and so on.

2. *SSR assumptions.* These are assumptions about the Strategy Selection Regimes (SSRs) employed by the agents, or players, in the game. Classical game theory makes two kinds of SSR assumptions, which typically apply to all players [203, 287]:

   (a) *Ideal rationality assumptions.* It is normally assumed that agents are "rational" and that Rational Choice Theory in some form (e.g., Savage's Subjective Expected Utility theory) characterizes this kind of (ideal) rationality. Roughly, agents are assumed to have utilities and to be maximizers of their utilities.

   (b) *Knowledge assumptions.* It is normally assumed that agents are omniscient with respect to the game. The agents know everything about the game, common knowledge obtains among all the players, and all agents have unlimited computational/ratiocination powers.

This chapter reports on a series of experimental investigations that examine play in games under non-standard SSR assumptions, at least as judged by the classical game theory literature. We investigate a series of games that are well recognized in the classical literature and that have been extensively studied. Our game—that is, Strategic Regime—assumptions are conventional, although we focus on iterated games.

It is, and has always been, recognized that the classical SSR assumptions (as we call them) are unrealistic. The original experimental work on Prisoner's Dilemma, among other games [99], was motivated by such concerns. Even so, they—and the consequences they engender—are interesting. The assumptions often afford tractability, allowing games to be "solved." Because they capture the notion of a certain plausible kind of ideal rationality, it is interesting to determine how well they describe actual human behavior. Even if they are inaccurate, they have value as a normative benchmark. And given the considerable powers of human cognition and institutions, it is not prima facie implausible that classical SSR assumptions will often yield accurate predictions.

This is all well and good, but the story is not over. There are certain puzzles or anomalies associated with the classical SSR assumptions. Famously in the Prisoner's Dilemma game, and in other games, the Nash equilibrium (NE) outcome is not Pareto efficient. Classical theory sees the NE as the solution to the game, yet many observers find it anomalous and experiments with human subjects often indicate support for these observers [203, 253]. Further, the NE need not be unique, posing thereby a challenge to the classical theory, which often struggles, or has to be stretched, to predict equilibrium outcomes that seem natural and that are reached by human subjects easily. In short, the classical theory has often proved to be a poor—weak and inaccurate— predictor of human behavior [263].

Besides the well-known puzzles and anomalies, there is another category of reasons to study games under variations of the classical SSR assumptions. Rational Choice Theory and omniscience may be plausible assumptions for experienced humans in certain favorable institutional settings (e.g., well-established markets). They are often not plausible assumptions for games played by birds, bees, monkeys up in trees, bacteria, and other similarly less cognitively well-endowed creatures. It is, simply put, scientifically interesting to investigate the play and outcomes in games in which the SSR assumptions of classical game theory are relaxed sufficiently to be capable of describing these kinds of more limited agents. Equally so, this is interesting from a practical, applications-oriented perspective. Adaptive artificial agents, e.g., fielded for purposes of electronic commerce, will inevitably resemble the lower animals more than their creators, at least in their cognitive powers.

With these motivations principally in mind, we investigated iterated play by simple, adaptive agents in a number of well-known games. Any such investigation, however, faces an immediate and urgent theoretical problem: There are indefinitely many ways to relax the classical SSR assumptions; how does one justify a particular alternative? We choose with a number of criteria in mind.

1. *Simple.* There are few ways to be ideally rational and indefinitely many ways not to be. In examining alternatives it is wise to begin with simple models and complexify as subsequent evidence and modeling ambition require.

2. *New.* Much has been learned about non-ideally rational agents through studies of the replicator dynamic (see [124] for a review). These investigations, however, see *populations* as evolving, rather than individual agents adapting. The individuals are typically modeled as naked, unchanging strategies, rather than adaptive agents, which proliferate or go extinct during the course of continuing play. Agents in some 'spatialized', cellular automata-style games have been given certain powers of state change and adaptation, but these have on the whole been limited in scope (e.g., [93, 135]). Experimenting with game-playing agents that are using reinforcement learning is a comparatively under-developed area.

3. *Theoretically motivated.* Reinforcement learning as it has developed as a field of computational study has been directly and intentionally modeled on learning theories from psychology, where there is an extensive supporting literature. This important class of learning model is a natural first choice for modeling agents in games, because it appears to apply broadly to other areas of learning, because its theoretical properties have been well investigated, and because it has achieved a wide scope of application in multiple domains.

4. *Adaptive.* Agents should be responsive to their environments and be able to learn effective modes of play.

5. *Exploring.* Agents should be able actively to probe their environments and undertake exploration in the service of adaptation; agents face the exploration-exploitation tradeoff and engage in both.

In addition, the SSRs should be realizable in sense that they specify definite procedures that simple agents could actually undertake. It is here, perhaps, that the present approach, which we label *procedural game theory*, differs most markedly from classical game theory and its assumption of ideal rationality, irrespective of realizability constraints.

Now to a discussion of the elements of reinforcement learning needed as background for our experiments.

6.2   Reinforcement Learning

6.2.1   Simple Q-Learning

Our experimental agents used a simple form of Q-learning, itself a variety of reinforcement learning. Detailed description of Q-learning is easily found in the open literature (e.g., [304, 328, 329]). We limit ourselves here to a minimal summary for the purposes at hand.

The Q-learning algorithm works by estimating the values of state-action pairs. The value $Q(s, a)$ is defined to be the expected discounted sum of future payoffs obtained by taking action a in state s and following an optimal policy thereafter. Once these values have been learned, the optimal action from any state is the one with the highest Q-value. The standard procedure for Q-learning is as follows. Assume that $Q(s, a)$ is represented by a lookup table containing a value for every possible state-action pair, and that the table entries are initialized to arbitrary values. Then the procedure for estimating $Q(s, a)$ is to repeat the following loop until a termination criterion is met:

1. Given the current state s choose an action a. This will result in receipt of an immediate reward r, and transition to a next state s'. (We discuss below the policy used by the agent to pick particular actions, called the exploration strategy.)

2. Update $Q(s, a)$ according to the following equation:

$$Q(s, a) = Q(s, a) + \alpha[r + \gamma \max_b Q(s', b) - Q(s, a)] \qquad (6.1)$$

where α is the learning rate parameter and $Q(s, a)$ on the left is the new, updated value of $Q(s, a)$.

In the context of iterated games, a reinforcement learning (Q-learning) player explores the environment (its opponent and the game structure) by taking some risk in choosing actions that might not be optimal, as estimated in step 1. In step 2 the action that leads to higher reward will strengthen the Q-value for that state-action pair. The above procedure is guaranteed to converge to the correct Q-values for stationary Markov decision processes.

In practice, the exploration policy in step 1 (i.e., the action-picking policy) is usually chosen so that it will ensure sufficient exploration while still favoring actions with higher value estimates in given state. A variety of methods may be used. A simple method is to behave greedily most of the time, but with small probability, ε, choose an available action at random from those that do not have the highest Q-value. For obvious reasons, this action selection method is called *epsilon-greedy* (see [304]; equivalently, ε-greedy). Softmax is another commonly used action selection method. Here again, actions with higher values are more likely to be chosen in given state. The most common form for the probability of choosing action a is

$$\frac{e^{Q_t(a)/\tau}}{\sum_{b=1}^{n} e^{Q_t(b)/\tau}} \tag{6.2}$$

where τ is a positive parameter and decreases over time. It is typically called the temperature, by analogy with annealing. In the limit as $\tau \to 0$, Softmax action selection becomes greedy action selection. In our experiment we investigated both epsilon-greedy and Softmax action selection.

6.2.2 Implementation of Q-Learning for 2×2 Games

A Q-learning agent does not require a model of its environment and can be used on-line. Therefore, it is quite suited for iterated games against an unknown co-player (especially an adaptive, exploring co-player). Here, we will focus on certain iterated 2×2 games, in which there are two players each having two possible plays/actions at each stage of the game. It is natural to represent the state of play, for a given player, as the outcome of the previous game played. We say in this case that the player has memory length of one. The number of states for a 2×2 game is thus four and for each state there are two actions (the pure strategies) from which the player can choose for current game. We also conducted the experiments for the case that players have memory length of two (the number of states will be 16) and obtained broadly similar results. The immediate reward a player gets is specified by the payoff matrix.

For the Softmax action selection method, we set the decreasing rate of the parameter τ as follows.

$$\tau = T * \Theta^n \tag{6.3}$$

T is a proportionality constant, n is number of games played so far. Θ, called the annealing factor, is a positive constant that is less than 1. In the implementation, when n becomes large enough, τ is close to zero and the player stops exploring. We use Softmax, but in order to avoid cessation of exploration, our agents start using epsilon-greedy exploration once the Softmax progresses to a point (discussed below) after which exploration is minimal.

6.3 Experiments

6.3.1 Motivation

Iterated 2×2 games are the simplest of settings for strategic interactions and are a good starting point to investigate how outcomes arise under a regime of exploring rationality versus the ideal rationality of classical game theory.

The Definitely Iterated Prisoner's Dilemma, involving a fixed number of iterations of the underlying game, is a useful example. Classical game theory, using a backward induction argument, predicts that both players will defect on each play [203]. If, on the other hand, a player accepts the risk of cooperating, hoping perhaps to induce cooperation later from its counter-player, it is entirely possible that both players discover the benefits of mutual cooperation. Even if both players suffer losses early on, subsequent sustained mutual cooperation may well reward exploration at the early stages.

Motivated by this intuition, we selected eight games and parameterized their payoffs. The players are modeled as Q-learners in each iterated game. In five of the games the Pareto optimal (socially superior, i.e., maximal in the sum of its payoffs) outcome does not coincide with a Nash Equilibrium. The remaining three games, which we included to address the multi-equilibrium selection issue, each have two pure-strategy NEs.

6.3.2   The Games and the Parameterization

We parameterized each of our eight games via a single parameter, δ, in their payoff matrices. In the payoff matrices below, the first number is the payoff to the row player and the second is the payoff to the column player. We mark the Nash equilibria with # and the Pareto efficient outcomes with *. Resource optimal (socially superior, aka: Hicks optimal) outcomes are labeled with **. C and D are the actions or pure strategies that players can take on any single round of play. The row player always comes first in our notation. Thus, CD means that the row player chose pure strategy C and the column player chose pure strategy D. So there are four possible outcomes of one round of play: CC, CD, DC, and DD.

The first two games are versions of Prisoner's Dilemma (PD). The value of δ ranges from 0 to 3. When its value is 2 (see Table 6.1), it corresponds to the most common payoff matrix in the Prisoner's Dilemma literature.

	C	D
C	$(3, 3)^{**}$	$(0,3+\delta)^*$
D	$(3+\delta, 0)^*$	$(3-\delta, 3-\delta)\#$

TABLE 6.1: Prisoner's Dilemma, pattern 1.

While the Prisoner's Dilemma, in its usual form, is a symmetric game (see Tables 6.1 and 6.2), the following three games, adapted from [253], are asymmetric. The value of δ ranges from 0 to 3 in our experiments with these games. Note that as in Prisoner's Dilemma, in Games #47, #48, and #57 (Tables 6.3—6.5) the Nash equilibrium does not coincide with the Pareto optimal outcome.

	C	D
C	(3,3)**	(0, 3+δ)*
D	(3+δ, 0)*	(δ, δ)#

TABLE 6.2: Prisoner's Dilemma, pattern 2.

	C	D
C	(0.2 ,0.3)#	0.3+δ, 0.1*
D	0.1, 0.2	(0.2+δ, 0.3+δ)**

TABLE 6.3: Game #47.

	C	D
C	(0.2, 0.2)#	(0.3+δ, 0.1)*
D	(0.1, 0.3)	(0.2+δ, 0.3+δ)**

TABLE 6.4: Game #48.

For games with two NE, the central question is which equilibrium (if any) is most likely to be selected as the outcome. We choose three examples from this class of game. The game of Stag Hunt has a Pareto optimal solution as one of its two NE. The game of Chicken and the game of Battle of Sexes are coordination games. In Battle of the Sexes the two coordination outcomes (CC and DD) are NEs and are Pareto optimal. In Chicken, the coordination outcomes (CD and DC) may or may not be NEs, depending on δ. The value of δ ranges in our experiments from 0 to 3 for Stag Hunt and Battle of the Sexes. For Chicken, the range is from 0 to 2.

6.3.3 Settings for the Experiments

We set the parameters for Q-learning as follows. Learning rate, $\alpha = 0.2$ and discount factor, $\gamma = 0.95$. We ran the experiment with both Softmax action selection and epsilon-greedy action selection. For Softmax action selection, T is set to 5 and the annealing factor $\Theta = 0.9999$. When τ is less than 0.01, we began using epsilon-greedy action section. We set ε to 0.01. We note that these parameter values are typical and resemble those used by other studies (e.g., [273]). Also, our results are robust to changes in these settings.

Each game was iterated 200,000 times in order to give the players enough time to explore and learn. For each setting of the payoff parameter δ, we ran the iterated game 100 times. We recorded the frequencies of the four outcomes

	C	D
C	(0.2 ,0.3)#	(0.3+δ, 0.2)*
D	(0.1, 0.1)	(0.2+δ, 0.3+δ)**

TABLE 6.5: Game #57.

	C	D
C	(5,5)**#	(0,3)
D	(3,0)	(δ, δ)#

TABLE 6.6: Stag Hunt.

	C	D
C	(δ, 3-δ)**#	(0,0)
D	(0,0)	(3-δ, δ)**#

TABLE 6.7: Battle of the Sexes.

	C	D
C	(2,2)*	(δ, 2+δ)*#
D	(2+δ, δ)*#	(0, 0)

TABLE 6.8: Chicken. $0 \leq \delta < 2$. CC is ** for $\delta \leq 1$. CD and DC are ** for $\delta \geq 1$.

(CC, CD, DC, and DD) every 100 iterations. The numbers usually become stable within 50,000 iterations, so we took frequencies of the outcomes in the last 100 iterations over the 200,000 iterations to report, unless noted otherwise.

The summary of results tables, below, all share a similar layout. In the middle column is the payoff parameter δ. On its left are the results for epsilon-greedy action selection. The results for Softmax action selection are on the right. Again, the numbers are frequencies of the four outcomes (CC, CD, DC, and DD) in the last 100 iterations, averaged over 100 runs.

6.3.4 Results

It is generally recognized as disturbing or at least anomalous when classical game theory predicts that a Pareto inferior Nash equilibrium will be the outcome, rather than a Pareto optimal solution ([99, 203] and ever since). This is exactly what happens in our first five games, in which the unique subgame perfect Nash equilibrium is never the Pareto optimal (or even a Pareto efficient!) outcome. Will the outcomes be different if agents use adaptive, exploring SSRs, such as reinforcement learning? More specifically, can players learn to achieve a Pareto optimal solution that is not a Nash equilibrium? Among competing NEs, will players find the Pareto efficient outcome? Our results indicate a broadly positive answer to these questions.

Consider Table 6.9 (summarizing results for the parameterized PD game in Table 6.1). If δ is close to zero, the two players choose to defect most of the time. (That is, see above, during the final 100 rounds of 200,000 iterations, they mostly play DD. The entries in Table 6.9, and in similar tables report counts out of 100 rounds × 100 runs = 10,000 plays.) We note, by way of explanation, that there is not much difference in rewards between mutual defection and mutual cooperation: 3-δ and 3, with δ small. The Pareto optimal outcome does not appear to provide enough incentive for these players to risk cooperation. But as δ gets larger, we see more cases of mutual cooperation. The last row in Table 6.9 has an interesting interpretation: The players have incentive to induce each other's cooperation so as to take advantage of it by defecting. This is always the case in Prisoner's Dilemma, but exacerbated here (final row of Table 6.9) because the temptation for defection in the presence of cooperation is unusually large. Consequently, we see many CDs and DCs, but less mutual cooperation (CC). Notice that CC is maximized and DD minimized somewhere in the range of [1.75, 2] for δ. (Softmax and epsilon-greedy results are, here and elsewhere, in essential agreement.) When δ is low the benefit of mutual cooperation is too low for the agents to find the Pareto optimal outcome. When δ is very high, so is the benefit of defection in the face of cooperation, and again the agents fail to cooperate jointly. In the middle, particularly in the [1.75, 2] range, the benefits of mutual cooperation are high enough and the temptation to defection is low enough that substantial cooperation occurs.

Further insight is available by considering Table 6.10, the Wealth Extrac-

epsilon-greedy action selection				δ	Softmax action selection			
CC	CD	DC	DD		CC	CD	DC	DD
3	87	82	9828	0.05	0	106	101	9793
0	92	105	9803	0.5	0	90	94	9816
52	110	111	9727	1	1	111	111	9777
51	110	93	9746	1.25	2475	338	358	6829
1136	160	198	8506	1.5	3119	526	483	5872
1776	245	381	7598	1.75	4252	653	666	4429
3526	547	413	5514	2	789	883	869	7549
848	766	779	7607	2.5	496	2276	2368	4860
544	2313	2306	4837	2.95	539	2821	2112	4528

TABLE 6.9: Results summary for Prisoner's Dilemma, pattern 1.

tion Report for Table 6.9. The Total Wealth Extracted (WE) by an agent is simply the number of points it obtained in playing a game. Table 6.10 presents the Total WE for the row chooser in PD, Pattern 1. (Results are similar for the column chooser; this is a symmetric game played by identically endowed agents.) WE-Q:Pmax is the ratio (quotient, Q) of (a) Total WE and (b) 100 iterations × 100 runs × Pmax, the maximum number of points row chooser could get from outcomes on the Pareto frontier. Pmax $= (3 + \delta)$ and is realized when DC is played. WE-Q:Pgmax is the ratio of (a) Total WE and (b) 100 iterations × 100 runs × Pgmax, the maximum number of points row chooser could get from outcomes on the Pareto frontier whose total rewards are maximal (among Pareto efficient outcomes). Here, Pgmax $= 3$ and is realized when CC is played. WE-Q:Pgmax might be called the "wealth extraction quotient for socially optimal outcomes." Each of these measures declines as δ increases. Our agents have progressively more difficulty extracting available wealth. This hardly seems surprising, for at $\delta = 0.05$ the game is hardly a PD at all and the reward 3 for mutual cooperation is a paltry improvement over the "penalty" for mutual defection, 2.95. As δ increases, however, strategy selection becomes more and more of a dilemma and the agents become less and less successful in extracting wealth from the system. Note that these trends are more or less monotonic (see Table 6.10), while the actual outcomes change rather dramatically (see Table 6.9). From the perspective of classical game theory, changes in δ should not matter. Each of these games is a PD and should produce identical outcomes, all DD. Note that had the players played DD uniformly when $\delta = 2.95$, the row (and similarly the column) player would have extracted a total wealth of 10,000 × 0.05 = 500. In this light, extracting 14,410 is a considerable achievement.

Consider now the parameterized family of Prisoner's Dilemma Pattern 2 games (see Table 6.2). Here, the players stand to lose almost nothing by trying to cooperate when δ is close to zero. Exploration seems to help players reach

Softmax		(DC)		(CC)	
δ	Total WE	Pmax	WE-Q:Pmax	Pgmax	WE-Q:Pgmax
0.05	29197	3.05	0.957	3	0.973
0.50	24869	3.50	0.711	3	0.829
1.00	20001	4.00	0.500	3	0.667
1.25	20897	4.25	0.492	3	0.697
1.50	20339	4.50	0.452	3	0.678
1.75	21456	4.75	0.452	3	0.715
2.00	14261	5.00	0.285	3	0.475
2.50	16942	5.50	0.308	3	0.565
2.95	14410	5.95	0.242	3	0.480

TABLE 6.10: Row chooser's total wealth extracted in Prisoner's Dilemma, pattern 1.

the superior ("socially superior") Pareto optimal outcome (CC) and as we can see from Table 6.11, mutual cooperation happens 94% of time. Consider the scenario with δ close to 3. Note first, there is not much incentive to shift from the Nash equilibrium (DD) to the socially superior Pareto outcome (CC), since there is not much difference in payoffs; second, the danger of being exploited by the other player and getting zero payoff is much higher. Indeed, the players learn to defect most of the time (98%).

ε-greedy action selection					Softmax action selection			
CC	CD	DC	DD	δ	CC	CD	DC	DD
9422	218	183	177	0.05	9334	302	285	79
9036	399	388	150	0.5	9346	294	220	140
5691	738	678	2693	1	7537	954	1267	242
3506	179	275	6040	1.25	8203	542	994	261
1181	184	116	8519	1.5	7818	767	775	640
2	98	103	9797	1.75	4685	270	422	4623
97	114	91	9698	2	1820	217	220	7743
0	100	92	9808	2.5	0	77	117	9806
2	96	94	9808	2.95	0	90	114	9796

TABLE 6.11: Results summary for Prisoner's Dilemma, pattern 2.

The Wealth Extraction Report, Table 6.12, for Pattern 2 corresponds to Table 6.10 for Pattern 1. We see that our row chooser is able to extract a roughly constant amount of wealth from the game, even as δ and the strategy choices vary drastically. Note further that WE-Q:Pgmax is approximately constant (mostly over 90%) even though the players are mostly not playing CC at all.

Softmax		(DC)		(CC)	
δ	Total WE	Pmax	WE-Q:Pmax	Pgmax	WE-Q:Pgmax
0.05	28875	3.05	0.947	3	0.963
0.50	28878	3.50	0.825	3	0.963
1.00	27921	4.00	0.698	3	0.931
1.25	29160	4.25	0.686	3	0.972
1.50	27902	4.50	0.620	3	0.930
1.75	24150	4.75	0.508	3	0.805
2.00	22046	5.00	0.441	3	0.735
2.50	25159	5.50	0.457	3	0.839
2.95	29577	5.95	0.497	3	0.986

TABLE 6.12: Row chooser's total wealth extracted in Prisoner's Dilemma, pattern 2.

We now turn to games #47, #48, and #57, which are asymmetric games having a common feature: the row player has a dominant strategy C. Thus a fully rational row player will never choose D. What will happen if players are able to explore and learn? Tables 6.13–6.15 tell us that it depends on the payoffs. If δ is close to zero, the outcome will be the Nash equilibrium (CC) almost always. As δ increases, however, the incentives favoring the socially superior Pareto outcome (CC) concomitantly increase, drawing the players away from CC (Nash) to DD (socially superior Pareto). We note that row chooser would prefer CD to DD, yet in all three games (see Tables 6.13–6.15) we see a similar pattern of CD play as δ increases.

ε-greedy action selection					Softmax action selection			
CC	CD	DC	DD	δ	CC	CD	DC	DD
9790	101	101	8	0	9808	94	98	0
4147	137	156	5560	0.1	9812	94	93	1
3019	123	165	6693	0.15	9799	95	104	2
2188	141	132	7539	0.2	8934	85	109	872
185	355	130	9330	0.5	730	284	208	8778
131	309	135	9425	1	120	532	138	9210
138	288	99	9475	1.5	77	471	103	9349
99	321	131	9449	2	88	441	126	9345
126	172	88	9614	3	64	366	92	9478

TABLE 6.13: Summary of results for game #47.

The Wealth Extraction Reports for game #47 are also useful for understanding the row versus column power relationship in these games (Tables 6.16–6.17). Notice that at the Nash equilibrium (CC) total WE for row is

ε-greedy action selection				δ	Softmax action selection			
CC	CD	DC	DD		CC	CD	DC	DD
9789	102	107	2	0	9787	106	105	2
3173	515	173	6139	0.1	9811	86	101	2
2832	457	207	6504	0.15	8127	256	137	1480
1227	348	141	8284	0.2	2986	755	230	6029
109	627	143	9121	0.5	143	631	146	9080
90	492	139	9279	1	79	1320	126	8475
88	318	134	9460	1.5	117	1076	128	8679
241	236	119	9404	2	62	473	126	9339
76	284	139	9501	3	64	277	128	9531

TABLE 6.14: Summary of results for game #48.

ε-greedy action selection				δ	Softmax action selection			
CC	CD	DC	DD		CC	CD	DC	DD
9767	119	107	7	0	9764	131	105	0
1684	587	175	7554	0.1	9794	106	98	2
531	518	191	8760	0.15	9550	105	105	240
238	543	159	9060	0.2	1048	497	257	8198
126	307	121	9446	0.5	224	852	152	8772
118	520	114	9248	1	113	753	119	9015
104	526	125	9245	1.5	74	538	117	9271
66	225	102	9607	2	57	569	123	9251
123	296	116	9465	3	61	302	125	9512

TABLE 6.15: Summary of results for game #57.

2/3 of that for column. See Tables 6.16–6.17 for $\delta = 0$. As δ increases and CC play decreases both players uniformly increase their WE. At the same time, their WE becomes more and more equal, and by the time $\delta = 2$ row chooser is extracting more wealth from the game than column chooser. This occurs even though in more than 92% of the games the play is DD and column chooser extracts more wealth than row chooser! The difference is due to the occasional "defection" by row chooser to play C. Finally, we note that DC is neither Nash nor Pareto in these games. Our agents play DC at a rate that is low and essentially invariant with δ. That rate may be interpreted as a cost consequence of exploration.

Finally, it is instructive to note that when $\delta = 3$ the expected value for column playing D is $3.3 - 3.2p$, if row plays C with probability p. Similarly the expected value of playing C is $0.2 + 0.1p$. Consequently, column should play D so long as $p < \frac{31}{33}$. These considerations lead us to wonder whether our row chooser agents have not learned to be sufficiently exploitive. They may be too generous to column chooser, although column chooser is not without recourse. However, the fact that C and D for row chooser are so close in value, given that column chooser plays D, may impute stability in this stochastic, noisy, learning context. Note that in all three games CD is more rare when $\delta = 3$ than when $\delta = 2$.

Softmax	#47	(CD)		(DD)	
δ	Total WE	Pmax	WE-Q:Pmax	Pgmax	WE-Q:Pgmax
0	2000	0.30	0.667	0.20	1.000
0.1	2010	0.40	0.502	0.30	0.670
0.15	2014	0.45	0.447	0.35	0.575
0.2	2189	0.50	0.438	0.40	0.547
0.5	6539	0.80	0.817	0.70	0.934
1	11781	1.30	0.906	1.20	0.982
1.5	16767	1.80	0.931	1.70	0.986
2	21604	2.30	0.939	2.20	0.982
3	31559	3.30	0.956	3.20	0.986

TABLE 6.16: Row chooser's total wealth extracted in game #47.

In PD and games #47, #48, and #57, the Nash equilibrium is not on the Pareto frontier. The Stag Hunt game is thus interesting because its Pareto optimal solution is also one of its two pure strategy NEs. But which one, or which mixture, will be sustained remains a challenging problem for classical game theory. A mixed strategy seems natural in this iterated game for classical game theory. Table 6.18 shows that the outcomes for our reinforcement learning agents do not conform to the prediction of a mixed strategy. Say, for example, when δ is equal to 1, the mixed strategy for both players will be choosing action C with probability 1/3 and D with probability 2/3. (Let p

Column Chooser:					
Softmax	#47	(DD)		(DD)	
δ	Total WE	Pmax	WE-Q:Pmax	Pgmax	WE-Q:Pgmax
0	2962	0.30	0.987	0.30	0.987
0.1	2963	0.40	0.741	0.40	0.741
0.15	2961	0.45	0.658	0.45	0.658
0.2	3136	0.50	0.627	0.50	0.627
0.5	7291	0.80	0.911	0.80	0.911
1	12076	1.30	0.929	1.30	0.929
1.5	16909	1.80	0.939	1.80	0.939
2	21577	2.30	0.938	2.30	0.938
3	31342	3.30	0.950	3.30	0.950

TABLE 6.17: Column Chooser's Total Wealth Extracted in Game #47.

be the probability of playing C, then at $5p + 0p = 3p + (1 - p)$ the players are indifferent between playing C or D. This happens at $p = \frac{1}{3}$.) We should expect to see CC with a frequency less than 33%, while Table 6.15 shows CC happening at a rate of 88%.

ε-greedy action selection					Softmax action selection			
CC	CD	DC	DD	δ	CC	CD	DC	DD
9390	126	122	362	0	9715	108	109	68
9546	91	108	255	0.5	9681	120	121	78
9211	112	125	552	0.75	9669	111	101	119
8864	119	110	907	1	9666	98	102	134
8634	115	132	1119	1.25	9598	139	134	129
7914	122	130	1834	1.5	9465	99	109	327
7822	122	104	1952	2	9452	126	126	296
5936	87	101	3876	2.5	8592	116	89	1203
5266	121	106	4507	3	3524	111	115	6250

TABLE 6.18: Summary of results for Stag Hunt.

In Stag Hunt, CC is Pareto optimal but risky, while DD is riskless (on the downside) but Pareto dominated. As δ increases from 0 to 3.0 the risk/reward balance increasingly favors DD. Our agents respond by favoring DD at the expense of CC and in consequence they extract a decreasing amount of wealth. It is as if they were operating with a risk premium, yet we know they are not.

The remaining two games are coordination games. We are concerned not only with which NEs are to be selected, but also with a larger question: Is the Nash equilibrium concept apt for describing what happens in these games? The later concern arises as we observe different behavior in human experi-

Row Chooser:			
Softmax	Stag Hunt	(CC)	WE-Q:Pmax
δ	Total WE	Pgmax	WE-Q:Pgmax
0.0	48902	5	0.978
0.5	48807	5	0.976
0.8	48737	5	0.975
1.0	48770	5	0.975
1.3	48553	5	0.971
1.5	48143	5	0.963
2.0	48230	5	0.965
2.5	46235	5	0.925
3.0	36805	5	0.736

TABLE 6.19: Row chooser's total wealth extracted in Stag Hunt.

ments. Rapport et al. [253] reported a majority of subjects quickly settling into an alternating strategy, with the outcome changing back and forth between the two Nash coordination points (CD and DC) when playing the game of Chicken.

From Table 6.20 we can see these two NEs (and coordination points) in Battle of the Sexes are equally likely to be the outcome in most cases since the game is symmetric and these two outcomes are superior to other two, which give both players a zero payoff. In the game of Chicken (Table 6.21) we see that if the incentive for coordinating is too small (i.e., δ is close to zero), the players learn to be conservative and land on the non-NE (CC) since they cannot afford the loss resulting from DD (getting zero). As δ increases, the game ends up more and more in one of the Nash coordination points (CD or DC).

ε-greedy action selection					Softmax action selection			
CC	CD	DC	DD	δ	CC	CD	DC	DD
2641	63	4571	2725	0	2872	73	4477	2578
3842	135	1626	4397	0.1	4615	101	1732	3552
5140	102	90	4668	0.5	4772	102	162	4964
4828	107	94	4971	1	4862	88	89	4961
4122	101	109	5668	1.5	4642	85	102	5171
4983	100	97	4820	2	4623	97	87	5193
3814	111	96	5979	2.5	5139	102	99	4660
4015	1388	107	4490	2.9	4303	1794	118	3785
2653	4921	70	2356	3	2593	4776	58	2573

TABLE 6.20: Summary of results for Battle of the Sexes.

ε-greedy action selection					Softmax action selection			
CC	CD	DC	DD	δ	CC	CD	DC	DD
9276	227	347	150	0	9509	165	222	104
9587	143	135	135	0.25	9119	428	320	133
9346	209	223	222	0.5	9375	220	225	180
6485	1491	1858	166	0.75	8759	424	632	185
1663	3532	4706	99	1	1339	4903	3662	96
385	4161	5342	112	1.25	158	5416	4323	103
113	4488	5274	125	1.5	115	4700	5099	86
111	4301	5504	84	1.75	100	4704	5083	113
100	4853	4953	94	2	94	4772	5044	90

TABLE 6.21: Summary of results for Chicken.

The Wealth Extraction Report for Chicken, Table 6.22, is particularly revealing. When δ is small, play is overwhelmingly CC. CC is non-Nash and Pareto and for $\delta \leq 1.0$ CC is socially superior Pareto. C is less risky for both players than D (both CD and DC are Pareto and Nash outcomes), so when δ is small it stands to reason that our agents should stick with CC. Note in this regard that if the players exactly alternate the CD and DC outcomes, they each will receive a payoff of $1 + \delta$ on average. See the column labeled "WE if Perfect Alternation" in Table 6.22. We see that when $\delta < 1.0$ (i.e., when CC is socially superior), CC play is preponderant and Total WE is greater, often substantially greater, than WE if Perfect Alternation. For $\delta \geq 1.0$, CD and DC are socially superior Pareto. In the neighborhood of 1.0, play transitions from predominantly CC to predominantly CD and DC. Note that Total WE increases uniformly with δ (except for a slight decline in the neighborhood of $\delta = 1.0$, which we attribute to transition-induced error). As δ ranges from 1.0 to 2.0, Total WE closely approximates WE if Perfect Alternation. In short, the agents are impressively effective at extracting wealth. Outcomes are Nash (for the most part) if and only if there is not more money to be made elsewhere.

In order to see if players can learn alternating strategies, as observed in human subject experiments, we conducted another 100 trials for these two games with δ set to 1 and with Softmax action selection. For most of the trials the outcomes converge (i.e., settle, [85, 86]) to one of the Pareto superior outcomes. But we did observe patterns showing alternating strategies for both games. These patterns are quite stable and can recover quickly from small random disturbances. For the Battle of the Sexes, we observed only one alternating pattern: the players playing the two Nash equilibria alternately, in sequence. This pattern occurred in 11 out of 100 trials. For Chicken, we observed other kinds of patterns and have summarized their frequencies in Table 6.23.

At 23% (10 + 13 of 100), the proportion of alternating patterns cannot be

Row chooser: Softmax		(DC)	Chicken			WE if Perfect
δ	Total WE	Pmax	WE-Q:Pmax	Pgmax	WE-Q:Pgmax	Alternation
0.0	19482	2.00	0.974	1.0	1.948	10000
0.3	19065	2.25	0.847	1.3	1.525	12500
0.5	19423	2.50	0.777	1.5	1.295	15000
0.8	19574	2.75	0.712	1.8	1.119	17500
1.0	18567	3.00	0.619	2.0	0.928	20000
1.3	21136	3.25	0.650	2.3	0.939	22500
1.5	25127	3.50	0.718	2.5	1.005	25000
1.8	27493	3.75	0.733	2.8	1.000	27500
2.0	29908	4.00	0.748	3.0	0.997	30000

Note: Pgmax assumes perfect alternation of CD and DC.

TABLE 6.22: Row chooser's total wealth extracted in Chicken.

The outcomes	Frequency in 100 trials
Alternating between CD and DC	10
Cycle through CD-DC-CC or CD-CC-DC	13
Converge to one of the three: CC, CD, or DC	76
No obvious pattern	1

TABLE 6.23: Frequencies of different patterns of outcome in the game of Chicken.

said to be large. Note first that we have used payoffs different from Rapoport et al. [253] and this may influence the incentive to form alternating strategies. Second, our players do not explicitly know about the payoff matrix and can only learn about it implicitly through play. Finally, there certainly are some features of human adaptive strategic behavior that are not captured in our current Q-learning model but that are important for human subjects to learn such alternating strategies. The main point, however, is how irrelevant the Nash equilibrium concept seems for describing the outcomes of the iterated coordination games—Chicken and Battle of the Sexes—as played by our agents.

6.4 Summary

Wealth extracted (WE) is the proper measure of an agent's performance in a game. When the game is an iterated one, it may well be to an agent's advantage to explore, taking different actions in essentially identical contexts. Our simple reinforcement learning agents do exactly this. They present perhaps the simplest case of an adaptive, exploring rationality. In utter ignorance of the game and their co-players, they merely seek to maximize their WE by collecting information on the consequences of their actions, and playing what appears to be best at any given moment. This is tempered by a tendency to explore by occasionally making what appear to be inferior moves.

Remarkably, when agents so constituted play each other and NEs are distinct from more rewarding Pareto outcomes (Prisoner's Dilemma, games #47, #48, and #57, Chicken with $\delta < 1$), Pareto wins. The drive to maximize WE succeeds. Similarly, a Pareto superior Nash equilibrium will trump a Pareto inferior NE (Stag Hunt). Finally, in the presence of Pareto outcomes that are socially superior but unequally advantageous to the players, the players learn to extract an amount of wealth close to the maximum available (Battle of the Sexes, Chicken).

Outcomes that are neither Pareto efficient nor Nash equilibria are rarely settled upon. Nash outcomes give way to Pareto superior outcomes when it pays to do so. A bit more carefully, in the case that there is one sub-game perfect NE, these results violate that as a prediction. In the case that the iterated games are seen as open-ended, there are (viz., the Folk Theorem) a very large number of NEs, but there is also insufficient theory to predict which will in fact occur. Again, our agents defy this as a prediction: they rather effectively maximize their Total WE. To sloganize, "It's not Nash that drives the results of iterated play, it's Pareto."

6.5   For Exploration

Q-learning is just one of many extant procedures that may be applied to learning in contexts of strategic interaction. Camerer's review of behavior game theory discusses the main behavioral models of learning in games as of the time of its writing [54, Chapter 6]. Different models have appeared in the economics literature, e.g., [43, 44] and especially [45, 46].

Learning theory is a staple of psychology in its behavioristic incarnation. Many of the models developed for behavioral experiments (strategic or not) might be pursued for strategic interactions [53, 148, 218]. See [117] for a comprehensive treatment from the perspective of a cognitive scientist. Mainstream game theory is represented in [113], although the procedures they investigate are, while adaptive, not exploratory.

This literature may serve as a starting point for investigating definite procedures for learning in games. Compare and contrast the performance of the various procedures.

6.6   Concluding Notes

Reinforcement learning in games has become an active area of investigation. A systematic treatment of the literature would require a rather lengthy review paper of its own. Instead, we shall confine ourselves to brief discussions of certain especially apt works. We begin with several papers describing investigations into reinforcement learning in games by artificial agents.

Hu and Wellman [156] essay a theoretical treatment of general-sum games under reinforcement learning. They prove that a simple Q-learning algorithm will converge for an agent to a Nash equilibrium under certain conditions, including uniqueness of the equilibrium. When these conditions obtain the Nash equilibrium is, in effect, also Pareto optimal or dominant for the agent.

Claus and Boutilier [64] investigate reinforcement learning (Q-learning) agents in coordination (aka: common interest) games. (Claus and Boutilier refer to these as cooperative games, which are not to be confused with cooperative game theory; the games played here are non-cooperative.) The paper studies factors that influence the convergence to Nash equilibrium under the setting of iterated play when using Q-learning. The empirical results show that whether the agent learns the action values jointly or individually may not be critical for convergence and that convergence may not be generally obtainable for more complicated games. The paper also proposes use of a myopic heuristic for exploration, which seems promising to help convergence to optimal (Pareto dominant) equilibrium. However, because the games tested in the paper are restricted to two particular coordination games, the results are somewhat limited in scope.

Bearden [25] examines two Stag Hunt games, one with "high" risk and one with "low" using reinforcement learning and a genetic algorithm to discover parameter values for the agents' learning schedules. His results are not easily comparable with ours, since his two games are effectively parameterized differently than our series of games (as δ changes). Broadly, however, our results are in agreement. Bearden's "high" risk game is closest to our game with $\delta = 2$ or 2.5, while his "low" risk game roughly corresponds to our case with $\delta = 0.75$ or 1. In both studies, there is considerably more joint stag hunting (cooperation) in the "low" risk case and considerably more joint hare hunting in the "high" risk case.

Mukherjee and Sen [228] explore play by reinforcement learning agents in four carefully designed 3×3 games, in which the "greedy" (i.e., Nash) outcome is Pareto inferior to the "desired" (by the authors) outcome. Besides the different games, the experimental treatment involves comparison of two play revelation schemes (by one or both players) with straight reinforcement learning. It is found, roughly, that when the "desired" outcome is also a Nash equilibrium (NE) the revelation schemes are effective in promoting it. This kind of investigation, in which the effects of institutions upon play are explored, is, we think, very much in order, especially in conjunction with further investigation of learning regimes.

Reinforcement learning, in a related sense, has become popular in behavioral economics. A rather extensive series of results finds that reinforcement learning models, often combined with other information, perform well in describing human subject behavior in games. See Camerer [54] for an extensive review. In part as a consequence of the experimental results, there has been theoretical interest by economists in reinforcement learning in games. Burgos [52], for example, tries to use reinforcement learning models to explain subjects' risk attitudes, which are one aspect of choice theory. The setting is pairwise choices between risky prospects with the same expected value. Two models are used for the simulation; one is from Roth and Erev [263] and Erev and Roth [94], the other from Börgers and Sarin [38, 39]. The papers demonstrate a possible explanation of risk aversion as a side effect of the learning regime. This raises the important question of whether risk aversion, risk seeking, and even individual utilities could be emergent phenomena, arising from simple underlying learning processes.

Finally, Bendor et al. [27] study long-run outcomes when two players repeatedly play an arbitrary finite action game using a simple reinforcement learning model. The model resembles that in Erev and Roth (1998). A distinguishing feature of the model is the adjustable aspiration level, which is used as an adaptive reference point to evaluate payoffs. Aspirations are adjusted across rounds (each round consists of a large number of plays). They define and characterize what they call Pure Steady States (either Pareto efficient or Protected Nash equilibrium of the stage game), and the convergence to such states is established under certain conditions. The model limits itself to selection of particular action, thus does not allow mixed strategy or trigger strategy such as "Tit for Tat" in Prisoner's Dilemma. In this simple, but general case, the authors prove that *"convergence to non-Nash outcomes is possible under reinforcement learning in repeated interaction settings"* (emphasis in original).

The results original to this paper are consistent with and complementary to the results reported in the above papers and other extant work. Further analytic and simulation results can only be welcomed. The experimental technique, however, has allowed us to discover hypotheses that merit continued investigation. In particular, our suggestion is that for agents playing games, and learning, wealth extraction (or some variant of it) is a key indicator for understanding system performance. Agents, we suggest, respond to rewards, but do so imperfectly and in a noisy context. If the reward signals are sufficiently clear, the agents will largely achieve Pareto optimal outcomes. If the signals are less clear, the outcomes obtained represent a balance between risk and reward. In either case, it is far from clear what causal contribution, if any, is made by the Nash equilibrium. Resolution of these issues awaits much more extensive investigation.

This material is based upon work supported by, or in part by, NSF grant number SES-9709548. An earlier version of this paper was published as [175]. Permission to republish this material is gratefully acknowledged, to Springer. I wish to thank an anonymous referee for a number of very helpful suggestions.

Chapter 7

Affording Cooperation

7.1 Introduction

The problem of cooperation is to explain and predict cooperative behavior. When does it occur and why? When does cooperation fail to occur and why? We also wish to support intervention in order to promote or discourage cooperation, as suits our interests. This raises the *institutional design problem*: How should we design and manage institutions in order to achieve optimal levels of cooperation? To these we can add the philosophers' perennial issue of characterizing the subject: Just what is cooperation? How does altruism, surely a form of cooperation, differ from other kinds of pro-social behavior? And so on. All of these problems can generate multiple subproblems and varieties.

What makes the *subject* of cooperation forcefully problematic is (a) there seems to be so much of it, and (b) there often seems to be so little of it. These facts are not fully explained. Tipping and helping strangers are just two of many well-known, routine behaviors that appear to be highly cooperative, even altruistic. Testimony is ready to hand from many sources as to the prevalence of such puzzling behavior. To pick one, Michael Tomasello on altruism:

> In the contemporary study of human behavioral evolution, the central problem is altruism, specifically, how it came to be. [SOK: And we might add, how come it has stayed around?] There is no widely accepted solution to that question, but there is no shortage of proposals either. [312, page 51]

I want to lengthen the list of proposals on the origin and maintenance of cooperation, including altruism and other forms of pro-social (helping) behavior. I do not aim to offer a settled explanation of any specific cooperative behavior. Rather, I wish to add a new, or at least revised, account to the existing roster. What I want to add is a conceptual tool, as a resource for empirical science. Uses and tests of the tool must perforce follow later, although I shall in passing present some evidence.

The argument I want to make unfolds in several steps, beginning in the next section. First, two preparatory issues quickly.

7.1.1   Definitions

What should count as cooperation, altruism, pro-social behavior, helping, and so forth? The terms in fact are commonly used with varying senses in the literature. I note two examples. First, Bronstein [47, pages 186–7] writes that

> Diverse terms are currently in use to define interspecific coopera-
> tive interactions. Some of these terms have well-accepted alterna-
> tive meanings, however. Given such inconsistencies, the terminol-
> ogy we choose to employ is more than a strictly semantic matter:
> it can determine whether we are all trying to explain the same
> phenomena. ...
>
> I use the term *mutualism* to refer to all mutually beneficial, in-
> terspecific interactions, regardless of their specificity, intimacy, or
> evolutionary history. ...
>
> Whereas mutualism [as Bronstein will use the term] denotes two-
> species beneficial interactions, *cooperation* has usually been used
> somewhat more vaguely to denote benefits in a within-species con-
> text. Some researchers have used the terms mutualism and co-
> operation interchangeably ...[she goes on to note a plethora of
> conflicting uses and senses]

Second, in a footnote to the first occurrence of the term cooperation, Richer-son, Boyd, and Henrich [255, page 358] say:

> "Cooperation" has a broad and a narrow definition. The broad
> definition includes all forms of mutually beneficial joint action by
> two or more individuals. The narrow definition is restricted to sit-
> uations in which joint action poses a dilemma for at least one in-
> dividual such that, at least in the short run, that individual would
> be better off not cooperating. ...The "cooperate" vs. "defect"
> strategies in the Prisoner's Dilemma and Commons games anchor
> our concept of cooperation, making it more or less equivalent to
> the term "altruism" in evolutionary biology. Thus, we distinguish
> "coordination" (joint interactions that are "self-policing" because
> payoffs are highest if everyone does the same thing) and division
> of labor (joint action in which payoffs are highest if individuals do
> different things) from cooperation.

Note as well that even an everyday word such as *exploit* can be prob-lematic because it has (at least) the senses of "put to good use" and "take unfair or immoral advantage of."[1] Bronstein (see above) writes of "reciprocal exploitation (i.e., mutualism)" [47, page 191] which she contrasts with "uni-lateral exploitation," which itself may or may not be particularly harmful to the exploited party.

---

[1]WordNet identifies three distinct senses for the word as a verb.

There is another obstacle to any attempt to finalize the relevant definitions: the list of relevant terms is growing, with no apparent end in view. We need also to consider *mutualism, reciprocity, pseudoreciprocity, indirect reciprocity, by-product mutualism, kin selection,* and *reciprocal altruism.*

It has been suggested that perhaps our language is inadequate to the phenomena, and that we would be best advised to focus on describing kinds of phenomena that seem to be relevant, worrying later about what to call them (e.g., [192]). I have considerable sympathy for this advice and aim to follow it not only here, but generally. Even so, some rough characterization of certain terms will be useful. Here we go.

As commonly described, altruism is "selfless concern for the welfare of others." While that is fine as far as it goes, I would prefer to put examples before us. At the entry level, helping strangers by giving directions and giving to charity we can take as counting (all things being equal). An agent freely provides—transfers—something of value without expectation of any immediate or very specific reward. I'll call this *altruism in the small.* At the advanced level, we have taken big risks to save someone else or even sacrificing one's life for another. I'll call this *altruism in the large.* Altruism, after all, may not itself be one thing (I'm inclined to think it isn't), and its nature is contested. Some theorize that apparent altruism is often (always?) discretely selfish. "Scratch an altruist and watch a hypocrite bleed." I want to move on and leave these questions for another time and place. Let us agree to take as altruistic any behavior that benefits someone else without any immediate prospect of commensurate benefit for the giver.

I want to use *cooperation* in the broad sense of [255, page 358, see above]. This I take to be essentially the same as mutualism in Bronstein's sense [47, page 186, see above], but with a relaxation to allow more than two individuals and to allow both inter- and intra-species interactions. I will use *pro-social* and *helping* interchangeably to include both cooperation and altruism, as well as anything else that is arguably similar but not similar enough to count as one or the other, such as strong reciprocity [95].

7.1.2 Existing Proposals

The second issue before we begin the argument directly is the list of existing proposals for explaining pro-social behavior. Many kinds of explanations for the establishment and maintenance of cooperation or altruism have been put forward. Called *evolutionary pathways to cooperation* [147, page 40], they include:

- kin selection

  Cooperation, and even altruism, among relatives is selected because relatives share the same genes (e.g., [332]).

- reciprocal altruism

Established reciprocity relationships among individuals lead to selection of cooperation and altruism (e.g., [332]).

- group selection

  A trait may be preserved in a population because the population in which the trait is prevalent out-competes populations in which it is not (e.g., [299]).

- culture

  Cultural institutions promote and enforce norms of cooperation (e.g., [254]).

- bounded rationality

  Cooperation, even altruism, is a side effect of heuristics that on balance are individually favorable and favored by selection (e.g., [122]).

- sexual selection

  Males compete for mating access to females, and can be selected because of their cooperative behaviors (e.g., [332]).

- social selection

  This is a generalization of sexual selection. Competition is for contribution to the entire reproductive process, not just mating ([265]).

- selective investment theory

  Cooperative behavior is seen as an investment. As such it may be advantageous on straightforward Darwinian grounds (e.g., [48]).

- pseudoreciprocity

  A helps *B* because *B*'s flourishing helps *A* (e.g., [81, page 49]).

- indirect reciprocity

  A helps *B* because it will lead to a third party *C* helping *A* (e.g., [239]).

- by-product mutualism

  Mutualism is maintained by selfish behavior that incidentally benefits others (e.g., [208, page 202]).

These possible causes of cooperation are not mutually exclusive. Determining the actual mix is an empirical question. Surely very much depends upon the particulars of the case.

Finally, we can distinguish between *proximal* and *distal* explanations of prosocial behavior. Some of the pathways are best understood as one more than the other. Reputation (a form of indirect reciprocity), social norms, ethnicity [147, page 32] and "identity economics" [2, 3, 4] are but proximal explanations of cooperative behavior. What explains why we rely on reputation effects to modify our behavior or why we value identities and rely on them to support cooperative behavior? These are questions that call for distal, presumably evolutionary and/or cultural explanations.

7.2 A Game That Affords Cooperation

Imagine a tribe of hunter-gathers. They are in a primitive state of development compared to modern humans. There is little or no supporting technology, or even cultural practice, *jointly* useful to individuals as they hunt for food. There is little in the way of *cooperation afforders* for the hunt, things that support cooperation in the hunt.[2] There is, in consequence, no joint gain to be made from cooperation. Everyone hunts hare, as it were. Then there arrives a new afforder with the property that it affords cooperation between two individuals if they *both* have and use it. Think of this afforder as perhaps a net that must be held by two people for catching a stag. The net must be assembled from two components, one brought to the field by each of the two hunters. Let us assume further that the cost of bringing a single component to the field is small but real. The afforder is cheap, not free. For the sake of definiteness, let the strategic situation be as in Figure 7.1.

	C		D	
		2		1
C	[PN]			
	2		0.99	
		0.99		1
D			[N]	
	1		1	

FIGURE 7.1: Cooperation Afforder game, canonical payoffs.

Think of the payoffs as measured in days of food harvested, or some other natural quantity. If an agent brings the afforder to the hunt (and uses it), this is to play the C or cooperative strategy; if not then it is playing the D or defecting strategy. Under the standard payoffs (Figure 7.1), two cooperating agents will each harvest 2 days of food. Two defecting agents will each harvest 1 day of food. When the counter-party to a cooperating agent is a defecting agent, the cooperator harvests a day of food minus a small cost of bringing

---

[2]I use the term afforder with allusion to Donald Norman's use *affordance* [236]. A design, in Norman's usage, provides certain affordances, things we can have or do because of the design. By the term *afforder,* I merely mean that which affords something, so a design is an afforder and what it affords is an affordance.

the afforder to the field, while the defecting agent gets the day of food without the extra cost.

In upshot we have a variety of Stag Hunt game. The philosopher Brian Skyrms has made a compelling case for attending to the Stag Hunt game. He has explored general Stag Hunt games and documented the problems of achieving sustained cooperation in them [295, 297].

	C	D
C	[PN] 3 / 3	2 / 0
D	0 / 2	[N] 1 / 1

FIGURE 7.2: Skyrms's standard Stag Hunt.

With the payoff regimes like those Skyrms has typically handled, e.g., Figure 7.2, it is indeed very possible for the replicator dynamics (see §4.4) to drive a population to all D, depending on initial conditions. For the game of Figure 7.2 and a start of Cs and Ds in equal numbers, a frequent result is conquest by Ds. If Cs are in the minority, it is very unlikely that the replicator dynamics will drive the population to fixation on C. Herein lies a puzzle for cooperation, as Skyrms has emphasized. How can cooperation get started?

	C	D
C	R / R	N / $N - c$
D	$N - c$ / N	N / N

FIGURE 7.3: Cooperation Afforder game, abstracted. R: reward for cooperation. N: payoff for no afforder. c: (small) cost of the afforder. $R > N > c > 0$.

The Cooperation Afforder game, as abstracted in Figure 7.3, is genuinely a Stag Hunt (or Assurance) game in the received sense. It is symmetric and $R > N > N - c$. According to classical game theory, only the ordering of R, N, and $N - c$ payoffs matter, not their magnitudes. For us it is different. Our concern is with evolution and other causal processes, not with what happens under some a-causal theory of ideal rationality (as in classical game theory). Payoff magnitudes matter. Under the replicator dynamics for the game in Figure 7.1, so long as the frequency of Cs is at or above about 2–3%, the population will reliably evolve to all Cs. With balking,[3] multiple rounds of play and reactive strategies, etc., starting frequencies of less than 2% for Cs can reliably result in conquest by cooperators. These are empirical claims, arrived at using ReplicatorDynamicsWithBalking.nlogo, but they are robust. Figure 7.4 shows ReplicatorDynamicsWithBalking.nlogo initialized and ready to run the Cooperation Afforder game. Only 3% of the agents are cooperators, shown as black small squares. The defectors are white small squares. After a number of generations, on the order of 60, the defectors (white squares) will be eliminated. The entire population will consist of cooperators.

These tests were with large populations, of about 95,000 individuals. With small populations it is quite plausible that strategy C could appear and *drift* (experience sampling error) to a frequency sufficient for its selective advantage to come into play and drive the population to all C. Such a population could expand at the expense of populations without the afforder.

There are four key properties of the Cooperation Afforder game that afford the evolution of cooperation.

1. There are substantial benefits to be had from cooperation.

   To illustrate, in the Cooperation Afforder game with standard payoffs, Figure 7.1, 2 days of food from mutual cooperation is substantially better than 1 day of food when not cooperating.

2. Cooperation is not very costly if unrequited.

   Although cooperation is not free, neither is it immediately devastating to offer cooperation and not have it reciprocated. In terms of our example, the fitness loss is low enough, compared to the benefits of cooperation, that even relatively rare encounters with fellow cooperators will pay for the cost of carrying the afforder.

3. Cooperation unrequited is not exploitable.

   If one agent cooperates and the other defects, the defecting agent gains no positive benefit of any size. There is a fitness loss by the cooperating agent, but the (relative) fitness gain is divided among all of the non-cooperators. The non-cooperating counter-party in question has no idiosyncratic gain. The replicator dynamics with the Cooperation Afforder game are simply a concrete, specific way to demonstrate this.

4. Fixation of cooperation in the population is stable.

   In fact fixation of non-cooperation is unstable. For the numerical example just given, a 3% drift away from universal non-cooperation will reliably lead to fixation of cooperation. It would take at least a 97% drift away from universal cooperation to lead to fixation of non-cooperation. This is extremely unlikely in a population of any significant size.

---

[3]Balking occurs when one player in a game refuses to play. ReplicatorDynamicsWith-Balking.nlogo supports balking models. See its online documentation (Information tab) for details.

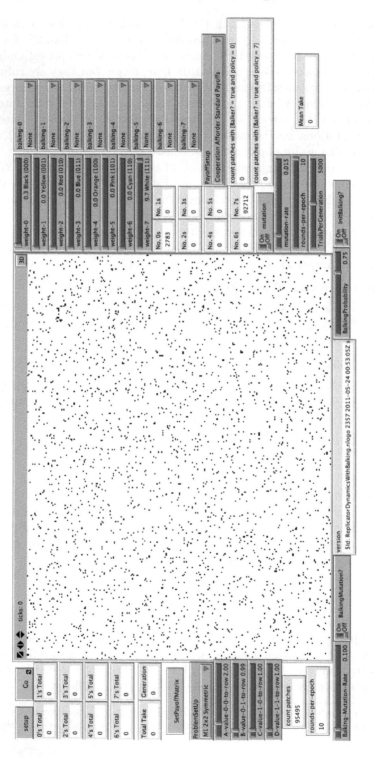

FIGURE 7.4: ReplicatorDynamicsWithBalking.nlogo set up for a run of the Cooperation Afforder game, starting with 3% cooperators, balking and mutation off.

Under these general conditions, which need not be specific to the Cooperation Afforder game, cooperation should often be favored by evolution and/or by learning. Put otherwise, when these conditions obtain, we are able to answer the question of how cooperation can get started and then be maintained. Broadly speaking, these conditions suffice for cooperation.

An analytic argument will serve to bolster the case and support the findings of the replicator dynamics model in ReplicatorDynamicsWithBalking.nlogo. Suppose that the frequency of the cooperation afforder is x. Then the expected return from having it is:

$$x \cdot R + (1 - x) \cdot S \tag{7.1}$$

where R is the reward from mutual cooperation and $S = P - \varepsilon$ is the "sucker's payoff," the payoff from carrying the cooperation afforder and not having it reciprocated. The expected return from not having the cooperation afforder is

$$x \cdot P + (1 - x) \cdot P = P \tag{7.2}$$

where P is the "penalty" for lack of cooperation.

Having the cooperation afforder has a higher expected value if

$$x \cdot R + (1 - x) \cdot S > P \tag{7.3}$$

Fixing P at 1 and S at 0.99, this yields

$$x \cdot R + (1 - x) \cdot 0.99 > 1 \tag{7.4}$$

Simplifying, we get

$$x \cdot (R - 0.99) > 0.01 \tag{7.5}$$

Now plot

$$x \cdot (R - 0.99) - 0.01 = 0 \tag{7.6}$$

For

$$x \cdot (R - 0.99) - 0.01 > 0 \tag{7.7}$$

the expected reproductive value of the cooperation afforder is larger than that for not having it. At $x = 0.02$ and $R = 2$ (we used 0.03 in our computational example, above) the value of the afforder is

$$0.02 \cdot (2 - 0.99) - 0.01 = 0.0102 > 0 \tag{7.8}$$

Rearranging expression (7.6)

$$x \cdot (R - 0.99) - 0.01 = 0 \tag{7.9}$$

we get R as a function of x:

$$R = \frac{0.01 + 0.99 \cdot x}{x} \tag{7.10}$$

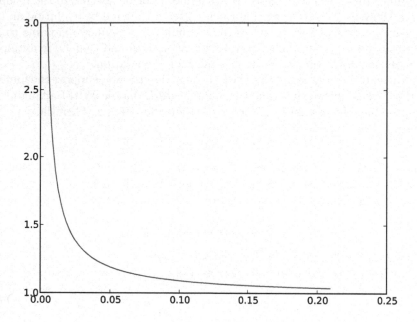

FIGURE 7.5: Cooperation affording region (above the curve) for R as a function of x.

Plotting R on the vertical as a function of x on the horizontal with these values gives us Figure 7.5.

For these numbers, we can see that the region of cooperation is quite large in the space of R and x. Any situation reasonably close to the Cooperation Afforder game standard payoffs will be driven by evolution in the direction of fixation of cooperation.

7.3   What's That Technology?

The upshot of the previous section is that there is no genuine impediment to understanding how cooperation could conquer a population and a species via something like the Cooperation Afforder game. This naturally raises questions regarding what afforder or afforders could, or even did, play this role. If evolution has to find anew every form of cooperation, at best we have made scant progress. To motivate my general answer to the issue, I shall take a brief digression.

7.3.1 Hi-Lo

Consider the simple, no-conflict, even trivial game represented in strategic form in Figure 7.6 (after [20]). For obvious reasons, the game is called Hi-

	H	L
H	2 / 2 [NP]	0 / 0
L	0 / 0	1 / 1 [N]

	H	L
H	A / A [NP]	0 / 0
L	0 / 0	B / B [N]

FIGURE 7.6: Hi-Lo. Assumed: $A > B$.

Lo. It is an example of an *inessential* game, a type normally passed over with dismissal because it is so trivial [253, page 22]. (We dismissed similar games in Chapter 4 and may want to reconsider them now.) Notably and commendably, Bacharach [20] disagrees and discusses the game in depth.[4] Let us review his position.

Any unmitigated equilibrium account of Hi-Lo (or indeed Stag Hunt), as in classical game theory, will fail for definiteness in the presence of multiple equilibria, and the classical account fails for that reason here.[5]

> Nash equilibrium fails to confer determinacy anyway, so there is no need to dispute the principle in order to show that standard theory [that is, classical game theory] fails for Hi-Lo. [20, page 61]

Things get worse. The theoretical problem of multiple equilibria is exacerbated in the case of Hi-Lo because the practical problem is absolutely, utterly unproblematic. People's behavior and all our intuitions agree that playing H is the rational thing to do. The theoretical challenge is to explain why— descriptively, in practice—it *is* unproblematic to play high, H. Bacharach puts it nicely.

> You are to play Hi-Lo, and it is common knowledge that you and your coplayer are intelligent people. It seems quite obvious that you should choose A [H in our Figure 7.6]. However, the question why it seems obvious, and the related question of why people almost always do choose A [H], have turned out to be anything but easy to answer. [20, page 35]

---

[4]While I agree with much of his analysis, I would depart considerably from elements of it. These matters, however, do not matter for the present.

[5]Bacharach is aware of [142] and related work on equilibrium refinement, which neither he nor I thinks succeeds in eliminating the paradox. See [20] for a discussion, which would take us too far afield.

Such is the Hi-Lo paradox of classical game theory. Game theory should, one would think, easily be able to offer a satisfying account of why players will overwhelmingly coordinate on Hi in Hi-Lo. It does not. Game theory should, one would think, also easily be able to offer a satisfying prescriptive account of why players should play Hi in Hi-Lo. Again it does not. We now turn to Bacharach's alternative account.

Bacharach observes that classical game theory presumes individual thinking, in which each of the players asks of itself: What is it best for me to do in this situation? Unfortunately for classical game theory, there is no determinate answer to this question via the theory. That is the thrust of the Hi-Lo paradox in Bacharach's view. As a way out of the paradox Bacharach proposes a different framing: team thinking. A team thinking agent, confronted by a strategic decision problem asks instead: What is it best for my team to do and what is my role in achieving it? Having answered these questions, the agent follows through by enacting its role in what on balance it judges is best for the team to do.[6]

Credible examples of team reasoning occur in team sports. Two players perceive an opportunity for the team and perceive that the other shares the perception. They spontaneously coordinate their actions. Simultaneously, one throws the ball to a spot and the other runs toward the spot. This happens— it has to happen—too quickly for the players to signal each other and agree. They act autonomously on what they understand and very often will achieve effective coordination. Ordinary life is rich in activities that will often be best understood as evidencing team reasoning by their members. Of course in quotidian interactions it can hardly require the skills of professional sports players. Examples include group hunting, rowing, dancing, singing, military activity, and various forms of work [223]. On the last, McNeill thinks we see it not only in gangs of laborers, but also in the efforts of modern corporations to instill company spirit through propaganda and (in Japan) calisthenics.

Bacharach's view is that coordination in Hi-Lo works similarly. The agents see the game (perceive the situation), are cued for team thinking, reason— trivially—that what is good for the team is for each player to play H, and end by ... playing H. (Wouldn't you?)

If the players are in fact engaging in team reasoning (or something like it), it is a virtue of Bacharach's account that it can easily explain and predict what should be obvious—coordination on H—but is unexplained entirely by individual thinking and classical game theory. This ends the digression.

---

[6]There are many complications possible which for present purposes need not detain us. For example, what if the agent thinks that some of the other team members cannot discern or enact their roles? See [20] for discussion of some of the related nuances.

7.3.2   Pro-Social Framing

My thesis, the explanatory path to cooperation I wish to put on the table, employs a version of Bacharach's account of team reasoning. There are several parts to the thesis. Here are the first few.

1. **Causation.** Pro-social frames, or framing, once present in a population could, by the argument with regard to the Cooperation Afforder game, evolve to domination, even fixation, in a population.[7]

2. **Behavior.** Distinct alternative frames of thinking about strategic interactions exist. Among them, prominently, are a selfish, self-centered frame and a pro-social frame. Individuals using their pro-social frames will tend to behave cooperatively, even altruistically. In the team thinking frame, individuals identify with the team and condition their actions on their estimates of what they should do for the sake of the team.

In short, *pro-social framing is an afforder whose proliferation may be explained straightforwardly on evolutionary grounds.* This is a candidate for explaining, or at least contributing to the explanation of, many cooperative phenomena. Let us call it the *cooperation afforder hypothesis with framing.* Cooperation that comes about and is maintained in virtue of the process hypothesized may be called *afforder-framing cooperation.*

The existence point—that there are both selfish and pro-social frames of thinking—is hardly a new idea. It goes by different names with different authors.[8] The idea has been in the air. Here is a recent representative passage.

> In shared cooperative activities, we have a joint goal that creates an interdependence among us—indeed, creates an "us." If we are carrying a table to the bedroom, I cannot simply drop it and run off without hurting us and our goal. In shared cooperative activities, my individual rationality—I want to transport the table to the bedroom so I should do X—is transformed into a social rationality

---

[7]Frames for Bacharach and in the other literature I discuss are alternate ways of organizing experience. One might speak of them as collections of presuppositions, which are used to direct action, arrive at new beliefs, and so on. So we might think of frames as collections of beliefs and a framing effect as something believed or done in consequence of a particular collection of beliefs. But this is not necessary. Frames may organize experience without recourse to anything directly propositional, for example by using a procedure to interpret sensory signals. The account I am giving here, and the accounts given by others employing framing, are mostly neutral with regard to how frames are built and to how framing effects result from them. This said, it is unlikely that frames are fundamentally propositional or doxastic. In the end, the issue is an empirical one whose resolution is of peripheral consequences for the case I am making.

[8]The authors may or may not cite one another. Tomasello [312] cites philosophers, [42], [123], [230], [281], and [313] who earlier introduced the terms I-mode thinking and we-mode thinking, corresponding roughly to individual and team thinking. On the other hand, Tomasello does not cite economists and game theorists, e.g., [20], Sugden as in [224, 303], and Colman as in [66]. Nor do they cite the philosophers or the biologists.

of interdependence: *we* want to transport the table to the bedroom, so I should do X and you should do Y. [312, page 41]

The case I am making is not specific regarding the details of the pro-social afforder and associated framing. Intendedly so, since I am making a basic point. There is much remaining to do by way of modeling and empirical investigation. I note, however, several features commonly identified with pro-social framing (we-mode thinking, team thinking, etc.):

1. **Shared goals.** The agent directs its activities to fulfilling the goals of the pertinent team, group, or society.

2. **Imitation.** Insofar as feasible, the agent emulates and conforms to characteristic behaviors of members of the group.

3. **Enforcement.** The agent undertakes to enforce norms of conformance, and recognizes an obligation, or at least need, to punish defectors from the group.

I wish to submit another afforder of cooperation, one that complements and abets cooperative framing and team reasoning. That afforder is communication.[9] The ability to communicate, even in a rudimentary way, will plausibly often have scope to instantiate these three key properties (see page 139 for the list of four) of the Cooperation Afforder game:

1. There are substantial benefits to be had from cooperation.

   The potential benefits of cooperation have to be provided exogenously by a combination of the environment and the agents' needs and capabilities. If they are such that cooperation abetted by communication would in fact reward the players amply, then cooperation may well ensue by the Cooperation Afforder game argument, above.

2. Cooperation is not very costly if unrequited.

   It is not safe to assume that the act, or the capacity, of communication is free, especially considering the necessary apparatus for being able to communicate. Given the apparatus, however, the energetic costs of signaling are surely low, or often can be.

3. Cooperation unrequited is not exploitable.

   If the addressee of the communicative act does not reciprocate with cooperation, there is, or can be, very little strategic cost to the sender. This may happen because the target fails to understand or in some other way fails to have uptake of the message. The target might have perfectly adequate uptake, but choose to ignore the message. What is here being excluded, for the Cooperation Afforder game to apply, is that the target

---

[9]Again Brian Skyrms has broken important ground in exploring the subject [298].

of the communication is able to use the message to its advantage and to the disadvantage of the sender, say by signaling a third party and effectively denying the original sender access to the resource. Of course it can happen that unrequited cooperation is exploitable and is exploited. The suggestion here is that this can be a rare enough eventuality that the conditions for the Cooperation Afforder game do in fact approximately apply.

Note in this regard Grice on communication and the important role of the cooperation principle [134]. Communication in Grice's analysis presumes cooperation and to violate the rule of cooperation is to abuse the institution. Could it be that a propensity for pro-social framing afforded the development of communication?

7.4   What about Evidence?

I will be brief, given my goal of (mainly) presenting an evolutionary pathway to cooperation.

The cooperation afforder hypothesis can explain, or can help explain, the presence of something on the order of team thinking, we-mode thinking, and shared cooperative activity thinking. Whether it is uniquely qualified to do so is beyond the scope of the present discussion. Little or nothing really rides on that issue. The more urgent question, I submit, is whether there is in fact evidence for something on the order of team thinking, we-mode thinking, shared cooperative activity thinking. Do we indeed have these two distinct modes of thinking, the one pro-individual, the other pro-social?

We do. There is both direct and indirect evidence. On the indirect side, the presence of framing and framing effects in human psychology is very well established. Richard Gregory [132, 133] has made us well aware of visual illusions. The Necker cube, Figure 7.7, is but one of very many examples of so-called optical illusions. These teach us that what we see depends on how we see, how we frame things.

The effect occurs outside of vision and even perception. We see it in our conceptualizing of the world. This is, I take it, a main point the philosopher Ludwig Wittgenstein was after in his duck-rabbit discussions.[10] The observation is an ancient one. Is Antigone's opposition to Creon a defiance of law and duty to her fellow citizens or is it a noble display of fidelity to and honoring of her dead brother? The frames, the presuppositions, of Antigone and Creon conflict with tragic upshot in Sophocles's play *Antigone*. The effect is often present in moral dilemmas, but has been established in many contexts and is

---

[10]The duck-rabbit is a standard optical illusion. Wittgenstein's discussion of it and related matters is to be found in [334].

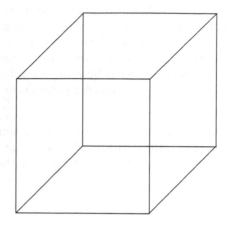

FIGURE 7.7: Necker cube.

well accepted in the psychology literature [317]. It is also often on display in language-based humor, a wonderful genre of which collects newspaper headlines. A recent random example I found: "District judges receive sex-assault guidance" —a headline in *The Philadelphia Inquirer* B3, Friday, March 18, 2011. Then there is "Hot tubs may be safer than thought" from *The Ann Arbor News*. The *Columbia Journalism Review* publishes a page of multiply-interpretable headlines every month. The indirect point is that we see framing effects all over and should not be surprised to find another one.

Experimental evidence for team reasoning in Hi-Lo and related games is also available, although only a few controlled experiments have been reported. Team reasoning in Hi-Lo is especially difficult to explore experimentally because there is no variation to work with: everyone always goes with H. Similar games can be designed, however, that do not universally elicit successful team coordination. Subjects may then be cued with team-promoting or individual-promoting messages. We would expect, if the team reasoning hypothesis is correct, that team-cued subjects would achieve coordination more often than individual-cued subjects. In experiments of this kind, Colman and others [66, 67] convincingly do obtain support for the team reasoning hypothesis.

More or less direct evidence of another sort is available from the literature on friendship, which has been reviewed recently by Silk [290]. Her money quote is: "Friendship is not mutualism" [290, page 47]. Silk's concept of mutualism is, unlike Bronstein's, tied to near-term reciprocity, requiring close bookkeeping that is absent in friendships. She puts her point this way.

> Taken together these experiments provide empirical support for the distinction between exchange and communal relationships. More importantly, they support the hypothesis that communal relationships are not based on strict TIT FOR TAT reciprocity. People

use TIT FOR TAT reciprocity as a diagnostic criteria for the existence of close friendships; when benefits are balanced directly, relationships are assumed to casual and ephemeral [i.e., exchange rather than communal]. People seem to make concerted efforts to obscure the accounting of costs and benefits among their friends—in joint tasks, they hide their own contributions and avoiding [sic] monitoring their friends' contributions. [290, page 46]

Continuing, she writes,

...the exchange-communal distinction implies that the process that preserves the balance in these two different kind [sic] of relationships differs. In exchange relationships, help is given with the explicit expectation that it will be reciprocated. In communal relationships, help is given because it is needed or desired; when both partners have the same communal orientation, benefits will flow back and forth, but they will not be strictly contingent on expectations of future benefits. [290, page 46]

Communal thinking is, it would seem, much the same thing as team thinking. Perhaps it is a generalization. In any case, communal thinking is a form of framing. It is liable to escape by experience, just not by comparatively small deviations from strict reciprocity. Team thinking and communal thinking make us more patient with our counter-parties.

The existence of framing and framing effects is reasonably well established. We should view the existence of something on the order of team reasoning, we-mode thinking, communal thinking, etc. as a credible hypothesis, one to be investigated further.

Before beginning to generalize the argument in the next section, it may be helpful to rehearse in compressed form the argument, or perhaps narrative is a better term, so far.

1. The Cooperation Afforder game is a model. It may be useful in explaining the origin and maintenance of cooperation in some cases. Calculations with the model demonstrate how a low-cost, reasonably high-benefit afforder of mutual cooperation could evolve to fixation in a population, after appearing even in very small numbers.

2. Given the model and its interesting property of explaining the fixation of cooperation, we would like to know how accurately it tracks relevant aspects of the real world. That is a very large question, one that I am not attempting to dispose of here. Other important questions include:

   (a) Are there credible candidates for the role of a *general* cooperation afforder?

   (b) If there are credible candidates, what does this tell us about further consequences of the narrative?

3. I take up question (2b) in §7.5, and address question (2a) immediately below.

4. Regarding question (2a), candidates that are very specific to particular circumstances are not very credible as sources for the high level of cooperativeness we find about us. The Cooperation Afforder game may explain the fixation of several afforders, but it would be a stretch to ask it to explain hundreds of them (e.g., one for hunting animals, another for hunting fish, another for farming, and so on).

5. Uncooperative behavior, the propensity to act in terms of one's perceived selfish interests, without regard to the interests of one's group, is a behavior that has not been seen to be problematic from an evolutionary perspective.

6. The Cooperation Afforder model teaches us that a propensity to act in terms of the interests of one's group (of two so far) is, under certain circumstances (not too costly, etc.), also not problematic from an evolutionary perspective.

7. So there are (at least) two propensities to attend to. One is for selfish behavior, the other for cooperative behavior.

   (a) Do we need both to explain the phenomena?

   (b) If so, how do they work and what are their properties?

8. Regarding (7a), the question of whether we need both propensities, that we do has prima facie support from the evident fact, very widely accepted, that there is quite a lot of prima facie selfish behavior and quite a lot of prima facie cooperative behavior. We can of course try to eliminate the need for one or the other, and this is what some of the paths to cooperation (above, §7.1.2) try to do. The exercise here is, in part, one of offering a kind of explanation that would serve to buttress the judgment that both selfish and cooperative propensities exist. We should recall as well Bacharach's argument, discussed above in §7.3.1, that the behavior universally found for the Hi-Lo game *cannot* be explained with the assumed selfish propensity of contemporary game theory.

9. Regarding (7b), the question of how these propensities work, I submit that they are, or operate like, the cognitive frames we are familiar with and for which we have received considerable evidential support quite independently of the issues immediately before us. If not quite evidence, this is an attractive property of the suggestion.

10. Cognitive frames are cued by experience and once in place the agent operates with them as defeasible presuppositions. Additional experience may cause a frame in use to be replaced by another. Sometimes, as with some optical illusions, frames are easily displaced. Most people can make

the Necker cube flip left and right at will. In other cases, such as with friendships, frames are not so easily displaced. They may be said to be sticky.

Given the evolutionary argument for the fixation and maintenance of a mutualistic propensity for cooperative behavior, and the hypothesis that the cognitive basis for this relies on reasonably sticky framing, what else can we explain?

7.5   Ramifications beyond Cooperation?

The Cooperation Afforder game models the capacity for two-player mutualistic cooperation spreading and becoming established in a population, once appearing in low frequency from exogenous causes. How do we get from this to an explanation of altruistic behavior, of contributing to public goods, and in general of defaulting to cooperative, even altruistic behavior? It is these behaviors that are indeed most in need of explaining.

If what evolves or is learned to support cooperation is specific to place, context, goals, and so on, it is indeed very difficult to see how the Cooperator Afforder game process would generalize to other circumstances. If what we acquire is a particular ability to cooperate in hunting stag, why would it apply elsewhere? There is little reason to expect altruistic donation to the Red Cross in consequence of cooperation in Stag Hunt.

The first thing to notice is that there is nothing in the Cooperation Afforder game account that requires cooperation be limited to *two* agents. If teams of agents, of communicating, communal-reasoning agents, find opportunities for cooperative behavior in which the costs of offering cooperation are low and cannot be exploited, then we can expect groups of cooperators to thrive. Perhaps three can bag a stag better than two.

I want to add another part of the cooperation afforder with framing hypothesis. I call it the *principle of imperfect specificity*.

1. **Causation.** Pro-social frames, or framing, once present in a population could, by the argument above with regard to the Cooperation Afforder game, evolve to domination or fixation in a population.

2. **Behavior.** There exist frames of thinking, among them, prominently, a selfish, self-centered frame and a pro-social frame. Individuals using their pro-social frames will tend to behave cooperatively, even altruistically.

3. **Imperfect specificity.** Individuals with pro-social frames will compete with (and among) agents having pro-individual frames. They may, in many contexts, prove to be competitively superior to individuals playing only pro-individual frames.

Recall the explanation for play in Hi-Lo. The player sees and understands the game, goes into team reasoning mode, reasons that its job is to play H, and plays H. The pattern is:

1. Perceive a situation calling for action.

2. Have cued pro-social thinking.

3. Decide on an action based on the presumed framework and assumptions of the pro-social thinking.

What I suggest is that given a frame, ordinary thinking is all that is required for action, but different frames will lead to different actions because different assumptions are being made. Once a frame is available that supports collaborative activity, cooperation, et cetera, the agent may acquire cues for that frame that are different from those that originally supported it. Because cues are always imperfect, there will be a tendency to apply the frame in new situations. In short, a pro-social frame, if available, will compete to be cued with a pro-individual frame. And if the rewards are superior it will tend to be used.

What could make pro-social frames out-compete pro-individual frames in contexts beyond simple mutualism? Sexual selection is one obvious source.[11] It is not much of a stretch to imagine that in our evolutionary history a male conspicuously exhibiting pro-social behavior would be especially attractive to females because of the prospect of contributions to raising the offspring.[12]

Punishment of asocial or antisocial behavior is a second possible source of competitive advantage for pro-social framing. If a pro-social frame includes a propensity to imitate others in the group, to conform to the group's behavior, and to punish those who do not, we should not be surprised if this frame pervades a given group. If in addition the cooperation so induced is productive, the group may be expected to grow and to out-compete groups without the frame. Support for this kind of story may be found originally in [40]; see [147] for a recent treatment.

One other source (there are others I will not discuss): transaction costs. Part of the reason there is imperfect specificity is surely that it is too expensive (or even impossible) to maintain an apparatus that is highly accurate in discriminating when to and when not to engage in communal thinking and communication. Deciding when to engage one's cooperation afforder itself has costs. In consequence we may often engage in seemingly altruistic behavior—tipping in places not to be visited again, helping strangers, and so on—as a default, which is not overridden by careful scrutiny and calculation. If the opportunity cost is low (Just how much richer would you be if you always refused to give directions when asked by a stranger?) and you live in a society

---

[11] Or social selection in Roughgarden's sense [265].

[12] Roughgarden [265] makes a similar point about sexual selection in birds. I do not believe she envisions any generalized pro-social framing, however; it is not necessary for contributing to brood support.

in which others mostly have made similar cost-benefit tradeoffs, then neither learning nor natural selection may disfavor it. The alternative is simply too costly for the available benefit.

This kind of argument may go far to account for altruism in the small. As for altruism in the large, I would expect to see less of it. Even so, kin selection plus imperfect specificity may account for some cases. Among non-kin, catastrophic errors of altruism that occur rarely may be governed by imperfect specificity because the alternatives are too expensive.

7.6 Summary and Conclusion

Here is the mode of explanation I would offer for explaining certain forms of pro-social, helping behavior.

1. The Cooperation Afforder game (in conjunction with the replicator dynamics or any similar evolutionary model) teaches us that a low-cost afforder supporting cooperation and avoiding exploitation could, once it appeared even in small numbers (say due to mutation), evolve to dominance and even fixation in a population.

2. If the afforder acquired from such a process were a generalized ability to engage in pro-social thinking (and perhaps communication), this afforder would inevitably be tried in circumstances beyond those originally occasioning it.

   (Recall the saying: "To someone with a hammer, everything looks like a nail.")

3. There are plausible circumstances in which pro-social thinking would out-compete pro-individual thinking, and in consequence be reinforced by evolution and/or learning.

4. A cooperation afforder (such as a generalized ability to engage in pro-social thinking and communication) could also grow and be maintained in a population due to its uses in situations that are (somewhat) exploitable, if they are infrequent enough and the cost of discerning them is high.

   Although agents might often have incentive to defect from pro-social thinking, the monitoring and discrimination costs may well be prohibitive and/or the frequency of exploitation by counterparts may be low.

5. This process may explain much altruistic behavior. In the small, agents are simply relying on an imperfect heuristic—pro-social thinking—that

is usually successful in the circumstances as perceived. In the large, catastrophic mistakes in very unusual circumstances are also subject to the logic of cost-benefit tradeoffs for monitoring and discrimination.

(Recall Franklin's adage: "Honesty is the best policy.")

In the briefest summary, pro-social framing with communication is a cooperation afforder that in many circumstances can be directly favored by selection and/or learning. It is a heuristic, one that I believe has been supported by evolution and learning. In consequence, humans pervasively adopt it as the default stance in social encounters, although we can, and do, learn to recognize exceptions. Often, what looks like (or simply is?) altruism is an application of a pro-social framing heuristic to which its carrier has not formed an exception.

This picture fits nicely with several well-established bodies of experimental findings.

1. Human infants, as early as 14 months old, spontaneously exhibit helping behavior. The effects are remarkable, and have been documented by Tomasello and his group (see [312] for an overview).[13]

2. Humans as adults robustly display *strong reciprocity,* that is "people willingly repay gifts and punish the violation of cooperation and fairness norms even in anonymous one-shot encounters with genetically unrelated strangers" (see [95, page 55], who also provide a useful review of the data).

3. In the well-known Dictator game, one of two agents is given the power to decide on a split of an amount of money. This is played as a one-shot game between anonymous counterparts. Typically, the dictator allocates a non-trivial amount of money to the counterpart, who must accept the dictator's decision. Under the assumptions of classical game theory and economics, the dictator should always appropriate the entire amount, leaving nothing for the counterpart. Interpreted in light of the cooperation afforder hypothesis, allocating a non-trivial amount of money to the counterpart indicates the presence by default of a pro-social framing heuristic and measures its strength or resistance to being accepted.

4. In the well-known Ultimatum game, one of two agents (the proposer) is given the power to propose a split of an amount of money and the other agent (the disposer) is given the power to accept or reject the proposal. If the disposer accepts the proposal it is implemented and the game is over; if the disposer rejects the proposal, the game is over and neither player receives any portion of the money. Ultimatum is also played as a one-shot game between anonymous counterparts. Typically,

---

[13]Perhaps the reader will be struck, as I have been, with the incisive insight revealed in the lyrics of the song "You've Got To Be Carefully Taught" from the musical *South Pacific.* It would not be the first time that a work of art has done this.

proposers propose a split that allocates a substantial portion (50% or even more, but more usually in the 20–40% range) of the available fund. Also typically, disposers reject portions amounting to less than, say, 20% of the fund.

Under the assumptions of classical game theory and economics, the proposer should always appropriate almost the entire amount, leaving next to nothing for the counterpart, and the counterpart should always accept such a proposal. Interpreted in light of the cooperation afforder hypothesis, the disposer's rejection of ungenerous proposals indicates the presence by default of a pro-social framing heuristic and measures its strength or resistance to being excepted. The proposer's allocating a substantial portion of the fund to the disposer both indicates the presence by default of a pro-social framing heuristic and measures its strength or resistance to being accepted, and indicates the proposer's understanding of the disposer's likely frame of thought.[14]

With the cooperation afforder hypothesis on the table as an evolutionary pathway to cooperation, we may hope that systematic empirical and modeling investigations will explore it and its ramifications, including those pertaining to social policy and institutional design.

It is appropriate to remind the reader that cooperation is normatively ambiguous. If a finite common resource requires cooperation to exploit it, the account here explains how cooperation could be supported by evolution with the resulting cooperation leading to a tragedy of the commons [141]. This is the common-pool resource problem. Concretely, exploitation of fossil fuels requires extensive cooperative action, yet it may well lead to destruction of the earth's ability to support our species.

7.7 For Exploration

1. Classify each of the pathways to cooperation listed in §7.1.2 as either distal or proximate causes of pro-social behavior.

2. For each of the proximate causes identified in the previous question, discuss whether and if so how they can be explained distally (that is, by biological evolution or learning).

3. We use the term *commodity* very generally, for goods, for stuff, for anything that has value. A commodity can be concrete or abstract, physical

---

[14]The literature on both the Dictator game and the Ultimatum game is voluminous. [54] provides a good overview and summary of these games, and of much else that is relevant to the cooperation afforder hypothesis.

or social. It may be material thing (stuff on the shelf of a store), a service, an obligation, a right, and so on.

In economic analysis it is standard to classify commodities with regard to *excludability*. A commodity has high excludability if it is easy for the owner of the commodity to restrict access and use of the commodity by others (and low excludability if this is difficult). One's personal property, such as clothes, food, place of residence, and so on, normally has high excludability. One's knowledge and one's job are less excludable. National defense and the earth's atmosphere are standard examples of highly non-excludable goods.

Subtractability is a second standard property used to classify commodities. For a given commodity, does using it, taking advantage of it, lead to a high reduction in its value to others or to a low reduction? One's personal property is normally highly subtractable. If someone else is wearing your clothes, you cannot wear them yourself at the same time. Highly subtractable commodities are also said to be *rivalrous*. The atmosphere, national defense, information, knowledge, and know-how are, normally, non-rivalrous, having low subtractability. Public infrastructure, such as roads, rail lines, and bridges, will often have low subtractability when they are not overcrowded.

With two dimensions to hand, we have framework, the *excludability-subtractability framework*. See Figure 7.8 (after [242, page 7] and [241]).

		Subtractability	
		Low	High
Excludability	Easy	Toll goods	Private goods
	Difficult	Public goods	Common-pool resources

FIGURE 7.8: Excludability-subtractability framework.

Discuss and comment upon the following statement:

> In view of the excludability-subtractability framework, it would appear that commodities that are public goods or common-pool resources are subject to market failure. Common-pool resources will be over-exploited and public goods will be under-provided. To remedy these kinds of market failures it is necessary either to impose social order (which Hobbes advocated, calling it Leviathan) or develop and maintain cooperative behavior sufficient to the situation.

4. Following up on the previous questions, assess the degree to which and the conditions under which the various pathways to cooperation listed

in §7.1.2 as well as the Cooperation Afforder hypothesis might produce sufficient pro-social behavior to adequately solve problems of provision of public goods and sustainment of common-pool resources.

7.8 Concluding Notes

The edited volume by Hammerstein [140] is rich in biological examples of various forms of cooperation. Frans de Waal's Tanner Lectures, *Primates and Philosophers: How Morality Evolved* [74], are an excellent source of information as well as careful and stimulating thought on cooperation.

> It is fine to describe animals (and humans) as the product of evolutionary forces that promote self-interests so long as one realizes that this by no means precludes the evolution of altruistic and sympathetic tendencies. [74, page 14]

Exactly.

An earlier version of this chapter was presented at the 2010 Philosophy of Social Science Roundtable meeting in St. Louis. I am grateful for the ensuing discussion and comments. I am also very grateful to Andrew J. I. Jones for many stimulating conversations on the subject of communication; they have had much uptake in my thinking for this chapter.

Part III

Markets and Applications

Chapter 8

Competitive Markets

8.1   Competitive Markets

Imagine a small market in which gadgets, coming in discrete units, are bought and sold. Three firms supply gadgets to the market. Firm 1 can supply up to three gadgets. For this firm the cost of producing the first gadget is 1 taler (the taler being our conventional currency). Producing a second gadget costs more, 3.1 talers. The third gadget, if produced, costs 11.2 talers. We can represent firm 1's supply position with a *quantity-cost* table:

Cost	1	3.1	11.2
Quantity	1	2	3

Let us be clear on interpreting the table properly. The cost to firm 1 of producing two gadgets is 4.1 talers, this is the cost of producing the first gadget, 1 taler, plus the cost of producing the second gadget, 3.1 talers.

There are two other firms, firms 2 and 3, supplying the market. Their quantity-cost tables are, respectively, as follows:

Cost	3.1	3.1	4.4	11.2
Quantity	1	2	3	4

and

Cost	4.4	6.4	6.5	7.7	8.3
Quantity	1	2	3	4	5

Taking the suppliers all together, the production (per time period) maxes out at 12 gadgets. Of course, depending upon demand, fewer could be produced. Combining the several quantity-cost tables of the gadget suppliers ("summing them") we get a total quantity-cost table for the suppliers, or *supply schedule* for the market, as shown in Table 8.1. Although the table indicates which supplier is associated with each unit of gadget, this is often unneeded information, and I shall normally suppress it.[1]

---

[1] Note as well an inevitable arbitrariness in the table. Suppliers 1 and 2, for example, both produce one unit of gadgets at 11.2 talers, yet supplier 2 is scheduled as the provider of the eleventh gadget while supplier 1 is the twelfth.

Cost	1	3.1	3.1	3.1	4.4	4.4	6.4	6.5	7.7	8.3	11.2	11.2
Quantity	1	2	3	4	5	6	7	8	9	10	11	12
Supplier	1	1	2	2	2	3	3	3	3	3	2	1

TABLE 8.1: Example supply schedule for the gadgets market.

On the demand side let us suppose that there are a number of buyers who have various maximum prices they are willing to pay for a gadget. Combining ("summing") the buyers' demands for gadgets and suppressing their identities, let us suppose we arrive at the following *demand schedule*, Table 8.2. This table

Price	12	6	6	5	5	4	4	3	3	3	2	2
Quantity	1	2	3	4	5	6	7	8	9	10	11	12

TABLE 8.2: Example demand schedule for the gadgets market.

tells us that if there is only one gadget available for sale in the market, the highest price that a buyer will pay for it is 12 talers. If there are two gadgets available, there is a buyer for the first one at (at most) 12 and a buyer for the second one at (at most) 6 talers. And so on. The demand schedule represents the most that buyers are willing to pay for the nth unit of gadgets. The supply schedule represents the least that sellers are willing to require to part with the nth unit of gadgets. By definition, in a *competitive market* the supply schedule represents the true costs of the suppliers and the demand schedule represents the true values to the buyers. Both buyers and sellers are said to be *price takers*. If it comes down to selling at (or just above) their costs of production, or not selling (and not producing) at all, the sellers will take the price and agree to sell. Similarly, if it comes down to buying at (or just below) their true values for the gadgets, or going without, the buyers will take the price and agree to buy. This is simply what is meant by a competitive market.

Figure 8.1 (page 163) shows the supply and demand schedules (for a discrete commodity) plotted in the conventional fashion. Connecting the supply dots we get what is called the *supply curve,* and connecting the demand dots we get the *demand curve.* This terminology will be useful to us as we proceed.

Indeed, the point of telling this story about a competitive market for gadgets is to present concepts that generalize usefully for our purposes. There is nothing peculiar to the gadgets market I just described that prevents a straightforward generalization of the concepts to any market for a commodity presented in discrete units. As for commodities presenting more or less continuously (think: water, oil, things coming in large numbers of small units), abstracting to continuous functions for both the supply and demand curves is in fact now straightforward. See Figure 8.2. As in the figure, the continuous

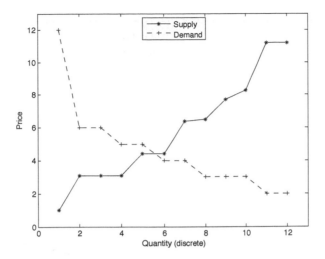

FIGURE 8.1: Example supply and demand curves for a commodity in discrete quantities.

case with linear functions is the usual one in the literature. We shall for now be carrying on in that tradition.

8.2 Competitive Story 1: The Standard Account

Textbooks in economics will tell a story—our Competitive Story 1—about *competitive markets*. Characteristically, as discussed in §8.1, both buyers and sellers simply accept the prevailing price and make their economic decisions accordingly, so as to maximize their economic statuses. They *take the price as given*. Neither buyers nor sellers can influence the price in any significant way (e.g., [322, page 289]).

Buyers, for example, cannot reduce the prevailing price by refusing to purchase for a time in the market, nor do they attempt to do so. Accepting the prevailing price, buyers buy at the level that straightforwardly suits them. They act myopically (naïvely and greedily) and accept any offer of trade that makes them better off. Any buyer deviating from this behavior would have only a minuscule effect on the overall market price and, having foregone the benefit of purchasing at the present price, would lose more than could be gained by waiting. Think of a customer who prefers to have lunch at a restaurant each day, but prepares his own meal today in hopes of getting a price reduction tomorrow. The competitive market story (§8.1) assumes that this

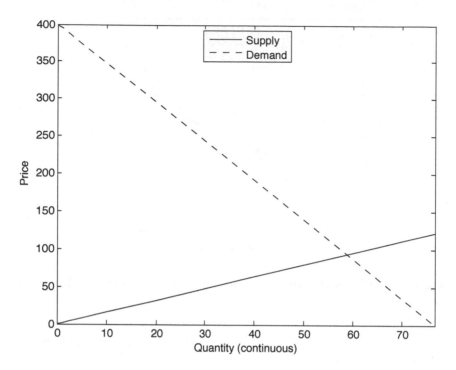

FIGURE 8.2: Example supply and demand curves for a commodity in continuous quantities.

is not profitable for our customer, who would not be so foolish as to engage in such unprofitable behavior.

Similarly, sellers cannot increase the prevailing price by withholding supply when it would be immediately profitable to sell in the market at the prevailing price. They, too, act myopically, accepting any offer of trade that makes them immediately better off. Any attempt to do otherwise would be self-defeating because any consequent increase in price would be too small and too temporary to overcome the foregone income from selling now at the prevailing price. Think of a lunch counter that decides to close one day a week in hopes of being able to raise its prices. The competitive market story assumes that this is not profitable for our producer and that it will not be so foolish as to engage in such unprofitable behavior.

Our Story 1 for competitive markets is a story about what happens in them. As we saw in §8.1, there are three principal elements besides the buyers and sellers in this story:

1. A single commodity which is subject to trade between buyers and sellers.

2. A *demand curve* (or *demand schedule*) that describes the the aggregate demand by the buyers for the commodity, at a given price.

3. A *supply curve* (or *supply schedule*) that describes the aggregate supply that will be provided by the sellers, at a given price.

In short, Competitive Story 1 assumes a competitive market and tells us what will happen in it. The upshot of the story is that the market will go to *equilibrium* of the demand and supply schedules.

Now let use see what all this means. First, we may think of the single commodity as presenting itself as a continuously varying quantity, like water or oil, in which case we might call it *widget* and refer to so many liters or kilograms of widget (think: barrels of oil, liters of water, kilograms of coal, etc.). Or we may think of the commodity as presenting itself as discrete individuals (think: cars and computers), in which case we might say *gadgets* and refer to so many gadgets, or so many hundred gadgets. Linguistically, *widget* is a *mass term* and *gadgets* is a *count term*. Story 1 (for competitive markets) works best for continuously varying commodities, so we shall employ the mass term version, speaking of widget. The story is plausibly extended to gadgets (count term commodities) providing they are small enough in relation to the total size of the market that our mass-term story provides a good approximation, as often it will.

The demand schedule for a commodity tells us, for a given price, P, the quantity, Q, that would be purchased by the buyers taken all together. This is most conveniently represented as a function, and a linear one at that, having the form:

$$Q_D(P) = a - b \cdot P \tag{8.1}$$

($Q_D(P)$ is "the quantity demanded by the buyers as a function of the price.") We will assume that a, b, $P \geq 0$ and so demand falls as price increases. Notice that our focus on continuously varying commodities (hence our use of the mass term widget) coheres nicely with our expression of the demand schedule (and later the supply schedule) as a linear function of a real-valued variable.

The supply schedule tells us how much, for a given price, the total quantity that would be produced by the sellers taken all together. This, too, is most conveniently represented as a function (again, a linear one for convenience) having the form:

$$Q_S(P) = c + d \cdot P \tag{8.2}$$

($Q_S(P)$ is "the quantity supplied by the providers as a function of the price.") We assume that c, d, $P \geq 0$ and so supply increases as price increases.

At equilibrium (of the supply and demand schedules)

$$Q_S(P) = Q_D(P) \tag{8.3}$$

Verbally, this says that at equilibrium the quantity supplied equals the quantity demanded, with both quantities depending on a common price, P. Of course equilibrium does *not* occur at every price, P, so we need to find the price or prices at which it does occur. Given our setup the calculation is trivial. We set supply equal to demand, as in expression (8.3). Substituting in the demand schedule (expression (8.1)) and the supply schedule (expression (8.2)) we get:

$$c + d \cdot P = a - b \cdot P \qquad (8.4)$$

Solving for P we get

$$P^* = \frac{a - c}{b + d} \qquad (8.5)$$

Verbally put, the value of the price at equilibrium, P^*, is $\frac{a-c}{b+d}$. The value of the quantity at equilibrium, Q^*, is then easily determined:

$$Q^* = c + d \cdot P^* = c + d \cdot \frac{a - c}{b + d} = a - b \cdot P^* = a - b \cdot \frac{a - c}{b + d} = \frac{ad + bc}{b + d} \qquad (8.6)$$

Let us look at a specific example algebraically and geometrically. Following traditional conventions, we will present the supply and demand curves as prices that are functions of quantities.[2] Let

$$P_D(Q) = DemandPriceIntercept - DemandSlope \cdot Q \qquad (8.7)$$

($P_D(Q)$ is "the price demanded by buyers as a function of quantity supplied") and

$$P_S(Q) = SupplySlope \cdot Q \qquad (8.8)$$

($P_S(Q)$ is "the price required by sellers as a function of quantity supplied"). At equilibrium

$$P_D(Q) = P_S(Q) \qquad (8.9)$$

Solving for Q we find that

$$Q^* = QuantityEquilibrium = \frac{DemandPriceIntercept}{(SupplySlope + DemandSlope)} \qquad (8.10)$$

and

$$P^* = PriceEquilibrium = (SupplySlope \cdot QuantityEquilibrium) \qquad (8.11)$$

and

$$P^* = (DemandPriceIntercept - DemandSlope \cdot QuantityEquilibrium) \qquad (8.12)$$

Further specifying the example, let us say *Price* for P and let *SupplySlope* = 1.59, *DemandSlope* = −5.22, and *DemandPriceIntercept* = 400. Figure 8.3 uses this terminology and draws the two curves, as does Figure 8.2. Our example plausibly has it that when *Price* = 0 no widget is supplied.

The equilibrium point, (*QuantityEquilibrium*, *PriceEquilibrium*), is at the intersection of the supply and demand curves. We can see this geometrically in Figure 8.3, or in terms of our prior discussion, the equilibrium value of $Q^* =$

---

[2]Technically, these are the so-called *inverse demand function* and *inverse supply function* forms, minutiae that need not detain us.

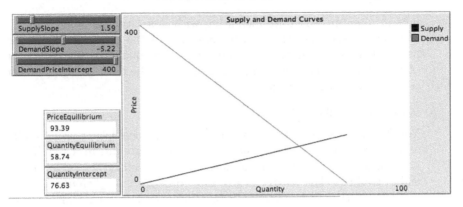

FIGURE 8.3: Supply and demand curves: Example, from CompetitiveMarket.nlogo.

QuantityEquilibrium is the value of Q that solves expression (8.9). There is a single point of equilibrium because the "curves" are straight lines and they are not parallel.

To see why this is an equilibrium, consider first a producer who contemplates supplying one more quantity of widget to the market. The effect will be to reduce the price realized (for the entire quantity of widget) from 93.39 (in our numerical example) to $93.39 - 5.22 = 87.17$, but by assumption, no producer finds this profitable. Conversely, any producer cutting supply from the equilibrium amount would forego the opportunity to make an attractive sale. Again, this is what is assumed or represented in the supply curve. So, no producer has incentive either to increase or decrease the quantity supplied to the market.

On the buyer side, the demand curve represents the most that buyers are willing to pay for a given quantity. Buying more violates this assumption. Buying less entails failing to make a trade (of money for widget) that the buyer in fact finds attractive. In sum, then, neither buyers nor sellers have incentive to depart from the equilibrium, once they are at it.

In consequence of a competitive market being at equilibrium, several properties are often asserted to obtain, including the following.

1. Does a single price prevail in the market?

   Yes. There is a single price that prevails in the market, and it is the equilibrium price. Corresponding to that price is a unique quantity of commodity produced.

2. What is the prevailing price?

   The prevailing equilibrium price equals what is called the *marginal cost of production*. That is, the cost to manufacture the last quantum of widget sold is equal to (or just below) the equilibrium price, and since unit or quantum costs are increasing with quantity, the next quantum of widget costs more to produce than any of the buyers is willing to pay. Put more carefully, the equilibrium price equals the marginal cost of production of the marginal producer, that is the most costly producer still making any sales at all.

3. Is the market efficient?

Yes. The market is said to *clear*, by which it is meant that no mutually attractive trades (between a buyer and a seller) go untransacted. There is, to employ the standard metaphor, "no money left on the table" at the end of trading. The market is also said to be *Pareto efficient* or *Pareto optimal*, in that it is impossible to make someone better off without reducing the welfare of someone else. To see this, notice that since the market has cleared, no mutually profitable trades remain to be made. So we have to seek Pareto improvements by breaking and rematching at least two consummated trades, but since all trades occur at the same price, there are no benefits to be had by switching trading partners.

4. Are there economic rents supported?

Yes, as seller's surplus. The standard view is that in the long run there is no *economic rent* to be had by the sellers. The following passage is typical.

> Any amount of money earned in excess of payments to the factors of production is a pure economic profit [i.e., constitutes economic rent]. But whenever someone finds a pure economic profit, other people will try to enter the industry and acquire some of that profit for themselves. It is this attempt to capture economic profits that eventually drives them to zero in a competitive industry with free entry. [322, page 406]

The claim for no economic rent in a competitive market relies on the "eventually" caveat, as is clearly expressed in the passage just quoted. Recalling Figure 8.3, note that if all trades are at the equilibrium price, then essentially all of the producers are obtaining economic rent, since the price at which they are willing to trade (their costs of production) for a quantum of widget is below the equilibrium price. The equilibrium price is set by the highest-cost producer whose costs are at the intersection with the downward-sloping demand curve. Put otherwise, the equilibrium price represents the coincidence of the buyer who values the commodity the least and the seller who requires the most resources to produce it.[3] All other sellers, selling at this price, receive economic rents, at least in the short term. In the long term, if other sellers can find

---

[3] Among all feasible buyers and sellers, that is, buyers whose values would permit them to trade.

a way to produce at less than the equilibrium price, we might expect them to enter the market and reduce the equilibrium price. Any sellers who continue to produce at less than the new, reduced equilibrium price will continue to enjoy economic rents, albeit reduced ones. Notice as well that this is a two-sided affair. Nearly all of the buyers are obtaining their quanta of widget at a price *lower* than what they would ultimately be willing to pay. Realistically, it is this result of mutual (economic) profit or rent that drives the market. How many dollars for quanta of widget trades would there be if both sides were indifferent as to the consequences of trading or not trading? In the lingo, there is a *buyer's surplus* and a *seller's surplus* generated. Buyer's surplus: the amount of money a buyer is willing to pay for something, minus the amount actually paid for it. Similarly for seller's surplus. It is these surpluses that drive the market.

5. Is the market fair in the sense that economically identical agents at the start of trade are economically identical at the conclusion of trade?

The outcome is broadly speaking fair in the sense that two equally endowed agents at the start of trade will be in economically identical positions at the conclusion of trade. Two sellers with identical reservation prices on their quanta of widget will sell their quanta at the same price. At the end of trade they will experience identical gains from trade.

Note: a market that is not fair in this sense is said to have *horizontal inequality*.

6. How is the surplus from trading distributed?

At a point C (cost of production) on the supply curve the surplus from trade received by the seller is $P - C$, the price minus the cost of production. At point V (value of the commodity) on the demand curve the surplus from trade received by the buyer is $V - P$. The total surplus generated by the trade is $V - C$. Who gets what portion of that surplus depends entirely on the price P. Assuming that P is the equilibrium price still leaves the distribution underdetermined. What determines the overall distribution of surplus between buyers and sellers are the actual supply and demand curves.

Such, in a nutshell, is Story 1 for competitive markets. What are we to make of it? Given that a market is competitive in the sense to hand, will it in fact reach equilibrium or something close to it? Economists typically think it will. Here is a representative passage:

It is at least conceivable that at any given time peoples' demands and supplies are not compatible [not in equilibrium], and hence something must be changing. These changes may take a long time to work themselves out, and, even worse, they may induce other changes that might "destabilize" the whole system.

This kind of thing can happen ... but it usually doesn't. ... we
typically see a fairly stable ... price from month to month. [322,
page 3]

Whether equilibrium is in fact normally obtained in markets we would call
competitive and if so whether the list of properties above actually does apply
to such markets is, of course, a matter for empirical investigation. On the
face of it, there are considerable grounds for skepticism. At the very least, we
should like to investigate alternative models that cohere with departures from
Story 1. This will hardly resolve all issues, but presentation and comparison
of alternative accounts constitutes a necessary starting step.

Notice as well that Story 1 merely asserts that equilibrium is reached. It
does not tell us how. There *is* a story for how the competitive equilibrium is
reached, but it invokes a nonexistent "Walrasian auctioneer" and as such is
hardly credible as an actual mechanism. (See [93, Chaper IV] for an accessible
account and a compelling critique. See any microeconomics textbook for a
description.) Let us continue, then, and look at other stories for competitive
markets that do offer rather definite mechanisms for describing trade. We shall
be interested in whether these mechanisms produce equilibria and if not, what
they do produce.

8.3   Competitive Story 2: Random, Bargaining

Story 2 is just like Story 1 (competitive market, single quantity, demand
and supply curves) except that instead of positing an outcome (equilibrium) it
posits a trading regime. It is this trading regime that produces the outcome.
The Story 2 trading regime is quite simple. A buyer and a seller are randomly drawn and paired. Each negotiates the sale price based on its private
reservation price (point on the supply or demand curve, as the case may be).
The buyer's reservation price, the maximum it will pay, is obtained by randomly drawing a quantity and finding the corresponding point on the demand
curve. The seller's reservation price is obtained similarly, but with a different,
independently drawn quantity. We assume that buyer and seller are equally
good negotiators and end up at a price midway between their two reservation
prices. Further, trades are only made if the drawn quantities for the buyer and
seller are both less than or equal to the equilibrium quantity, since neither,
we presume in Story 2, is interested in trading beyond that point, given the
position of its counterpart.

Re-describing the Story 2 trading regime more plainly, buyers and sellers
trade for a quantum of widget. They are arrayed in order of their reservation
prices, for buying or selling as the case may be. These arrays constitute the
supply and demand curves shown in Figure 8.3. Each point on the supply

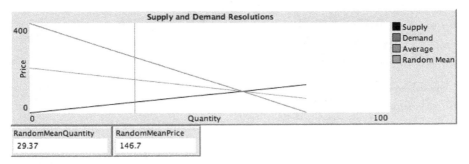

FIGURE 8.4: Supply and demand curves: Example, from CompetitiveMarket.nlogo.

curve corresponds to a producer who is willing to sell a quantum of widget providing the price is at least equal to the price at that point of the supply curve. Similarly, each point on the demand curve corresponds to a buyer who is willing to buy a quantum of widget providing the price is at most equal to the price at that point on the demand curve. We pick a random seller from the supply curve and a random buyer from the demand curve. They trade for a quantum of widget at the price midway between their reservation prices. We repeat this process many times and observe the behavior of the market.

Figure 8.4 is helpful for understanding the resulting behavior. The figure is like Figure 8.3, but with two additional lines drawn. The middle line, labeled Average, is the mean of the supply curve and the demand curve. The vertical line, labeled Random Mean, intersects the Quantity axis at exactly one-half of the equilibrium quantity. In our example, this is at 29.37, as shown in Figure 8.4. The Random Mean and Average lines intersect at the Price of 146.7 (and quantity 29.37) in our example. Let us call this point (quantity 29.37, price 146.7) the *expected random outcome*. It should approximate the average outcome of many trades under the Story 2 trading regime.

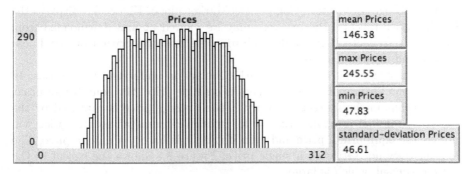

FIGURE 8.5: Supply and Demand Curves: Example, from CompetitiveMarket.nlogo.

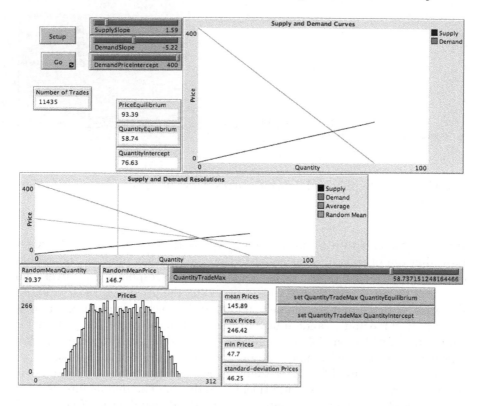

FIGURE 8.6: CompetitiveMarket.nlogo Interface tab.

Figure 8.5 confirms this conjecture. The histogram in the figure is for realized prices under the Story 2 trading regime, after about 13,000 trades. We see that the realized mean price of 146.38 is in fact close to our calculated value of 146.7. All three figures, 8.3–8.5, are taken from the Interface tab of the NetLogo application CompetitiveMarket.nlogo. Figure 8.6 displays the full Interface tab and the program may be viewed at and downloaded from the book's Web site. The problem affords experimenting with arbitrary linear supply and demand curves.

It is instructive to compare market behavior under the two stories. In Story 1, equilibrium obtains and there is a single price, the equilibrium price, that prevails in the market. In Story 2, the average price is captured by the expected random outcome. Except for special circumstances this price will not be the equilibrium price and even if it is, it is hardly a unique price. The market does not settle on a single price; Figure 8.5 shows the extent of the price dispersal, which is large.

In our example, the expected random price is higher than the equilibrium price. This is a consequence of the fact that in the example the absolute value of the slope of the supply curve (=1.59) is lower than the absolute value of the slope of the demand curve (5.22). I invite the reader to experiment in this regard with CompetitiveMarket.nlogo.

We may summarize for purposes of comparison the main characteristics of the competitive market under Story 2.

1. Does a single price prevail in the market?

   No. As we have seen there is a distribution of prices at which trades occur.

2. Is the equilibrium price approximated on average? If not, what is the prevailing price?

   No, the equilibrium price is not approximated on average. The average prevailing price may be higher or lower than the equilibrium price, depending on the slopes of the supply and demand curves.

3. Is the market efficient?

   Yes. By construction the market clears. Every feasible trade that appeared was transacted.

4. Are there economic rents supported?

   Yes, clearly. All sellers are selling at a price above their costs of production. As is the case in Story 1, buyer surplus and seller surplus are generated by the market.

5. Is the market fair in the sense that economically identical agents at the start of trade are economically identical at the conclusion of trade?

   No. It is quite possible for initially identical agents to have by chance quite different experiences in the market and end up different economically after trade.

6. How is the surplus from trading distributed?

   Compared to Story 1, the amount of surplus appropriated by the buyers is lower, since the average price is higher.

We turn now to our next story.

8.4 Competitive Story 3: ZI-C Agents

A well-known paper by Gode and Sunder is the source of our Story 3 [126]. Figure 8.1 and our story for competitive markets remain apt for the basic setup. The (artificial) markets Gode and Sunder explored involved discrete commodities, as in Figure 8.1. Now, however, the market is continuing.

Details of the Gode and Sunder setup are as follows.[4] Buyers and sellers (six of each in their experiments) are individually given endowments (for sellers) or requirements to buy (for buyers) of "one or more" items. The original description is incomplete:

> At the beginning of each period, each buyer was endowed with the right to buy one or more units of an unspecified commodity. The buyer was privately informed of the redemption value v_i of each unit i, and the buyer's profit from buying this unit at price p_i was given by $v_i - p_i$. Redemption values v_i, $i = 1, 2, \ldots, n$, defined the individual buyer's demand function for the commodity. Since redemption values for each buyer were private, the market demand function was unknown to the buyers. [126, page 122]

Once each buyer receives its information, an implicit market demand function results. What is unclear in this passage (and in the paper) is exactly how the market demand function is assigned to individual buyers. Let us assume that the market demand function is given exogenously and has positive prices for at least 6 units of the commodity. Further, we assume that units on the market demand function are assigned randomly, in round-robin fashion, to the buyers. For example, if there are 9 units of demand, the six buyers each get one unit randomly assigned and the three remaining units go, one each, to randomly-chosen buyers. Thus, each buyer is endowed randomly with 1 or 2 units.

The sellers are endowed similarly:

> At the beginning of each period, sellers were endowed with the right to sell one or more units of the commodity, c_i being the cost of the ith unit to the seller. The seller's profit from selling the ith unit at price p_i was $p_i - c_i$. Sellers had no fixed costs and incurred costs only for units sold. The market supply function was unknown to the sellers. Every trader had to trade the ith unit before trading the $(i + 1)$th unit. [126, page 122]

As with the buyers, let us assume that the market supply function is given exogenously and that units from it are assigned randomly in round-robin fashion to the sellers.

Figure 8.7 presents in pseudocode the trading regime of Gode and Sunder. Points arising:

1. The trading regime is said to be a double auction because buyers post bids, sellers post asks, and the market mechanism effects the trade transactions when bids match or exceed asks.

---

[4]The original paper is not entirely clear regarding the setup of the artificial market, although the missing details do not seem likely to be material. I am attempting here a sympathetic description that is specific enough to be replicated independently. Also, I am describing only some of the results of the paper. See the original: [126].

1. Initialize: n_b (the number of buyers), n_s (the number of sellers), n_p (the number of periods of play), t_p (amount of time for trading in each period), S (the market supply schedule), D (the market demand schedule), v_{max} (the maximum value of a unit of commodity to any buyer).

2. Create n_b buyer agents, n_s seller agents.

3. Do for each of the n_p periods of play:

   (a) Randomly endow each buyer in round-robin fashion, exhausting the market demand schedule, D.

   (b) Randomly endow each seller in round-robin fashion, exhausting the market supply schedule, S.

   (c) Do until elapse of time t_p:

      i. In random order each seller agent with an unsold item posts a ZI-C ask for its least costly available item.

      ii. In random order each buyer agent with an unexercised right to buy an item posts a ZI-C bid for its most valuable outstanding right.

      iii. Transact any feasible trades, based on the bids and asks.

      iv. Update the agents, removing from their inventories any items sold or bought.

FIGURE 8.7: Pseudocode for ZI-C auction trading.

2. Regarding step (1) in Figure 8.7, in the Gode and Sunder work [126], there were always six buyers, six sellers, and six periods of play. t_p was set to 30 seconds. Five pairs of supply and demand schedules were explored. (They are presented only graphically in the paper.)

3. Regarding steps (3(c)i) and (3(c)ii) in Figure 8.7, we come to the point of the entire exercise. ZI-C stands for "zero-intelligence with constraint." A ZI-C buyer creates a bid for a unit of value v_i that is randomly (uniformly) drawn from the range $[1, v_i]$. A ZI-C seller creates an ask for a unit of cost c_i that is randomly (uniformly) drawn from the range $[c_i, v_{max}]$. (See step (1) in Figure 8.7 for the definition of v_{max}.)

   In a nutshell, ZI-C traders bid and ask randomly, provided they would not make a loss on the trade if effected.

4. Regarding step (3(c)iii) in Figure 8.7,

   A transaction canceled any unaccepted bids and offers. Fi-

nally, when a bid and ask crossed, the transaction price was
equal to the earlier of the two. [126, page 122]

What did Gode and Sunder find? How did their artificial markets of ZI-
C agents behave? We may summarize for purposes of comparison the main
characteristics of the competitive market under Story 3 as follows.

1. Does a single price prevail in the market?

   No. As is obvious from the description of the setup, many ZI-C trades
   take place at non-equilibrium prices. It is only at the end of the periods
   that the equilibrium price is approached.

2. Is the equilibrium price approximated on average? If not, what is the
   prevailing price?

   Yes, the equilibrium price approximated on average. In each market,
   averaged over the six periods of play the end-of-period commodity price
   closely approximated the equilibrium price as defined under Story 1.
   In this behavior, the ZI-C agents resembled human subjects trading in
   equivalent (or very similar) market arrangements, although the ZI-C
   agent prices were much more volatile than those of the humans.

3. Is the market efficient?

   Close to it. Most periods ended with "no money on the table." That
   is, it was generally *not* the case that a unit of value v_i remained unsold
   for which there was a buyer who valued it at at least that amount. The
   markets do not clear every period, but nearly always they do. With
   efficiency defined as the sum of all profits actually made divided by the
   total profit that could be made, in the five markets examined by Gode
   and Sunder the ZI-C agents had efficiencies ranging from 97.1 to 99.9.

4. Are there economic rents supported?

   Yes, clearly. All sellers are selling at a price above their costs of production, buyers are buying at prices below the redemption values of the
   commodity. As is the case in Story 1, buyer surplus and seller surplus
   are generated by the market.

5. Is the market fair in the sense that economically identical agents at the
   start of trade are economically identical at the conclusion of trade?

   No. Although Gode and Sunder do not address this question, clearly it
   is quite possible for initially identical agents to have by chance quite
   different experiences in the market and end up different economically
   after trade.

6. How is the surplus from trading distributed?

   Although Gode and Sunder do not address this question, clearly the
   distribution of surplus between buyers and sellers will depend upon the

actual supply and demand curves (as in Stories 1 and 2) and on the value of $v_{\max}$. Note that if we use a very much higher value the market will clear less often and when trades are made the sellers will generally do better.

8.5 Competitive Story 4: Trade on the Sugarscape

Epstein and Axtell report on a number of artificial markets, each constructed in the spirit of Kreps' envisioning of the competitive market story for realistic agents.

> ...we can imagine consumers wandering around a large market square, with all their possessions on their backs. They have chance meetings with each other, and when two consumers meet, they examine what each has to offer, to see if they can arrange a mutually agreeable trade. ...If an exchange is made, the two swap goods and wander around in search of more advantageous trades made at chance meetings. [[184, page 196], quoted in [93, page 101]]

The several artificial markets they explore share the following aspects.

1. Trade takes place on the "Sugarscape," a two-dimensional lattice. This is a variant of, and very like, the world in NetLogo, which we have discussed earlier. It is a gridscape (checkerboard) with patches (squares). Agents are located on patches.

2. Two commodities grow on the Sugarscape patches: sugar and spice.

3. Different patches support different levels of growth and abundance of sugar and spice. There are two hills of spice, in the northwest and southeast, and two of sugar, in the southwest and northeast. Hills are defined in terms of commodity richness, increasing with the height of the hill.

4. Agents need both sugar and spice in order to survive.

5. Agents are differentially endowed with vision, how many patches they can see in their neighborhood.

6. Agents are differentially endowed with metabolisms, how fast they burn sugar and spice.

I want now to focus our attention on one of these artificial markets, the base case example. I omit many details, which are readily available in the original article [93, Chapter IV].

During each episode of play, each agent surveys its neighborhood as far as its sight permits, then moves to the unoccupied visible patch offering the greatest welfare improvement for itself, based on the agent's sugar and spice inventories and their availabilities at the patch. Upon arriving at the chosen patch, the agent appropriates all available sugar and spice. After all agents have made their moves and consumed sugar and spice, adjacent agents (sitting on neighboring patches) consider trading sugar and spice. If a mutually advantageous exchange is possible it is transacted. During each episode, agents' stocks of sugar and spice are depleted according to their endowed metabolism rate. Agents exhausting either stock die and are removed from the Sugarscape.

What did Esptein and Axtell find? What happens on the Sugarscape? We may summarize as follows.

1. Does a single price prevail in the market?

   No. Prices at first are significantly scattered. Over time they converge to near the theoretical equilibrium level. Variation is never eliminated. In the base case it becomes small, but in other cases, with added realism, price variation may not converge very tightly.

2. Is the equilibrium price approximated on average? If not, what is the prevailing price?

   Yes, the equilibrium price approximated on average, after a period of convergence.

3. Is the market efficient?

   Epstein and Axtell note that their artificial market is not perfectly efficient, but they do not report a measure of its efficiency. In the base case, under discussion here, it would appear that the overall efficiency of the market approaches a reasonably high value. Epstein and Axtell report that the converged market tends to be locally Pareto optimal (and efficient), but "the resulting market has far from optimal welfare properties globally" [93, page 116].

4. Are there economic rents supported?

   Everyone is a hunter-gatherer in the Sugarscape model. Production prices are set to 0. Trades do not happen, of course, unless both buyer surplus and seller surplus are generated. Surplus is measured relative to the existing stocks of the agents. Their values are highly dependent upon context. The value to the agent of a unit of sugar depends upon how much spice the agent has, and vice versa.

5. Is the market fair in the sense that economically identical agents at the start of trade are economically identical at the conclusion of trade?

   No. Substantial horizontal inequality exists. Adding realistic features to the base case increases horizontal inequality.

6. How is the surplus from trading distributed?

Esptein and Axtell use a bargaining regime based on a multiplicative model of the agents' internal valuations of their positions. See [93, pages 102–4] for the technical details. What they do is sensible, if a bit ad hoc. The important thing to note is that any given agent may be a buyer or seller of sugar (or spice), depending on the condition of its stock of both sugar and spice. This condition may, and often does, change back and forth over time. The internal evaluation scheme, and hence the bargaining model, which determines the relative prices of sugar and spice, is entirely neutral between the two commodities.

8.6 Discussion

These stories taken together invite us to think about markets somewhat differently than we have heretofore presupposed. Recall our pictures of supply and demand curves for discrete commodities (Figure 8.1, page 163), and for continuous commodities (Figure 8.2, page 164). When a trade is made let us continue to assume that the buyer values the commodity more than the price paid and the seller values the money received more than the commodity given up. We can express this with formal notation—

$$V_B(P,Q) \geq V_S(P,Q) \tag{8.13}$$

—although it is hardly necessary. The simple idea is that the value to the buyer of a trade at price P of quantum Q is at least as great as the value to the seller; otherwise we would not expect the trade to happen.[5]

The *potential surplus* from a possible trade (in a given market) of Q quanta of commodity is simply the difference between the buyer's value and the seller's value for the amount of commodity:

$$S(Q) = V_B(Q) - V_S(Q) \tag{8.14}$$

Notice that the price is not involved in the calculation. The *market potential surplus*, for a given value of Q, $S_M(Q)$, is just the sum from 0 to Q of the potential surpluses for the associated trades. Geometrically, in terms of Figures 8.1 and 8.2, the potential surplus for a given Q is the area under the demand curve, from 0 to Q, minus the area under the supply curve, from 0 to Q.[6]

I use the term *potential surplus* by analogy with *potential energy* in physics, where it is defined roughly as energy stored in a system and having the potential to do work. Potential energy is measured in joules; it will tend to be

---

[5]I am simplifying. It would be better to say that these are expected or anticipated values. This raises issues that are important but beyond our present scope.

[6]I am using language appropriate for the continuous case. Adjusting for the discrete case is trivial.

released, perhaps to do work, until it is dissipated, at which time the system is in equilibrium. Similarly, before trading, if there is a positive potential surplus in a market, it will in the sequel tend to drive trading until it is dissipated and the market is in equilibrium.

A lesson of Story 1 is that *if* all trading is at the equilibrium price (determined by the intersection of the supply and demand curves), then the potential surplus will be entirely dissipated and, in consequence, the system will be at equilibrium. To see this, consider varying Q to maximize $S_M(Q)$, the market potential surplus. It is easy to see that Q^*, the equilibrium quantity, will maximize $S_M(Q)$.

Story 2 teaches us that the market can be in equilibrium (all market potential surplus is dissipated) without there being a single price governing all trades. Stories 3 and 4 show us how the equilibrium price might be approximated without the market being in equilibrium. What will actually happen, of course, has to be determined empirically.

8.7   For Exploration

1. We have reviewed four stories describing competitive markets. Story 1 is the textbook story of neoclassical economics. Summarize how the other three stories agree and disagree with Story 1. When two stories disagree, can you find plausible conditions in practice in which either or both might be correct?

2. Story 2, uniquely among our four stories, has an account of how the settled price in a competitive market might differ systematically from the equilibrium price. Can you think of other trading arrangements which would be realistic and would have a similar consequence?

3. If, contrary to the assumption of a competitive market, sellers could collude to raise prices above the competitive level, fewer trades would be made, and the marginal producers would fail to have sales. Why might the marginal producers go along with this and not offer prices that were below the collusion price and yet still at or above their costs of production?

4. For the supply and demand curves shown in Figure 8.1 (page 163), what is the equilibrium quantity? Price?

5. In context of the *discrete* supply and demand curves shown in Figure 8.1 (page 163), we said (on page 162):

   > The demand schedule represents the most that buyers are willing to pay for the nth unit of gadgets. The supply schedule

represents the least that sellers are willing to require to part with the nth unit of gadgets. By definition, in a *competitive market* the supply schedule represents the true costs of the suppliers and the demand schedule represents the true values to the buyers. Both buyers and sellers are said to be *price takers*. If it comes down to selling at (or just above) their costs of production, or not selling (and not producing) at all, the sellers will take the price and agree to sell. Similarly, if it comes down to buying at (or just below) their true values for the gadgets, or going without, the buyers will take the price and agree to buy. This is simply what is meant by a competitive market.

How should this passage be rephrased to describe a competitive market when the commodity is continuously presented?

6. Gode and Sunder also investigated human subjects and ZI-U agents, and compared them with the ZI-C agents. ZI-U (zero-intelligence unconstrained) agents were like ZI-C agents except that they bid randomly without regard to their reservation prices. Whereas a ZI-C seller would ask randomly in $[c_i, v_{max}]$ the ZI-U seller would ignore its cost of production and ask randomly in $[1, v_{max}]$. Similarly, while a ZI-C buyer would bid randomly in $[1, v_i]$, the ZI-U buyer would bid randomly in $[1, v_{max}]$.

Gode and Sunder write that

> The main question addressed in the article is, How much of the difference between the market outcomes with ZI-U traders and those with human traders is attributable to intelligence and profit motivation, and how much is attributable to the market discipline. . . ? [126, page 128]

They found that the ZI-C behavior was close to the human behavior, and far from the ZI-U behavior. From this they conclude that "market discipline," that is the structure imposed by the market institution, accounts for the main behavioral features of the market, not human intelligence and profit seeking.

Give a sympathetic summary of their argument and do so in more detail than I have presented here. Then evaluate the case they make.

7. Using NetLogo (or some other appropriate programming environment) implement and explore an artificial market for ZI-C, ZI-U, etc. traders. What do we learn from this?

8. Suppose that you were an agent trading in a double auction market otherwise entirely populated by ZI-C traders. Design one or more strategies that would allow you to do systematically better than the ZI-C traders.

Describe these strategies with sufficient precision that they could be implemented for artificial agents in an artificial market.

9. A market that is efficient (or Pareto efficient) is often said to be *socially optimal*. Discuss. In what sense is an efficient market socially optimal? Is it possible for a market to be efficient and not be socially optimal in some other sense? If so, what other sense or senses?

10. Comment on the following passage:

> [T]he putative case for laissez-faire economic policies is that, left to their own devices, market processes yield equilibrium prices. Individual (decentralized) utility maximization at these prices then induces Pareto optimal allocations of goods and services. But if not price equilibrium occurs, then the efficiency of the allocations achieved becomes an open question and the theoretical case for pure market solutions is weakened. [93, page 95]

11. Comment on the following passage:

> The case for laissez-faire economic policies does not rest on market equilibrium or any other theoretical property from economics. It rests on the fact that the alternatives to laissez-faire, all involving government intervention in the market, are guaranteed to be worse than any bad outcome produced without interference.

12. How in practice could you determine whether a given, seemingly competitive market was behaving in accordance with any of our four stories? In particular, upon observing a prevailing price in the market, how could you realistically determine whether it was the equilibrium price?

13. Must competitive markets have a unique equilibrium price? If not, provide a counter-example.

14. Comment on the following passage:

> Horizontal inequality does not imply unfairness. If, prior to trading, any two equally endowed agents have the same expected outcome, then the system is fair. We should not be concerned, as a matter of social policy, on unequal outcomes that are produced solely by chance.

15. Gode and Sunder [126, page 120] write that

> Standard economic theory is built on two specific assumptions: utility-maximizing behavior and the institution of Walrasian tâtonnement.

Varian [322, page 3] makes essentially the same point.[7]

> In much of economics we use a framework built on the following two simple principles.
>
> **The optimization principle:** People try to choose the best patterns of consumption that they can afford.
>
> **The equilibrium principle:** Prices adjust until the amount that people demand of something is equal to the amount that is supplied.

Story 1, the standard account, in fact relies on these two assumptions. Discuss the relationships between Stories 2–4 and these two principles or assumptions.

16. Story 1 encourages assessment of the outcome of a market on the basis of efficiency. Why is efficiency in this sense a good thing? If it were found that a market was highly inefficient, what might be done to improve its efficiency? Besides efficiency, what other aspects of a market might be valued, positively or negatively? How do all these values interact? Are they ever in conflict? If so, how should they be traded off against each other?

17. Suppose that the sellers in an efficient market were given a *truth detection machine* (TDM) that revealed to them the reservation price of any prospective buyer. How would this affect prices in the market? What if the shoe were on the other foot, and only the buyers had a TDM? What would happen if everyone had it and which of our stories would this situation most closely match? Now consider the consequences of an imperfect but reasonably reliable TDM. Is such a machine plausible?

18. Suppose that an actual competitive market has sustained variation in the prices realized. Given that everyone else is a price taker, devise a strategy by which you can profit.

19. Assuming a competitive market and Story 1, what do the supply and demand curves look like when there is very little supplier (buyer, buyer and seller) surplus to be generated by trade?

Suppose that the supply curve is such that there is little seller surplus

---

[7] "People try to choose..." is a bit misleading. The standard account in economics assumes, as Gode and Sunder say, "utility-maximizing behavior." This means, both here and generally in the literature, that agents *actually* maximize their utilities. Similarly, "optimizing" means actually finding an optimal solution. These are technical uses of the terms. In ordinary language, to say that an agent is maximizing or optimizing suggests, or allows, that the agent is merely *trying* to maximize or optimize, but may not be succeeding perfectly. This ordinary language usage is *not* what is employed in economics. Given the potential for confusion, I shall use such terms as *groping* and *heuristically groping* to describe agents that are (in some reasonable sense) seeking actually to maximize or optimize, but may or may not succeed.

that can be created by trade, according to Story 1, but the demand curve is such that there is potential for substantial buyer surplus. What do Stories 2–4 teach us about what will happen in such a market?

20. Suppose, in a competitive market, that the supply curve is flat; all the producers have the same variable cost of production. What, according to Story 1, should the suppler surplus be?

    Comment on whether this is realistic. How might evidence be collected that would bear on ascertaining the true situation in a given market?

    Consider a realistic variety of a TDM (truth detection machine). There is a competitive market for widgets with a large number of suppliers and a flat supply curve. Suppliers are constantly undertaking experiments regarding their prices, so that on a given day in a given store, the posted price is likely to be a little above or below the average prevailing price in the market. Suppliers keep careful sales records, and track closely the associations between sales volumes and prices. How is this likely, over time, to influence the prevailing average price in the market and, in particular, the sellers' surplus?

21. Consider Competitive Story 5, which is like Story 2 except that trading may occur past the equilibrium quantity. QuantityTradeMax, the maximum on the quantity axis, beyond which no trades occur, was set to the EquilibriumQuantity in Story 2. Now we want to consider setting it higher. Assume we set it to the point at which the demand curve intersects the quantity axis, QuantityIntercept. This is easily done in CompetitiveMarket.nlogo. See the Interface tab, shown in Figure 8.6. What will happen? Why? How might we empirically distinguish this situation from the equilibrium account, Story 1?

8.8   Concluding Notes

Standard microeconomics textbooks present our Story 1 as the received view. Good examples include [184, 234, 321, 322].

Testfatsion and Judd have edited a useful volume touching on many of the issues raised in this chapter, as well as issues raised in the sequel [311]. See also [213] and [308].

Gode and Sunder say more on the forces of market discipline in [127].

Chapter 9

Monopoly Stories

9.1 Monopoly Story 1: The Standard Account

Introductory textbooks in microeconomics will tell a story—our Story 1—roughly as follows. Suppose that there is a market demand, Q (think: Quantity demanded), for a particular product and that this demand is a linear function of the product's price, P:

$$Q(P) = Q = c - b \cdot P \tag{9.1}$$

Assume: $b, c > 0$. Here, c is a constant, representing the quantity demanded when the price is 0 and, since we are linear, the price point at which demand disappears when the price is too high (when $b \cdot P = c$). The symbol $Q(P)$ indicates that Q (demand) is a function of P (price). Q simply abbreviates $Q(P)$.[1]

Rearranging (9.1) we get:

$$P(Q) = P = a - dSlope \cdot Q \tag{9.2}$$

where $a = \frac{c}{b}$ and $dSlope = \frac{1}{b}$. (We assume $dSlope > 0$; mnemonic: downward Slope.) Expression (9.2) is the standard form of what is called the *inverse demand function* (or simply the *demand function*). The notation $P(Q)$ indicates explicitly that price, P, is a function of quantity, Q. $P(Q)$ is often abbreviated to P (price) without engendering confusion.

Into this market enters a monopolist, the sole supplier of the good in question. We assume the revenue obtained by the monopolist for selling Q units, $r(Q)$, is *unit price × number of units sold* $= P \cdot Q = (a - dSlope \cdot Q) \cdot Q = a \cdot Q - dSlope \cdot Q^2$. If we further assume that the production costs are proportional to the number of units sold, then the costs faced by the monopolist in selling Q units, $c(Q)$, is *cost per unit × number of units sold* $= k \cdot Q$ ($k \geq 0$). Setting profit, π, to *unit price − cost per unit*, the profit made by the monopolist is just revenue minus cost or

$$\pi = r(Q) - c(Q) = (a - dSlope \cdot Q) \cdot Q - k \cdot Q = (a - k) \cdot Q - (dSlope \cdot Q^2) \tag{9.3}$$

---

[1] The story does *not* depend in any significant way on the linearity of the demand function, although it simplifies things if we can assume that $Q(P)$ is monotonic and twice differentiable.

The monopolist will seek to set Q, the number of units made available to the market, so as to maximize π. The optimal value of Q (for the monopolist) can now be found by a simple exercise with the calculus.

$$\frac{d\pi}{dQ} = (a - k) - (2 \cdot dSlope \cdot Q) \qquad (9.4)$$

Setting $(a - k) - (2 \cdot dSlope \cdot Q)$ to zero and solving for Q yields

$$Q^* = \frac{(a - k)}{(2 \cdot dSlope)} \qquad (9.5)$$

Further, the second-order condition for maximization obtains (since we assume that $dSlope > 0$):

$$\frac{d^2\pi}{dQ^2} = -2 \cdot dSlope < 0 \qquad (9.6)$$

So, Q in (9.5) is Q^*, the optimal quantity for the monopolist to put on the market.[2]

Even though the model is exceedingly simple, it teaches valuable lessons. If having found Q^* the monopolist also finds that $\pi > 0$, the monopolist will cheerfully enter the market. In perfectly competitive markets $\pi = 0$ in the long run. Thus, the monopolist with sufficiently low production costs earns positive profits, also known as *monopoly rents*. Story 1 is an account of how and why firms operating as monopolies can (depending on production costs) be positively profitable, while firms operating in a perfectly competitive market are not (but that's another story; see Chapter 8).

Story 1 also offers cautionary lessons. Are consumers always worse off facing a monopolist rather than a perfectly competitive market? It is at least conceivable that k, the cost of production, is lower for the monopolist, resulting in lower prices for the consumer than would occur under full competition. This might occur, for example, if advertising and marketing costs were high under competition and low under monopoly. Is the monopolist's position socially optimal in the sense that the monopolist produces at an optimal scale, given the demand function? It is possible that the monopolist at its optimal position produces so as to minimize its unit costs. If this happens, however, it is due to coincidence, not to the factors operating in the market. The monopolist will maximize its profits, period. There is no particular reason why doing that should be globally efficient. Story 1, with its attendant model and analysis, teaches us that these are empirical questions.

---

[2]More generally, we assume that the total cost to the monopolist of producing Q units is $K(Q)$. Then we may characterize the monopolist's problem as

$$\max_{Q} P(Q) \cdot Q - K(Q) \qquad (9.7)$$

without assuming any specific functional forms for $P(\cdot)$ and $K(\cdot)$, other than for technical purposes.

9.2 Monopoly Story 2: Quantity Adjustment

To repeat, the classic account of monopoly (adumbrated as Story 1, above) offers much insight into the phenomena. Remarkably, it does this without appeal to any first-order (observational) data, relying on the presumed existence of a demand curve and assumption of utility maximization by the monopolist. With this it can explain the common observation that sellers seem to do better in, and seem to prefer, monopolistic markets over competitive markets with many suppliers. The story, however, is hardly a complete account of monopoly. As one author notes with regard to the standard—or "classical"—theory of monopoly,

> That's it. That's all there is to the standard theory of monopoly. We think of the monopoly being able to select the quantity it will supply or, equivalently, the price it will charge, and the resulting equilibrium price or quantity is read off the demand curve. But this simple classic theory is subject to a lot of questions, caveats, variations, and elaborations. [184, page 301]

An obvious way in which Story 1 is an incomplete account of monopoly is that the procedure it exhibits for finding the optimal quantity (or price) from the monopolist's perspective will usually be unrealistic. Story 1 is about the monopolist deciding what quantity to put on the market by *optimization*. Even if we assume that the monopolist knows its own costs precisely, it will rarely have access to an exact, twice differentiable, static demand function with which to calculate its optimal response. In short, optimization will often not be available, or accessible, to the monopolist.

The alternative to using an optimizing procedure is to use a *heuristic procedure* (often shortened to just *heuristic*). A heuristic, as I shall use the term, is a procedure that is not guaranteed to optimize but is warranted by experience to produce good results with good reliability. Someone who is using an optimizing procedure is said to *optimize* or to be *optimizing* (the representation of the problem). There is no established analog term for heuristics, so I will coin one and say that someone who is using a heuristic procedure *gropes with warrant* or simply *gropes* (on the representation of the problem). The word gropes (and its relatives) is perhaps not ideal for this purpose. The absence of an established alternative necessitates some neologism or other and *groping* has the virtue of not being otherwise employed. Unfortunately, in its ordinary English usage, *groping* suggests absence of intelligence and warrant. The reader should keep in mind, then, that I am stipulating a different meaning.

The standard account of monopoly (Story 1 will do) tells us what happens if the monopolist optimizes. Although optimization will often not be credible, because it is not available, the account is plausible because it is credible that a monopolist could use a heuristic procedure and grope its way to a position close to what could be achieved with optimization. With Story 2 we take a step toward cashing in that promissory note.

In Story 2 we assume, as before, that a stable, linear demand function exists and that the monopolist has zero marginal costs. Now, however, the monopolist is not given the demand function. This is something it must discover. A number of approaches are possible. For starters, let us assume that the monopolist begins with a rough idea of the demand function, which it uses to set what we will call its initial base (or anchor) quantity, or `initialQuantity` in the model. The monopolist operates under a PROBE AND ADJUST heuristic procedure. The market proceeds in discrete *episodes,* during which the monopolist offers a certain quantity and receives a price determined by the demand function. Although the demand function is quite real in this model, the only access the monopolist has to it is by seeing the prices that result from placing particular quantities on the market. The monopolist probes in each period by offering a quantity that is uniformly distributed in [`currentQuantity` $- \delta$, `currentQuantity` $+ \delta$], where `currentQuantity` is the current anchor quantity, and δ is an exploration parameter, typically small in relation to `currentQuantity`. After offering the probe, the monopolist observes its return—here a function of the price obtained—and records the return, classifying it as resulting either from an increase or a decrease, as the case may be, to the anchor quantity.

The monopolist continues in this fashion for a certain number of periods (or *episodes*), set by the model parameter `epochLength`, for which 30 would be a typical value. After this number of probes, the monopolist determines whether it has, on the whole, done better by offering quantities larger than the anchor or smaller. Having made this determination, the monopolist adjusts its anchor quantity by an amount ε (another parameter) in the indicated direction. This completes the current epoch. The next market period (episode) begins a new epoch and the process repeats itself. Figure 9.1, page 189, presents the basic PROBE AND ADJUST procedure in pseudocode.

The NetLogo program MonopolyProbeAndAdjust.nlogo may be used to illustrate this PROBE AND ADJUST scenario. We see the results of one run in Figure 9.2. Under the assumptions of the case the optimal quantity for the monopolist to offer is 100. The monopolist's initial hunch about the demand curve results in setting `initialQuantity` to 115.0. This value is read into the variable `currentQuantity` at initialization. The remaining parameters— δ, ε, `epochLength`—are set, respectively, to 3, 1, and 50. (MonopolyProbe-AndAdjust.nlogo is available on the book's Web site. The reader is invited to download and explore the program.)

The essential results of the run are displayed in the plot "Monopoly and Current Quantities." As this plot makes clear, PROBE AND ADJUST leads the monopolist rather promptly and directly to the vicinity of the optimal quantity, after which the monopolist's anchor quantity randomly adjusts in the neighborhood of the optimal quantity, within the limits set by δ and ε. Points arising:

1. Set parameters δ, ε, currentQuantity, epochLength
   (Typically, $\varepsilon < \delta \ll$ currentQuantity and epochLength ≈ 30.)

2. episodeCounter $\leftarrow 0$

3. returnsUp $\leftarrow$ []   (Initialize returnsUp to an empty list.)

4. returnsDown $\leftarrow$ []   (Initialize returnsDown to an empty list.)

5. Do forever:

6. episodeCounter $\leftarrow$ episodeCounter $+ 1$

7. bidQuantity $\sim U[\text{currentQuantity} - \delta, \text{currentQuantity} + \delta]$
   (The agent's bidQuantity is drawn from the uniform distribution within the range currentQuantity $\pm\delta$.)

8. return $\leftarrow$ *Return-of* bidQuantity
   (The agent receives return from bidding bidQuantity.)

9. If (return $\geq$ currentQuantity) then:
   returnsUp $\leftarrow$ *Append* return *to* returnsUp
   else:
   returnsDown $\leftarrow$ *Append* return *to* returnsDown

10. If (episodeCounter mod epochLength $= 0$) then:
    (Epoch is over. Adjust episodeCounter and reset accumulators.)

    (a) If (*mean-of* returnsUp $\geq$ *mean-of* returnsDown) then:
        currentQuantity $\leftarrow$ currentQuantity $+ \varepsilon$
        else:
        currentQuantity $\leftarrow$ currentQuantity $- \varepsilon$

    (b) returnsUp $\leftarrow$ []

    (c) returnsDown $\leftarrow$ []

11. Loop back to step 5.

FIGURE 9.1: Pseudocode for basic PROBE AND ADJUST.

1. PROBE AND ADJUST does not create a model of the demand curve. Instead, it is a form of direct reinforcement learning with a bit of patience in the form of the `epochLength` parameter.

2. A monopolist undertaking PROBE AND ADJUST would appear to behave in a manner generally consistent with the "law of effect" from behavioral psychology, according to which agents engage in randomized exploratory behavior and environmental "responses that are followed by pleasant or satisfying stimuli will be strengthened and will occur more often in the future" [218, page 365]. The parameters of PROBE AND ADJUST temper this behavior. In particular `epochLength` enforces a degree of patience.

3. PROBE AND ADJUST as described so far can never converge *exactly* to the optimal quantity, since it continues to explore within the range

   [`currentQuantity` $- \delta$, `currentQuantity` $+ \delta$].

   A more intelligent form of PROBE AND ADJUST could reduce δ over time, as well as ε. In principle, this could result in arbitrarily small deviations from optimality. The efficacy of this would, of course, depend on the actual form and stability of the demand function.

4. PROBE AND ADJUST is *computationally tractable*. It does not require substantial computational resources, whether in processing, memory, or storage.

5. PROBE AND ADJUST is hardly the only learning algorithm—or even the only tractable learning procedure—that could be used successfully by a monopolist in the face of a static, linear demand function. PROBE AND ADJUST serves to demonstrate the existence of at least one such algorithm and as such it may serve as a default for explanation and as a benchmark for other algorithms. The problem of discerning an empirically valid description of how monopolists learn to set quantities and prices remains an open question. Presumably, whatever methods are actually employed will perform for their users at least as well as PROBE AND ADJUST.

6. It should be evident that PROBE AND ADJUST does not require the demand function to be linear. Multi-modal demand functions (those with more than one local maximum) and demand functions with very flat regions will be problematic for PROBE AND ADJUST. Strictly monotonic demand functions will generally be handled well. Further, small departures from strict monotonicity may be overcome by the continuing exploration of PROBE AND ADJUST.

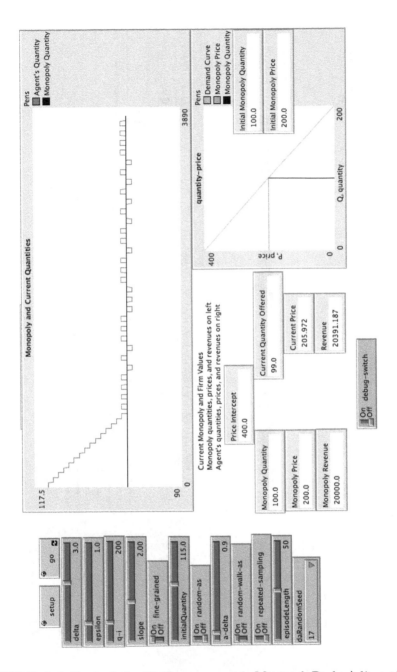

FIGURE 9.2: Run of the NetLogo program MonopolyProbeAdjust.nlogo. A monopolist uses PROBE AND ADJUST to learn in the presence of a fixed, linear demand.

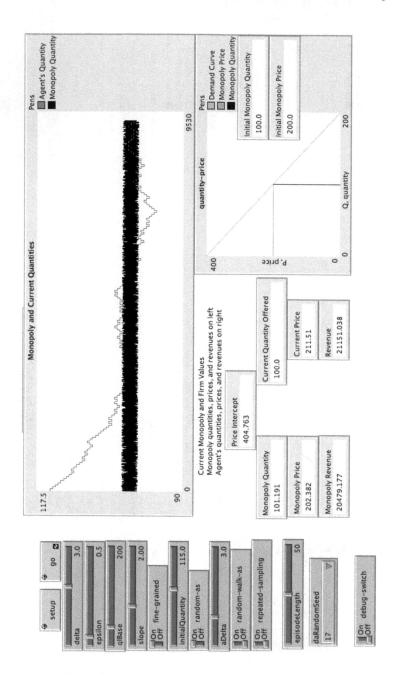

FIGURE 9.3: Run of the NetLogo program MonopolyProbeAdjust.nlogo. A monopolist uses PROBE AND ADJUST to learn in the presence of a randomly varying, linear demand.

9.3   Monopoly Story 3: Randomly Varying Demand

Assuming a purely static demand function may not be plausible in many circumstances. Even if the demand function is linear, that it should never change will surely be characteristic only of a few markets. If we are to examine monopoly behavior (or potential behavior) in the presence of varying demand, we will need a model of learning (informed quantity- or price-setting behavior), as before, and a model of varying demand. Monopoly Story 3 is the story of randomly varying demand of a certain sort and learning to set quantities via PROBE AND ADJUST.

Assume as above that the demand curve is linear: $P = a - dSlope \cdot Q$, where P is the unit price, Q is the quantity offered, $dSlope > 0$ is -1 times the slope, and a is the price intercept, the price when $Q = 0$. Given `quantityIntercept` as the value of Q (when $P = 0$), we can compute `priceIntercept` as the value of a.[3] All of this happens in Story 3 yielding the basic (inverse) demand function

$$P(Q) = P = \texttt{priceIntercept} - (\texttt{dSlope} \cdot Q) \tag{9.8}$$

In Story 3 we assume that `quantityIntercept` varies uniformly in [qIBase $\pm$ aDelta], where qIBase is the base (and initial) value of `quantityIntercept`. At the beginning of each period, the demand function is generated using dSlope as a constant and drawing a value for `quantityIntercept` uniformly from [qIBase $\pm$ aDelta], i.e.,

$$\texttt{quantityIntercept} \sim U[\texttt{qIBase} \pm \texttt{aDelta}]$$

How does a PROBE AND ADJUST monopolist fare in this environment? We can use the NetLogo program MonopolyProbeAdjust.nlogo to explore this question. Figure 9.3 illustrates. What we see in the "Monopoly and Current Quantities" plot is that the monopoly quantity fluctuates quite randomly within the range [98.5, 101.5]. As before, our monopolist starts out with a substantial overestimate of what the optimal quantity is, but proceeds rather directly to approach the neighborhood of the optimal value for quantity. The exact quantity, however, changes in every period and it is impossible for the monopolist to predict where in the range [98.5, 101.5] it will be. Under these particular circumstances, it would be heuristically best for the monopolist to settle on a quantity of 100. This would still entail a loss from actual optimality (guessing or otherwise getting it right each period). As it is, however, our PROBE AND ADJUST agent cannot learn what is heuristically best. Instead, it is buffeted by chance fluctuations. Although it settles for the most part in or near the (black) band of variation, the monopolist's offered quantity will drift high and low.

---

[3]Set $P = 0$ and $Q = \texttt{quantityIntercept}$, then solve $0 = a - (\texttt{dSlope} \cdot \texttt{quantityIntercept})$ for a and set `priceIntercept` to that value.

Notice that `epsilon`, the size of the adjustment step in PROBE AND AD-JUST, is set at 0.5, down from 1.0 in Figure 9.2. We can go further in tilting PROBE AND ADJUST toward exploration on the exploration-exploitation tradeoff. In Figure 9.4 we see the results of reducing `epsilon` to 0.2 and increasing `episodeLength` to 100. Now the initially bad starting point for the monopolist is more costly because it takes longer for PROBE AND ADJUST to find the neighborhood of the black band. Once there, however, PROBE AND ADJUST keeps the monopolist there nearly always. We may imagine even a modestly intelligent (or evolving) monopolist learning to adjust these parameters.

9.4   Monopoly Story 4: Demand as a Random Walk

In Story 3 demand variation was generated by uniformly disturbing the quantity intercept of a linear demand function. PROBE AND ADJUST was able to perform reasonably well, depending of course on its parameter settings. These, we noted, might be learned by a simple process or might evolve in a population of competing individuals.

In Story 4 we make the demand vary in a more challenging, and yet more realistic, way. Now the quantity intercept of the linear demand function will vary as a random walk. This capability can also be demonstrated in our NetLogo program: MonopolyProbeAndAdjust.nlogo. See Figure 9.5. If both the Interface tab switches `random-as` and `random-walk-as` are set to On, then in each period the current value of the quantity intercept (stored in `quantityIntercept`) is randomly perturbed. Specifically

$$\texttt{quantityIntercept}_{t+1} =$$

$$U[-\texttt{aDelta}, \texttt{aDelta}] + \texttt{quantityIntercept}_t \qquad (9.9)$$

The plot "Monopoly and Current Quantities" in Figure 9.5 presents results from a typical run of PROBE AND ADJUST. Notice that `epsilon` is back up at 1.0 and that `aDelta` is also set to 1.0. As usual, we have the monopolist start out too high. Now, however, the monopoly price (in black) drifts upward a bit and the monopolist soon tracks the demand reasonably well. In this stream of random numbers, however, there is a secular trend that reduces the optimal demand. During this (probabilistic) precipitous drop the PROBE AND ADJUST monopolist follows along but inevitably does less well. Later, the steep trend abates, the monopolist catches up and continues to track fairly well.

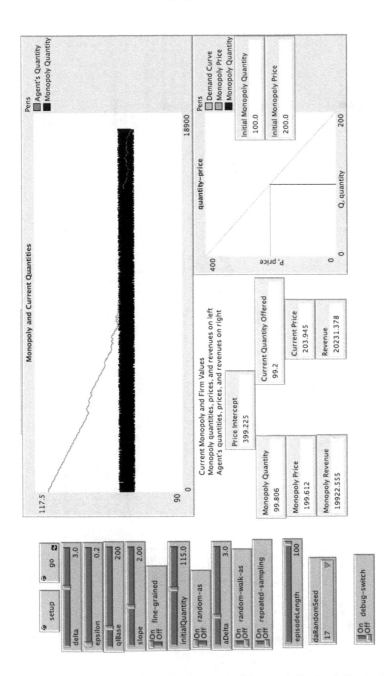

FIGURE 9.4: Run of the NetLogo program MonopolyProbeAdjust.nlogo. A monopolist uses PROBE AND ADJUST to learn in the presence of a randomly varying, linear demand.

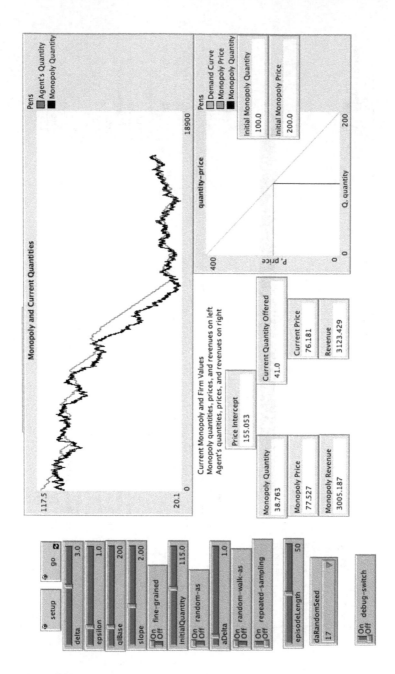

FIGURE 9.5: Run of the NetLogo program MonopolyProbeAdjust.nlogo. A monopolist uses PROBE AND ADJUST to learn in the presence of a random walk, linear demand.

9.5   Discussion

PROBE AND ADJUST is a simple, intuitive, computationally undemanding learning procedure for adjusting a continuous (or fine-grained discrete) quantity. Effective functioning of PROBE AND ADJUST in any given context of the sort examined here will depend upon (i) how well its parameters are set and (ii) the nature of the underlying random process that generates the demand curve.

On (i), I have noted that parameter values for PROBE AND ADJUST may themselves be learned, either by individuals or via selection in a population of competing individuals. Notice further that the parameters for PROBE AND ADJUST may be learned without having access to the demand function, other than by interaction with it. For example, long periods of unidirectional movement—as in the middle section of the "Monopoly and Current Quantities" in Figure 9.5 or the beginning of the plot in Figure 9.3—heuristically suggest increases in `delta` and `epsilon`. Conversely, substantial periods of fluctuation about a weak tread—as in the latter parts of the plot in Figure 9.3—heuristically suggest decreases in `delta` and `epsilon`. These remarks could be extended. The key point is that PROBE AND ADJUST is tunable in a computationally attractive fashion, using information available to the agent.

Regarding (ii), PROBE AND ADJUST may easily do worse than a learning or adaptive procedure that makes use of knowledge of the underlying process. In each case—static demand function, randomly oscillating demand function, and randomly walking demand function—it is possible to find a better procedure for that case. PROBE AND ADJUST, on the other hand, appears to be a reasonably good generalist, especially if it incorporates learning of parameter values.

The real significance of PROBE AND ADJUST is that it provides a demonstration of a computationally tractable and otherwise credible learning procedure by which a monopolist could effectively set quantity (or by extension, price), even in a stochastically changing demand environment. As such our discussion of PROBE AND ADJUST serves to validate and strengthen, and even to extend slightly, the classical theory of monopoly.

A final word on methodology. Classical behavioral theories of learning (stimulus response learning, operant conditioning, and related approaches [218]) emphasize learning of associations between (and among) categorized experiences. The motivating idea is captured (at the level of a slogan, but usefully nonetheless) in the "law of effect" mentioned above. Recent cognitive theories of learning (see, e.g., [117] for an overview) emphasize the role of "mental models," cognitive maps, broadly mental representations in learning and the use of knowledge. PROBE AND ADJUST may be seen as positioned between these two views. It is a procedure that balances exploration of the environment and exploitation of feedback. While it certainly responds to rewarded behavior with new behavior (implicitly presumed to be) like that rewarded, it does not do so slavishly. It is patient and it continues to explore systematically. On the other hand, it would be a stretch to describe PROBE AND ADJUST as representational. It does not proactively seek data, fit a model to the data, and make decisions based on the model. PROBE AND ADJUST is not statistical regression analysis or anything like it representationally.

The distinction between broadly behavioral, cognitive, and procedural accounts of learning will be useful to us. Before saying more, however, it is well to remember that decision making by the classical monopolist is purely parametric, not strategic. We now turn to duopoly, which is very much a strategic context.

9.6   For Exploration

1. Suppose that a monopolist has estimated precisely the monopoly quantity for the market. Why might the monopolist find it attractive to offer a larger quantity to the market?

2. Is it possible for a monopolist to offer the monopoly quantity to the market, receive the corresponding monopoly rent, and leave the customers better off than if the market were fully competitive? Explain.

3. Recall Competitive Market Story 2 from Chapter 8. If that trading regime occurred in a monopoly market, what would happen? What would be the average price? How would prices be distributed?

4. Recall Competitive Market Story 3 from Chapter 8. If that trading regime occurred in a monopoly market, what would happen? What would be the average price? How would prices be distributed?

5. Recall Competitive Market Story 4 from Chapter 8. If that trading regime occurred in a monopoly market, what would happen? What would be the average price? How would prices be distributed?

6. What would happen in a competitive market if buyers and sellers used PROBE AND ADJUST to set their bids and asks?

7. Besides PROBE AND ADJUST, what informationally and computationally credible means are there for monopolists to discuss the demand curve and respond to it heuristically well? Consider, as we did above, demand curves that vary randomly.

8. Using MonopolyProbeAndAdjust.nlogo, explore the range of conditions under which PROBE AND ADJUST tracks the stochastically varying demand curve well. What are the conditions under which it breaks down?

9.7   Concluding Notes

Standard microeconomics textbooks present our Story 1 as the received view. Good examples include [184, 234, 321, 322].

Chapter 10

Oligopoly: Cournot Competition

10.1 Introduction

Research on decision making in the firm includes two streams—from the economics literature and from the management literature—having different goals. The management literature looks at how managers should learn about their circumstances and make decisions. The purpose of this literature is to advise and to train managers who have to make decisions in the course of running an organization. When looking at interactions of firms in a market, economists generally look at the firm as an aggregate and posit a theory of full information, ideally rational decision making in which the firm actually maximizes profits. The stylized abstractions of economists have been useful for developing theories of the competitiveness of markets and of public policy remedies for market power, a classic concern of microeconomics. The managerial literature deals with making the myriad decisions managers face in organizations. There is less emphasis on the public policy concerns of market power and more on decisions that improve profits and observations on the behaviors of participants. Thus, these two streams of literature address complementary issues, yet treat the same entity, the firm in its environment.

Models of market power standardly make one of two assumptions. Either firms offer prices and the firm with the lowest price gets the whole market, the Bertrand game, or firms produce quantities under the assumption that the other firms do not adjust their quantities in response to the firm's decisions, the Cournot game. In Kagel and Roth [162] and elsewhere, one finds comments to the effect that the behaviors of the agents in these economic models are not verifiable and often run counter to the literature in cognitive psychology and behavioral economics. Keunne [187] points out that firms in an oligopoly form a community with community-based norms, values, and hierarchies.

The managerial literature addresses the issue of making decisions without full information because it is aimed at practical application and in practice the information available is always incomplete. A better theory of the firm and markets would take into account how managers make decisions and what they know and when they know it. Managers know less about the parameters for functions such as demand curves and probability distributions but know more about the other players in their market. By using people in simulated markets it is possible to observe the decision making and outcomes that get beyond simple abstractions of the decisions managers face. Selten, Mitzkewitz, and Uhlich [282] simulate outcomes in a duopoly market by having students program agents. They give students full information on both the supply and demand parameters, and software in which the students can embed their strategies to play against other students in duopoly games. Key to the clever design of the experiments is the revealing and testing of the decision rules used by the players. The authors find that smart agents can obtain higher profits than those in the classic Cournot equilibrium.

Here are several of their observations on the results. One strategy of the players is to try to forecast the actions of the other agents. This strategy is close to positing what is known as the conjectural variation of the other player in response to the first player's actions [187] and performs badly in the experiments. In the Cournot game the conjectural variation is presumed to be zero. The best performing strategies are the oligopoly equivalent of TIT FOR TAT, the winning strategy in Axelrod's tournament with Prisoner's Dilemma [15]. The agents in the Selten et al. experiments punished excessive competitiveness and rewarded cooperation, moving the solution from the non-cooperative (defecting) Cournot equilibrium toward the monopoly outcome, which is the Pareto outcome. In essence they are reinforcing a community norm of cooperation without explicitly setting production quotas. (We note that the Pareto outcome of mutual cooperation also results from repeated play in the Prisoner's Dilemma game with the TIT FOR TAT strategy and that under general conditions of mutual learning the Pareto outcome is often achieved or approximated [175].)

The behavior of the players in the Selten et al. experiments is strikingly at odds with what would be predicted by the classic Cournot theory. The Selten et al. players learned to collude tacitly and thereby achieved rewards in excess of those available from the Cournot outcomes. They did this, however, using explicitly given knowledge of the market (e.g., the demand curve) that is in fact not generally available in practice.

The foregoing forcefully raises the question of whether there are effective procedures, using realistically available information, that may be actually used by managers in oligopoly settings and that produce the Cournot-improving outcomes found in the experiments by Selten et al. This is the question we explore in what follows. We begin with an overview of the current literature.

10.2   Overview of the Literature

See Tesfatsion and Judd [310] for a collection of articles on the current status of and issues in agent-based modeling in economics. Tesfatsion [309] surveys the applications of "agent-based computational economics" (ACE) to specific industries and supply chains. Brenner [44, 45] summarizes the literature on learning both from cognitive psychology and artificial intelligence. Brenner [43] and Bruun [49] are useful collections of relevant papers. Duffy [80] looks at what intelligence is necessary for an agent, describing zero-intelligence agents, reinforcement learning and evolutionary algorithms. Pyka and Fagiolo [249] provide an overview of the methodological issues in agent-based economic models.

To deal with the institutional features of electricity markets, researchers have developed agent-based models to simulate the auctions they use. See Bunn and Oliviera [50], Entrikan and Wan [92], and Marks [213].

The rational expectations literature—discussing situations in which economic agents try to forecast the future—relates to the issues raised by Selten, Mitzkewitz, and Uhlich [282]. Hommes et al. [155] experiment with human subjects to forecast the clearing price for a market. The forecasted price sets the production quantity and the actual price results from the production level, resulting in the cobweb model. They show that people can find the market-clearing price and the stability of the market depends on the price sensitivity of supply.

Arifovic [10] shows that in an oligopoly that is modeled with a population of agents that evolves using a genetic algorithm, the solution converges to the competitive solution, not the Cournot equilibrium. Vriend [325] shows that with social learning, where each player sees the returns of every player and can adopt the strategies of the successful players, the solution converges to the competitive equilibrium, and with individual learning, where the player sees only its returns, the solution converges to the Cournot equilibrium. Riechman [256] finds it necessary to have more complicated agents to find a solution different from the competitive solution even in an oligopolistic market. Waltman and Kaymak [327] use Q-learning and find that under certain circumstances the agents move to solutions between the monopoly and Cournot equilibria. Arifovic and Maschek [11] find that Vriend's results are parameter driven and not robust. Alkemade, La Poutr, and Amman [6] show that modeling the agents as chromosomes (the agent has an assigned strategy) in a genetic algorithm instead of modeling the strategies as chromosomes (the agent can choose its strategy) can lead to premature convergence in a genetic algorithm. The parameter settings determine whether the Cournot or the competitive solution is reached when agents are strategies and the Cournot solution is reached when the agents can choose the strategy. Note that the definitions used in this article are different from the definitions used here in that we define a policy as a choice of measure of success and the price and quantity are operating decisions.

Huck et al. [157] have a model of agent behavior that, like ours, requires little intelligence or information about its environment. They find that when agents make simultaneous moves of the same step size in a duopoly, the players maximize total social welfare and divide the market equally even though they see only their own welfare. This is because either both players see either the marginal revenue function of the monopolist when both players move in the same direction or the trial price as the marginal revenue, as if the market were in perfect competition, when the players move in opposite directions. The effect of these perceptions is a sequence of steps that leads to equalizing production levels and movements in the same direction. The players converge to the monopoly solution because that is the marginal revenue function they see. Huck et al. also model the players moving sequentially and monitoring the effects of their own actions, resulting in the Nash/Cournot equilibrium. In this case the players follow the tâtonnement process used to explicate the Cournot equilibrium.

Barr and Saraceno [22] use neural networks to represent learning agents that learn about the environment rather than learn the optimal production level. They show that the agents find the Cournot equilibrium. Agents with simple neural networks find the equilibrium faster but more complicated networks develop a better demand representation and find a more accurate solution in the long run.

Marks and Midgley [212] build a simulation of a market with an oligopoly of coffee manufacturers and a retailer between the producers and customers that decides which coffee promotion to accept. They then simulate the outcomes of retailer strategies that range from zero information to sophisticated measurement of the market and find that a zero information retailer does quite well. This is an example of bringing the management literature into the economic models of markets. Midgley, Marks, and Cooper [226] used an earlier version of this model to look at breeding profit maximizing retailers to examine the frequency of promoting coffee specials. Their work uses point-of-sale information for comparing the model retailer to the actual retailer. Sallans et al. [268] breed firms that compete on production positioning in a market and the firms have to finance their businesses using agents modeled as financial firms. They are able to replicate many phenomena observed in retail markets.

The business literature on decision making, especially the practitioner books and articles, is relevant to agent-based modeling because the economic models should represent business decisions in the way business people make them. Since the focus of oligopoly models is on setting the price and/or quantity, the most relevant literature is in marketing on pricing and capacity expansion. We focus on pricing here as the same issues arise in setting quantities.

We use Nagle and Hogan [231] to illustrate current managerial thought on best practices in pricing products (whether by setting prices or quantities). The main point this book makes is that a firm should first do everything to avoid a focus on price, for example creating product distinctions that are real or only in the minds of customers. The discussion of demand elasticities in this book covers 3 pages in a 30-page chapter with the main discussion about customer perception of product attributes of the firm's product versus the attributes of competitor products and social norms. That is, the discussion focuses on the position of the product relative to the competition with the goal of segmenting the market to customers who are willing to pay a premium for the product's perceived attributes.

The entire approach advocated presumes—and is incomprehensible without assuming—that customers are not fully rational and/or do not have full information. The following passages are representative.

> Unless customers actually recognize the value that you create and ask them to pay for, value-based pricing will fail. ...

> The reality is that customers generally don't know the true value delivered by items they buy unless the seller informs them. That leaves the most differentiated and highest quality supplier vulnerable to competitors who offer a lower-price alternative possessing only those value components the customer recognizes, and who portrays additional value elements.... [231, page 81]

> Although cost should not drive the prices you charge, your prices can definitely affect the cost-to-serve customers and, therefore, your profitability. Many companies differentiate their offers with bundled services, even when demands on those services are subject to customer discretion and therefore are not proportional to the volume of sales. "Service abusers" can boost your average cost of sales while "service avoiders" drive up your average cost of sales by abandoning you in favor of cheaper, low-service competitors. ...

> ... The solution is to create "roughly right" cost allocation indexes and use them to build a "roughly right" relative profit index by account or segment. [231, pages 113–4]

Their discussion on elasticities looks at the elasticities of the firm's products and not the market and they note how elasticities differ depending on market share, because the different products in a market are positioned for different customer segments, and products age. In the chapter on estimating price response they state that "The low accuracy of many numerical estimates makes blind reliance on them very risky." They conclude the chapter with "Even when actual purchase data cannot provide conclusive answers, they can suggest relationships that can then be measured more reliably with other techniques."

What should be taken from this short discussion is that the management literature recommends that managers explore rather than optimize. The data are not completely clear and circumstances change.

10.3   Experimental Setup

We procedurally model agents playing a Cournot game. Using NetLogo as the programming environment, we name our program OligopolyPutQuantity.nlogo.

Abstractly, the agent we define has three features:

1. a measure of success,

2. a data stream to measure its success, and

3. the ability to do experiments or to learn how its actions affect its success.

These three properties are the minimal set of properties for an economic agent to improve its outcomes when operating in a situation without full information. To add an element of realism, the agents can be made to operate in a noisy environment where the demand parameters are a random walk.

One measure of success we use is the classic measure, firm profitability. We term this measure and the policy of using this measure as the objective function "Own Returns." Another measure is the profitability of the whole industry, termed "Market Returns." We allow for both measures because firms operate in a complex institutional environment and the leaders of these companies set up and fund institutions that represent the industry. Examples are the National Petroleum Refiners Association and the Iron and Steel Institute. That is, firms choose when to cooperate and when to compete (see Brandenberger and Naibuff [41]). A less mainstream example is Cosa Nostra, which acts as a chamber of commerce for crime families that mediates conflicts, reduces killing among the families, and works to protect the profitability of organized crime.

We allow the players to use combinations of objectives as a measure of success. In the first, an agent pursuing the "Mixed" policy, at the end of its epochs, uses a convex combination of both objectives. In the second, an agent pursuing the "Market Returns, Constrained by Own Returns" (MR-COR) policy, at the end of its epochs, looks at the mean quantity it produces versus the mean total quantity produced for the entire market (its and the other player(s)'s production). If its mean quantity produced plus epsilon is lower than the mean quantity produced for the market, the agent raises its baseline production by epsilon; otherwise, it uses the "Market Returns" policy. Here a firm pursues a hierarchical policy where it looks to get its share of the market and then looks to keep the market as profitable as possible. Equal shares is the outcome of the Cournot solution when players have equal costs. In most oligopolies the firms have different sizes because of differences in product attributes, unique access to high-quality resources, or the history of the firms and markets, including acquisitions. We view this more as a stylized form of maintaining a sense of fairness while taking the larger view of the industry as a community norm of an implicit willingness to cooperate up to a point as in Kuenne.

In the Cournot game the players make quantity decisions and the market sets the price. Each player starts with a base quantity that remains fixed for a given number of periods and randomly adjusts the quantity up or down in each period, running experiments to observe the effects of altering the base quantity. The number of periods for which the base quantity remains the same for a player is termed an *epoch*. We use a uniform distribution for the random adjustments around the base quantity. Different players can have different epoch lengths. The player is interested in knowing whether it should increase or decrease its production and records its returns and/or the market returns for the quantity increases and decreases separately. After each epoch, the players assess their returns using their measures of success. If the profits for a player are higher with the increases than with the decreases, then the base quantity is increased and vice versa. This begins a new epoch. Epochs are repeated in the simulation until the pattern of behavior stabilizes. Note that each player knows the outcome only in relation to its decisions and retains no information on the other players' decisions. We term the search/learning method the agents employ PROBE AND ADJUST. Technically, the method is in the family of line-search algorithms where the algorithm finds the direction of improvement, takes a step of a certain size in that direction, and then assesses the benefit of that move. See Winston [333] for an introduction to algorithms in this class. More importantly, in our context, this algorithm represents a situation in which managers adjust their production incrementally to learn the consequences of their actions, without making radical changes that could risk the business. Think of this as a form of muddling through. The algorithm approximates the behaviors of consumer products companies that phase in price increases and the capacity-expansion decisions in commodity businesses such as petroleum refining, where increases or decreases in capacity are incremental because of environmental concerns removing the ability to build a wholly new refinery in the US. It does not reflect the situation where capacity has to be added in large increments, such as a firm building a green-field integrated steel mill.

10.4   Reference Model

For clarity we present the underlying model and resulting key quantities that we refer to throughout, as well as the terminology we use. To begin, we assume a linear inverse demand function:

$$P = a - slope \times Q \tag{10.1}$$

P is the price realized in the market. Q is the total quantity of good supplied to the market. a is the price intercept and $slope > 0$, we assume. We also assume that negative prices are not permitted, so (10.2) is actually what is assumed.

$$P = \max\{a - slope \times Q, 0\} \tag{10.2}$$

We begin with the duopoly case and then generalize the results. Let the agents have unit costs, k_i, which can differ. In the duopoly case the profit of firm 1 is then

$$\pi_1 = P \cdot Q_1 - k_1 \cdot Q_1 = (a - slope \cdot (Q_1 + Q_2)) \cdot Q_1 - k_1 \cdot Q_1 \tag{10.3}$$

For firm 2 we have

$$\pi_2 = P \cdot Q_2 - k_2 \cdot Q_2 = (a - slope \cdot (Q_1 + Q_2)) \cdot Q_2 - k_2 \cdot Q_2 \tag{10.4}$$

Differentiating we get

$$\frac{d\pi_1}{dQ_1} = a - 2 \cdot slope \cdot Q_1 - slope \cdot Q_1 \cdot \frac{dQ_2}{dQ_1} - slope \cdot Q_2 - k_1 \tag{10.5}$$

$$\frac{d\pi_2}{dQ_2} = a - 2 \cdot slope \cdot Q_2 - slope \cdot Q_2 \cdot \frac{dQ_1}{dQ_2} - slope \cdot Q_1 - k_2 \tag{10.6}$$

Setting $\frac{dQ_2}{dQ_1}$ and $\frac{dQ_1}{dQ_2}$ to 0 as usual leads to

$$0 = a - 2 \cdot slope \cdot Q_1 - slope \cdot Q_2 - k_1 \tag{10.7}$$

$$0 = a - 2 \cdot slope \cdot Q_2 - slope \cdot Q_1 - k_2 \tag{10.8}$$

and then on to

$$Q_1 = \frac{a - slope \cdot Q_2 - k_1}{2 \cdot slope} \tag{10.9}$$

$$Q_2 = \frac{a - slope \cdot Q_1 - k_2}{2 \cdot slope} \tag{10.10}$$

which when solved yield

$$Q_1^C(2, [k_1, k_2]) = \frac{a - 2k_1 + k_2}{3 \cdot slope} \tag{10.11}$$

$$Q_2^C(2, [k_1, k_2]) = \frac{a + k_1 - 2k_2}{3 \cdot slope} \tag{10.12}$$

Notice that

$$Q^C(2, [k_1, k_2]) = Q_1^C(2, [k_1, k_2]) + Q_2^C(2, [k_1, k_2]) = \frac{2a - k_1 - k_2}{3 \cdot slope} \tag{10.13}$$

The formula generalizes. With n players having proportional costs $k_i \in \{1, 2, 3, \ldots, n\}$ (total cost = unit cost × quantity = $k_i \cdot Q_i$) we have expression (10.17).

The monopoly quantity, Q^M, may be arrived at as the special case of (10.18) when $n = 1$:

$$Q^M(k) = \frac{(a-k)}{(2 \cdot slope)} \tag{10.14}$$

Finally, the rivalrous (or competitive, but we've already used C) quantity, Q^R, obtaining in a fully competitive market occurs when price $(a - slope \times Q)$ equals marginal cost (k, assuming all firms have the same marginal cost). Equating them and solving yields Q^R.

$$Q^R(k) = \frac{(a-k)}{slope} \tag{10.15}$$

Now assume there are n firms in the market, $n \geq 1$. The quantity supplied by firm i (in a given round or episode) is Q_i. We stipulate

$$Q = \sum_{i=1}^{n} Q_i \tag{10.16}$$

Each firm i has a unit (marginal) cost of production of $k_i \geq 0$.

Given these conditions, then in the Cournot model the equilibrium Cournot quantity, Q^C, is the sum of the individual Q_i^Cs, and

$$Q^C(n, \vec{k}) = Q^C = \sum_{i=1}^{n} Q_i^C = \frac{na - \sum_{i=1}^{n} k_i}{(n+1) \cdot slope} \tag{10.17}$$

When all k_i are equal to k we write $Q^C(n, k)$ for $Q^C(n, \vec{k})$. That is,

$$Q^C(n, k) = Q^C = \sum_{i=1}^{n} Q_i^C = \frac{na - \sum_{i=1}^{n} k}{(n+1) \cdot slope} \tag{10.18}$$

and the individual firm Cournot quantities are

$$Q_i^C(n, k) = \frac{(a-k)}{(n+1) \cdot slope} \tag{10.19}$$

10.5 Results

Our results are obtained from the program OligopolyPutQuantity.nlogo. This program affords procedural modeling of quantity-bidding (or Cournot) agents ("firms") in an oligopoly. Agents are able to use any of several forms of PROBE AND ADJUST, a learning algorithm suitable for adjusting a continuous parameter, here the quantity an agent offers to the market during

1. Set parameters δ, ε, currentQuantity, epochLength
   (Typically, $\varepsilon < \delta \ll$ currentQuantity and epochLength ≈ 30.)

2. episodeCounter $\leftarrow 0$

3. returnsUp $\leftarrow$ []   (Initialize returnsUp to an empty list.)

4. returnsDown $\leftarrow$ []   (Initialize returnsDown to an empty list.)

5. Do forever:

6. episodeCounter $\leftarrow$ episodeCounter $+ 1$

7. bidQuantity $\sim U$[currentQuantity $- \delta$, currentQuantity $+ \delta$]
   (The agent's bidQuantity is drawn from the uniform distribution within the range currentQuantity $\pm\delta$.)

8. return $\leftarrow$ *Return-of* bidQuantity
   (The agent receives return from bidding bidQuantity.)

9. If (bidQuantity $\geq$ currentQuantity) then:
   returnsUp $\leftarrow$ *Append* return *to* returnsUp
   else:
   returnsDown $\leftarrow$ *Append* return *to* returnsDown

10. If (episodeCounter mod epochLength $= 0$) then:
    (Epoch is over. Adjust episodeCounter and reset accumulators.)

    (a) If (*mean-of* returnsUp $\geq$ *mean-of* returnsDown) then:
        currentQuantity $\leftarrow$ currentQuantity $+ \varepsilon$
        else:
        currentQuantity $\leftarrow$ currentQuantity $- \varepsilon$
    (b) returnsUp $\leftarrow$ []
    (c) returnsDown $\leftarrow$ []

11. Loop back to step 5.

FIGURE 10.1: Pseudocode for basic PROBE AND ADJUST.

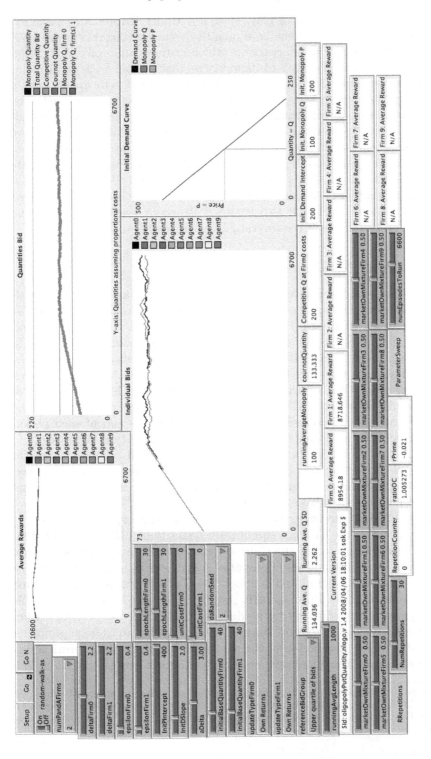

FIGURE 10.2: Results from `oligopolyPutQuantity.nlogo` with two quantity-offering players each using the Own Returns policy of play.

a single period (aka: round of play, episode). Under PROBE AND ADJUST, each agent maintains a current base quantity—currentQuantity—which it uses as the midpoint of an interval from which it draws uniformly each period to set its production quantity—bidQuantity—that period. Each agent maintains its currentQuantity for a number of episodes equal to its epochLength. When its current epoch is over (by count of periods or episodes played), the agent re-evaluates its currentQuantity, adjusting it up or down by its adjustment moiety, its epsilon, depending on whether production levels above its currentQuantity during the just-completed epoch have or have not been more profitable than those below its currentQuantity. Figure 10.1 presents the basic PROBE AND ADJUST algorithm in pseudocode.

Firms in oligopolyPutQuantity.nlogo may use any of several variations (or *policy versions*) of PROBE AND ADJUST. We now present results, focusing especially on the effects of the several policy versions available.

10.5.1   Profit Maximization

We begin with the classical objective function of maximizing firm profits. Consider first two agents—Firm0 and Firm1—who independently choose production quantities using PROBE AND ADJUST in the presence of a linear price function, $P(rice) = priceIntercept - dSlope \times Q(uantity) = 400 - 2Q$ in our examples below, that is unknown to them. In this section we consider the case in which each firm uses PROBE AND ADJUST and looks only to its own profits (rewards) when adjusting its currentQuantity. This is the policy of play labeled "Own Returns" in Figure 10.2. Specifically, a player using the Own Returns policy observes the market's price at the end of each period of play, calculates its profits (net of costs) for the quantity it produces (and sells), and uses this as its **return** in step 8 of Figure 10.1. There, we might label the *Return-of* function as *Own-Returns* in this case. Figure 10.2 shows simulation results from oligopolyPutQuantity.nlogo. Each firm has costs of $0 per unit and their base quantities—

(initialBaseQuantityFirm0 = initialBaseQuantityFirm1 = 40)

—start well below the Cournot equilibrium quantity of 133.33.[1] PROBE AND ADJUST leads them near to the Cournot solution with equal market shares. We see that after 6600 episodes (long after a stable settling has occurred) the average total quantity produced is about 134 (134.127 averaged over the last 1000 episodes of play). Each agent is producing about 67.1 at the end (this is a rough estimate, obtained by reading the Individual Bids charts, rather than computed exactly from the data). Firm0 is getting an average profit per play of about 8650 (averaged over the past 1000 episodes), while Firm1 is getting about 9004. The agents have essentially identical average profits overall

---

[1]The results we report are for long after the system has attained stability. Moreover, we find that initial base quantities do not affect either the location of, or the firm shares at, stability. For these two reasons we do not further discuss any starting point effects. There are none, except as noted below.

Firm	Average Reward	Standard Deviation of Average Reward
0	639.8964661413705	70.55142984995368
1	645.3864190695721	67.94933796315856
2	649.0329236965865	62.82872860151167
3	651.7672918703781	68.57175862731928
4	643.1174687232188	64.6236221246588
5	638.8300180138967	61.28529868477505
6	638.2949903490786	66.98362150327631
7	650.5939716430881	79.28907276367299
8	649.4548736132194	62.526625943544516
9	643.725796838893	65.38427951535789

TABLE 10.1: Results summary for 100 replications with 10 firms, all following the Own Returns policy of play.

(see Average Rewards chart). Thus, the cost of learning is quite small. Notice that the quantities fluctuate around the optimal quantity (from the Cournot perspective) because the agents are always randomly varying their base quantity. The standard deviation (`runningAverageBidSD` in the Figure) on total production is 2.847 units. We repeated the experiment 100 times, using the system clock to initialize the random number generator each time. Over the 100 trials the mean of the averages of the total quantity produced is 133.43 with a standard deviation of 1.57. Firm0's average of its average rewards was 8874.72 (standard deviation of 159.08) and Firm1's was 8863.70 (155.99). Note that the standard deviations of profits and production quantities are roughly 2 percent of profits and quantities produced.

We also repeated the experiment 100 times with 10 firms in the market (and using the system clock to initiate the random number generator). The Cournot quantity is now (with 10 firms) 181.818. Averaged across the 100 repetitions, the running average production was 182.08 with a standard deviation of 0.74. Table 10.1 reports the averaged rewards (and their standard deviations) for the 10 firms in this run of 100 replications.

Thus, we can say the model provides a good approximation to the Cournot solution, in spite of not having the full information assumed in the classical analysis and without each player giving the best response to the other player's plays. A simple explanation of these results is that since there is no correlation or coordination of the player's moves during an epoch, on average the players see the marginal revenue function of a Cournot player. Unlike Riechmann [256] then, we find that a pair of simple searching agents can find the Cournot solution. Figure 10.2 shows, and we observe in all other runs as well, a certain amount of oscillation about the Cournot quantity. The range of oscillation varies depending on the length of the epoch. Reducing the epoch length to 10 results in an average over five runs of the running average quantity to be 133.85

(using random number seeds 0 through 4), increasing the standard deviation on average to 3.52. After reducing `epochLengthFirm0` and `epochLengthFirm1` further to 5, the running average quantity averaged over five runs (seeds 0 through 4) is 132.8114 and the average standard deviation is 4.2004. At 2 the numbers are 133.7688 and 6.1852. The range on the production levels of the individual agents shows more fluctuations with a smaller epoch length than the total because if one firm increases production beyond the optimal quantity due to randomness, the other is more likely to respond with a decrease in the next epoch, a consequence of the shape of the objective function, which is quadratic and has a steeper slope and more curvature the further away from the optimum. With an epoch length of 1 the agents engage in what looks like a random walk with reflecting barriers at production levels of 0 for each player and total production at the competitive equilibrium with no profits for either player. This is because there is either an up or down value but not both in an epoch. Using the parameters of the previous runs, the average of the running averages is 87.45 (= mean of 170.612 22.841 65.555 125.666 52.594) and the average of the standard deviations is 9.75 (= mean of 7.725 8.299 7.784 14.988 9.934). Patience, in the form of a longer epoch length (a tilt toward exploration and away from exploitation), has its rewards for these players.

Increasing the number of players from two to three but keeping other conditions constant (and returning to epoch lengths of 30) increases the Cournot quantity to 150 and increases the standard deviation of the running average of the total quantity. The numbers are 150.30 (= mean of 148.284 151.984 153.597 149.392 148.221) and 3.6482 (= mean of 3.904 3.591 3.605 3.286 3.855). With 10 firms the Cournot quantity is 181.818 and the numbers are 181.811 (= mean of 180.851 180.708 182.882 182.504 182.11) and 6.33 (= mean of 6.212 6.673 6.233 6.442 6.112). In general, variation in total production increases with the number of firms in the market, but the overall picture remains otherwise accurate.

10.5.1.1   Detailed Statistical Analysis of the Model Results

Because it is possible to generate large amounts of data from the simulation results, we are able to avoid making the assumption of normality and use weaker nonparametric tests to examine the statistical validity of the results. Table 10.2 presents the relevant parameters and their default values (the "Table 10.2 settings") for the PROBE AND ADJUST model. We ran 100 repetitions of the `oligopolyPutQuantity.nlogo` model under the conditions described in Table 10.2. Each repetition produces an ending value of `runningAverageBid`, the total quantity averaged over the past `runningAverageLength` (=1000 in the Table 10.2 settings) episodes (i.e., in episodes 5601–6600, given the

Agent Parameters	
numPandAFirms	2
epochLengthFirm0	30
epochLengthFirm1	30
initialBaseQuantityFirm0	40
initialBaseQuantityFirm1	40
updateTypeFirm0	Own Returns
updateTypeFirm1	Own Returns
deltaFirm0	3.0
deltaFirm1	3.0
epsilonFirm0	0.7
epsilonFirm1	0.7
unitCostFirm0	0
unitCostFirm1	0
Environment Parameters	
InitPIntercept (Price intercept of demand function)	400
InitDSlope (Negative of the slope of the demand function)	2
numEpisodesToRun (Number of episodes in a run)	6600
runningAvgLength (Running average length)	1000
daRandomSeed (Random number seeded with)	system clock
NumRepetitions (Number of repetitions)	100

TABLE 10.2: Default settings of the principal parameters.

Table 10.2 settings). Here is a summary of the 100 values obtained for runningAverageBid.[2]

runningAverageBid summary					
Min.	1st Qu.	Median	Mean	3rd Qu.	Max.
129.5	132.3	133.3	133.3	134.1	138.6

Note that the Cournot value is 133.333. These data are symmetrically and rather tightly centered on or near the Cournot value.

Using the exact binomial test (binom.test in R) on the differences between the 100 runningAverageBid values and the Cournot value, in this particular run of 100 we actually got 50 values above the Cournot value and 50 below (there were no ties). Under the null hypothesis of $p = 0.5$, the p-value is 1 and the 95% confidence interval is $[0.3983, 0.6017]$. There is no reason to reject the null hypothesis, that PROBE AND ADJUST with Own Returns leads the agents to the Cournot solution.

---

[2]These and all subsequent statistical calculations were made in R, which we gratefully acknowledge [250].

At the end of each run `oligopolyPutQuantity.nlogo` reports the standard deviation of the `runningAverageBid` (see Figure 10.2; in the present case, for the last 1000 episodes). Here is a summary of the 100 values obtained for `runningAverageBidSd`.

runningAverageBidSd summary					
Min.	1st Qu.	Median	Mean	3rd Qu.	Max.
2.596	2.865	3.099	3.146	3.333	4.417

While some asymmetry is apparent, these values are reassuringly regular and concentrated.

Turning now to the rewards obtained by the two individual firms, we would expect them to have no systematic differences. That this is so is certainly suggested by the following summary data.

summary of mean rewards						
	Min.	1st Qu.	Median	Mean	3rd Qu.	Max.
`meanRewardFirm0`	8509	8784	8886	8868	8958	9414
`meanRewardFirm1`	8366	8766	8900	8895	9006	9276

The nonparametric Wilcoxon rank-sum test (wilcox.test in R) yields a p-value of 0.2026 for the null hypothesis of no difference between the two sets of returns. This coheres with the suggestion we drew from the table.

We investigated whether the default number of episodes, 6600, is sufficient for getting beyond any transitory effects from the initialization of the model. As the three tables of summary information in Table 10.3 indicate, it is.

We now consider some "parameter sweeping" experiments, first on epoch lengths. We conducted a full factorial experiment using five levels of epoch length (24, 26, 30, 34, 38) for each of two variables (`epochLengthFirm0` and `epochLengthFirm1`) with 30 repetitions (for a total of $750 = 30 \times 5^2$ runs. (Otherwise, the parameter settings are the default settings of Table 10.2.) From the Table 10.4 it is evident that on average there is no deviation from our basic findings for the default settings. Looking within these data Table 10.5 is a summary for when Firm0's epoch length was 24. Restricting our attention further, Table 10.6 is a summary for when `epochLengthFirm1` = 38. Evidently, our basic findings are robust for changes in epoch length between 24 and 38.

Finally, we ran a large full factorial experiment with $\texttt{deltaFirm}_i \in \{2.4, 3.0, 3.8\}$, $\texttt{epsilonFirm}_i \in \{0.4, 0.7, 1.0\}$, $\texttt{epochLengthFirm}_i \in \{24, 30, 36\}$, and $i \in \{0, 1\}$. There were thus 3^6 unique combinations of factors and since we conducted 10 replications, there were $7290 = 10 \times 3^6$ runs in all. See Table 10.7.

The Wilcoxon signed rank test (wilcox.test in R) for the mean of the running average quantities against the null hypothesis of 133.3333 (the Cournot value) produces a p-value of 0.1347, so we do not reject the null hypothesis that

summary with 6600 episodes			
runningAverageBid	runningAverageBidSd	meanRewardFirm0	meanRewardFirm1
Min. :129.5	Min. :2.596	Min. :8509	Min. :8366
1st Qu.:132.3	1st Qu.:2.865	1st Qu.:8784	1st Qu.:8766
Median :133.3	Median :3.099	Median :8886	Median :8900
Mean :133.3	Mean :3.146	Mean :8868	Mean :8895
3rd Qu.:134.1	3rd Qu.:3.333	3rd Qu.:8958	3rd Qu.:9006
Max. :138.6	Max. :4.417	Max. :9414	Max. :9276

summary with 6000 episodes			
runningAverageBid	runningAverageBidSd	meanRewardFirm0	meanRewardFirm1
Min. :129.0	Min. :2.583	Min. :8386	Min. :8445
1st Qu.:132.5	1st Qu.:2.911	1st Qu.:8754	1st Qu.:8734
Median :133.6	Median :3.137	Median :8866	Median :8843
Mean :133.6	Mean :3.181	Mean :8878	Mean :8837
3rd Qu.:134.8	3rd Qu.:3.346	3rd Qu.:9002	3rd Qu.:8939
Max. :137.9	Max. :4.529	Max. :9396	Max. :9250

summary with 7200 episodes			
runningAverageBid	runningAverageBidSd	meanRewardFirm0	meanRewardFirm1
Min. :128.5	Min. :2.718	Min. :8347	Min. :8425
1st Qu.:131.9	1st Qu.:2.917	1st Qu.:8777	1st Qu.:8751
Median :133.2	Median :3.069	Median :8870	Median :8880
Mean :133.3	Mean :3.122	Mean :8876	Mean :8876
3rd Qu.:134.9	3rd Qu.:3.281	3rd Qu.:8994	3rd Qu.:9016
Max. :137.4	Max. :3.889	Max. :9570	Max. :9450

TABLE 10.3: 6600 episodes suffices.

summary epoch length sweeps			
meanRewardFirm0	meanRewardFirm1	runningAverageBid	runningAverageBidSD
Min. :8360	Min. :8364	Min. :129.2	Min. :2.503
1st Qu.:8775	1st Qu.:8755	1st Qu.:132.3	1st Qu.:2.889
Median :8887	Median :8872	Median :133.3	Median :3.068
Mean :8883	Mean :8875	Mean :133.3	Mean :3.134
3rd Qu.:9002	3rd Qu.:8990	3rd Qu.:134.3	3rd Qu.:3.308
Max. :9452	Max. :9373	Max. :139.2	Max. :4.622

TABLE 10.4: Summary of epoch length sweeps.

| summary epoch length sweeps with epochLengthFirm0 = 24 | | | |
meanRewardFirm0	meanRewardFirm1	runningAverageBid	runningAverageBidSD
Min. :8365	Min. :8364	Min. :129.2	Min. :2.556
1st Qu.:8790	1st Qu.:8761	1st Qu.:132.4	1st Qu.:2.913
Median :8900	Median :8868	Median :133.2	Median :3.098
Mean :8889	Mean :8863	Mean :133.3	Mean :3.140
3rd Qu.:9005	3rd Qu.:8982	3rd Qu.:134.3	3rd Qu.:3.332
Max. :9435	Max. :9268	Max. :138.2	Max. :4.537

TABLE 10.5: Summary of epoch length sweeps with epochLengthFirm0 = 24.

summary with epochLengthFirm0 = 24 & epochLengthFirm1 = 38			
meanRewardFirm0	meanRewardFirm1	runningAverageBid	runningAverageBidSD
Min. :8580	Min. :8590	Min. :129.8	Min. :2.721
1st Qu.:8797	1st Qu.:8673	1st Qu.:132.6	1st Qu.:2.914
Median :8895	Median :8880	Median :133.1	Median :3.060
Mean :8911	Mean :8853	Mean :133.3	Mean :3.129
3rd Qu.:9028	3rd Qu.:8979	3rd Qu.:134.1	3rd Qu.:3.344
Max. :9233	Max. :9172	Max. :136.5	Max. :4.086

TABLE 10.6: Summary with epochLengthFirm0 = 24 and epochLengthFirm1 = 38.

meanRewardFirm0	meanRewardFirm1	runningAverageBid	runningAverageBidSD
Min. :8069	Min. :8072	Min. :127.5	Min. :2.068
1st Qu.:8756	1st Qu.:8754	1st Qu.:132.3	1st Qu.:2.883
Median :8877	Median :8876	Median :133.4	Median :3.170
Mean :8875	Mean :8872	Mean :133.4	Mean :3.219
3rd Qu.:8994	3rd Qu.:8993	3rd Qu.:134.4	3rd Qu.:3.511
Max. :9538	Max. :9742	Max. :138.8	Max. :6.087

TABLE 10.7: Summary data, full factorial own-own experiment.

the agents are settling on a quantity total equal to the Cournot value. (Recall that this is across a sample of 7,290 data points.) Applying the Wilcoxon test to the mean rewards of Firm0 and Firm1 yields p-value $= 0.5189$. We cannot reject the null hypothesis that each firm is obtaining, on average, an equal reward.

Given the symmetry of the factorial design, these results are as expected and serve primarily to increase our confidence in the implementation of the model and to provide a baseline for comparison. If we focus on the extreme case in which the factors for Firm0 are at their lowest (`epochLengthFirm1 = 24`, `deltaFirm0 = 2.4`, and `epsilonFirm0 = 0.4`) and the factors for Firm1 are at their highest (`epochLengthFirm1 = 36`, `deltaFirm0 = 3.8`, and `epsilonFirm0 = 1.0`) we get the summary results in Table 10.8.

It is evident from these data that the extreme ends of the parameter settings we examined do not perturb the basic findings. If we now use the Wilcoxon (rank sum) test (wilcox.test in R) to compare the mean rewards received by the two firms, we get a p-value of 0.9118, prohibiting us from rejecting the null hypothesis that the two firms are, on average, getting the same level of reward, despite using very different parameter values in PROBE AND ADJUST.

We regressed `meanRewardFirm0` on `epochLengthFirm0`, `deltaFirm0`, `epochLengthFirm1`, `deltaFirm1`, `epsilonFirm0`, and `epsilonFirm1`, including all interaction terms, using OLS. The residuals have a mean of $-1.965907e - 14$ and appear to be quite symmetrical about the mean. The Q-Q plot, Figure 10.3, indicates a reasonably good match with the normality assumption. Multiple R-Squared for the full model was 0.01412, with a p-value of 0.001058. *None* of the fitted coefficient values had an associated p-value below 0.3.

In summary, for this base case we find that the model settles reliably in the neighborhood of the Cournot outcome, with each player obtaining approximately equal returns. Further, these results are very stable to small to moderate changes in the model's parameter settings.

10.5.2   Maximizing Total Market Profits

In the previous section we discussed results obtained when all agents use the Own Returns policy. These agents, using PROBE AND ADJUST as their learning regime, set their **return** value at step 8 (Figure 10.1) to the profit (net of revenue and costs) they individually received during the episode. Agents following the Market Returns policy instead set their **return** values to the total profits of the industry (i.e., all players). This would seem a remarkably unselfish behavior. And so it is, yet it offers certain insights.

When both players maximize total market profits in a duopoly game, the total production settles into an oscillation around the monopoly solution of 100 (given our standard settings; see Figure 10.2). However, the players can, and normally do, have very different average profits depending on which player has

Firm0 parameters minimal, Firm1 parameters maximal			
meanRewardFirm0	meanRewardFirm1	runningAverageBid	runningAverageBidSD
Min. :8646	Min. :8667	Min. :130.2	Min. :2.899
1st Qu.:8847	1st Qu.:8875	1st Qu.:131.9	1st Qu.:2.961
Median :8936	Median :8907	Median :132.8	Median :3.195
Mean :8915	Mean :8954	Mean :132.5	Mean :3.182
3rd Qu.:9028	3rd Qu.:9012	3rd Qu.:133.2	3rd Qu.:3.295
Max. :9129	Max. :9389	Max. :134.1	Max. :3.733

TABLE 10.8: Summary data, extreme comparison, own-own experiment. epochLengthFirm0 = 24, epochLengthFirm1 = 36, deltaFirm0 = 2.4, deltaFirm1 = 3.8, epsilonFirm0 = 0.4, epsilonFirm1 = 1.0.

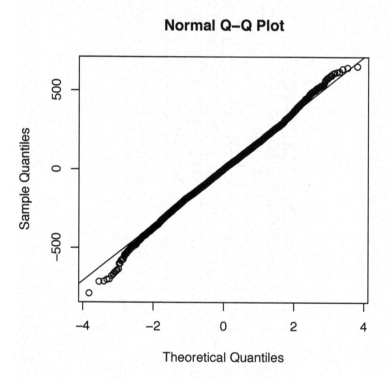

FIGURE 10.3: Q-Q plot of residuals from own-own regression.

the larger initial quantity. The reason is that the players see the same signals on total market profitability and tend to move in the same direction. In a simulation where we started one player at 60 and the other at 40, after 1,500 events the average profitability of the first player was around 11,500 and the second at 8,500. Also around 1,500 episodes (rounds of play), the first player was producing around 65 and the second 35. At 13,500 events the players were much closer, 55 and 45. There is random fluctuation in these runs. If the two players begin with the same initial base quantity, one typically produces more than the other. Which one is random, with small epoch lengths favoring more variation. Production levels can cross, even multiple times. Thus, although the total production stays close to the monopoly solution, the individual-player production levels are not identical. (This variation, as we shall see, averages out and may be deemed random.) A player starting out with a lower production level is at a strong disadvantage relative to the other player and, sometimes, relative to the Cournot solution.

Using the system clock to seed the random number generator we ran 100 repetitions of the standard configuration (above), with both players playing Market Returns. The average (over 100 repetitions) of the concluding running average for 6,600 episodes of play was 100.10 (sd 0.108). Firm0's overall average reward was 9,967.14 with a standard deviation of 628.92, which is very high compared to that for "Own Returns" (above). Firm1's results were similar: 10,019.57 (628.93).

Altruism has its limits. If one player maximizes market profits while the other maximizes its own returns, the total production settles on the monopoly solution. However, the first player is forced to exit the market and the second player settles on the monopoly quantity and reaps the profits of the monopoly solution. From these two simulations it is clear that being a good citizen without any regard to self-interest is a weak strategy.

10.5.2.1 Detailed Statistical Analysis of the Model Results

We conducted a full factorial experiment with 10 repetitions with the following factors and settings: numEpisodesToRun (6600, 7600), deltaFirm0 (2.4, 3.8), deltaFirm1 (2.4, 3.8), epsilonFirm0 (0.4, 1.0), epsilonFirm1 (0.4, 1.0), epochLengthFirm0 (24, 36), and epochLengthFirm1 (24, 36). The key summary information appears in Table 10.9.

Comparing Tables 10.9 and 10.7, we note several points:

1. The runningAverageBid in the market-market condition settles tightly and symmetrically about 100, the monopoly quantity, instead of the Cournot quantity of the own-own case (see Table 10.7).

2. The runningAverageBidSD value is discernibly lower in the market-market case.

3. Each firm in the market-market case obtains a mean reward in the neigh-

meanRewardFirm0	meanRewardFirm1	runningAverageBid	runningAverageBidSD
Min. : 6119	Min. : 6253	Min. : 99.40	Min. :1.927
1st Qu.: 9328	1st Qu.: 9348	1st Qu.: 99.93	1st Qu.:2.365
Median : 9986	Median :10000	Median :100.00	Median :2.675
Mean : 9986	Mean :10000	Mean :100.00	Mean :2.650
3rd Qu.:10639	3rd Qu.:10658	3rd Qu.:100.08	3rd Qu.:2.889
Max. :13728	Max. :13864	Max. :100.62	Max. :3.530

TABLE 10.9: Summary data for the full factorial market-market experiment.

borhood of 10,000, compared to 8,900 in the own-own case. On average both firms do better by adopting the "cooperative" or "altruistic" policy (acting so as to maximize industry returns, rather than individual returns), than by adopting the "selfish" or "best response" policy.

4. The variation in mean reward obtained is much higher in the market-market case than in the own-own case.

Comparing Table 10.10 with the analogous results for the own-own (Cournot) case in Table 10.8, we find some apparent differences for the players, depending on the players' parameter settings. In the own-own case, the minimized parameter player (Firm0 in Table 10.8) is at no apparent disadvantage. Table 10.10, however, shows Firm0 getting about 10% less on average than Firm1. Note that the mean quantity remains very near to the monopoly level. If we take the negative of these settings (with Firm0's parameters maximal and Firm1's minimal), we get the expected result: they settle near the monopoly quantity and Firm0 has the advantage. Evidently, mutual use of PROBE AND ADJUST is (under the current range of conditions) robust to parameter settings with respect to the total quantity, but individuals may gain comparative advantage through parameter settings. The nonparametric Wilcoxon rank-sum test (wilcox.test in R) yields a p-value of 0.0004871 for the null hypothesis of no difference between the two sets of returns. This coheres with the suggestion we drew from Table 10.10: there is a difference resulting from the two treatments.

10.5.3 Mixed Maximizing of Market and Own Returns

Between the extremes of maximizing one's own returns—Own Returns, §10.5.1—and maximizing on the market's overall returns—Market Returns, §10.5.2—there is an infinite number of weighted combinations. Under the policy of play of Mixture of Market and Own Returns each firm individually has a parameter, `marketOwnMixture` $\in [0, 1]$, by which it combines observed market returns and its own returns from each episode of play. Using PROBE AND ADJUST, the firm finds its own reward, `ownReward`, in each episode as well as the average reward for firms in the market, `averageReward`. The firm calculates its `mixedReward` as

$$(1 - \texttt{marketOwnMixture}) \times \texttt{ownReward} + \texttt{marketOwnMixture} \times \texttt{averageReward}$$

and records this value on the associated up or down list for returns for mixtures, depending on whether its production quantity is above or below its current quantity. (See Figure 10.1, especially step 9.) Running under the standard conditions, except as noted, we consider the case with two firms in the market each using Mixture of Market and Own Returns as its policy of play, each using 0.5 as its value for `marketOwnMixture`. In a representative run, after 6600 episodes the running average total production is 120.837 with a

Firm0 parameters minimal, Firm1 parameters maximal			
meanRewardFirm0	meanRewardFirm1	runningAverageBid	runningAverageBidSD
Min. : 7953	Min. : 9597	Min. : 99.75	Min. : 2.612
1st Qu.: 9029	1st Qu.:10310	1st Qu.: 99.84	1st Qu.:2.641
Median : 9488	Median :10498	Median : 99.97	Median :2.671
Mean : 9370	Mean :10616	Mean : 99.93	Mean :2.672
3rd Qu.: 9675	3rd Qu.:10956	3rd Qu.:100.04	3rd Qu.:2.701
Max. :10389	Max. :12033	Max. :100.06	Max. :2.730

TABLE 10.10: Summary data, extreme comparison, market-market experiment. epochLengthFirm0 = 24, epochLengthFirm1 = 36, deltaFirm0 = 2.4, deltaFirm1 = 3.8, epsilonFirm0 = 0.4, epsilonFirm1 = 1.0.

standard error of 2.983. Recall that the monopoly production is 100 and the Cournot production 133.333. Firm0's average reward (for episodes 6501–6600) is 9461.997 and Firm1's is 9651.893. We repeated the experiment 100 times, setting the mixture for both players to 0.1 market (and 0.9 own) returns, and using the system clock to initialize the random number generator each time. Over the 100 trials the mean of the averages of the total quantity produced is 131.31 with a standard deviation of 1.47. Firm0's average of its average rewards was 9020.88 (standard deviation of 191.22) and Firm1's was 8994.42 (162.64).

These results are not sensitive to the random seed used, except of course for which firm comes out slightly ahead. The results also extend in the obvious way to more than two firms (our program handles up to 10, but this is easily changed to an arbitrary number). By mixing consideration of their own returns and the market's returns the firms do better than the Cournot outcome but not as well as the monopoly position. The problem is that the Mixture of Market and Own Returns policy of play is exploitable in the same way as the Market Returns policy. In a representative run, with Firm0 employing Own Returns as its policy of play and Firm1 sticking with Mixture of Market and Own Returns with its `marketOwnMixture` set to 0.5, we got the following outcomes. The running average total production is 128.065 with a standard error of 3.185. Firm0's average reward is 11,024.902 while Firm1's is 7,379.554. (These general results are not sensitive to the random seed used.) Things are a bit more equal if firm 0 uses Mixture of Market and Own Returns but sets its `marketOwnMixture` to 0.4. The running average production comes in at 124.215 with a standard error of 2.946. Firm0's average reward is 9741.348 while Firm1's is 9,068.581. (Again, these results are not sensitive to the random seed used.)

At bottom, however, the mixture policy of play is vulnerable to exploitation.

10.5.4 Market Returns, Constrained by Own Returns

A player pursuing the policy of Market Returns, Constrained by Own Returns (MR-COR) operates as follows. At the end of its epochs it assesses whether the mean of its production quantities during the epoch plus its δ is less than the mean of all the production quantities during the epoch. (See Figure 10.1; δ is the search range on each side of the base quantity that a player uses each episode. The program also allows use of ε instead of δ.) If it is, the player increases its base quantity by ε. If it is not, then the player takes the market view and follows the Market Returns policy. The key feature of this policy is that the players have a sense of fair division of the market and if a player does not get what it perceives to be a fair share, it increases production. Two players on average split the market equally, the total production fluctuates around the monopoly total production of 100 and they each make an average profit of 10,000 (under the settings we are discussing).

The Market Returns, Constrained by Own Returns policy may be likened to TIT FOR TAT in Prisoner's Dilemma in that if one player tries to take too much market share, the other responds by matching the increase in production. This means that any random increase in production by one is matched by the other and their actual production levels track together, whereas with market returns only, the production levels of the two tend to look like mirror images around 50. It is interesting to see what happens when one player, Firm1, plays MR-COR against the other player, Firm0, playing Own Returns. Typical results under the standard conditions are that Firm1's average profit during the final 1,000 episodes is 8,833, while Firm0's is 8,977. In this type of case, the self-interest of Firm0, maximizing its own returns, drives the production to an oscillation around the Cournot equilibrium and has an average episode profit of about 9,000. The player that plays Market Returns, Constrained by Own Returns, Firm1, does slightly worse because it almost always produces less than the other player, as it has to be slightly forgiving on share to compensate for noise in the market-share results. Its profits are around 8,800, slightly below what it would achieve at the Cournot equilibrium playing myopically. Long-sighted behavior has its risks.

Firm	Mean of Running Averages of Rewards to Firm	SD of Running Averages
0	1994.2025916855077	69.987676171652
1	1988.0676679163748	62.37538356784939
2	1987.651615420313	57.6836838380063
3	1987.8337170456355	68.10393678546707
4	1995.5662569019985	61.40921846900001
5	1995.3897636449321	54.92891914419867
6	1982.9621113765784	48.40642783370098
7	1990.0869668751468	55.82401403170985
8	2004.7913882302219	73.96700484493032
9	1988.9208194082335	56.140148147371825

TABLE 10.11: Results for 100 repetitions of 10 firms each playing Market Returns, Constrained by Own Returns (MR-COR) under the standard conditions. The monopoly quantity is 100, the Cournot quantity is 181.818. Across the 100 repetitions the mean (standard deviation) of the running average of the production quantity is 101.9725498548145 (0.6607808254556182).

10.5.4.1 Detailed Statistical Analysis of the Model Results

We conducted a full factorial experiment with 10 repetitions with the following factors and settings: numEpisodesToRun (6600, 7600), deltaFirm0 (2.4, 3.8), deltaFirm1 (2.4, 3.8), epsilonFirm0 (0.4, 1.0), epsilonFirm1 (0.4,

1.0), `epochLengthFirm0` (24, 36), and `epochLengthFirm1` (24,36). Firm0 used the Own Returns update policy throughout, while Firm1 used Market Returns, Constrained by Own Returns. The key summary information appears in Table 10.12.

With this mixture of policies the agents revert to something near, but slightly below, the Cournot solution. The Wilcoxon signed rank test yields a p-value of $2.341e - 16$ for the null hypothesis that the average value is 133.33333, leading to its convincing rejection. The test fails to reject the null hypothesis that the mean value is the slightly lower number 133.1, showing how close the solution is to the Cournot solution. The Wilcoxon test also rejects the null hypothesis that the two firms are obtaining their rewards from the same distribution (p-value $< 2.2e - 16$). Firm0, using Own Returns, has a discernible advantage over Firm1, using Market Returns, Constrained by Own Returns, but is this Firm0's best policy?

Table 10.13 presents summary data for the full-factorial experiment (with 30 replications) when both players use the Market Returns, Constrained by Own Returns (MR-COR) policy of play. Notice that both players do very well, with mean rewards fully 1000 higher than in the Own Returns versus Market Returns, Constrained by Own Returns case, summarized in Table 10.12. Comparing the present case (MR-COR, Table 10.13) with the Market-Market (both altruistic) case, summarized in Table 10.9, it would seem that little or nothing is lost by playing Market Returns, Constrained by Own Returns. MR-COR is a robust policy of play. Agents have little to lose by using it and much to gain if everyone uses it. Table 10.14 presents summary information for extreme parameter settings, with Firm0 as usual having the low settings and Firm1 the high. From the table, it appears that Firm0 does somewhat better than Firm1. This is confirmed by the Wilcoxon rank sum test on `meanRewardFirm0` and `meanRewardFirm1`. The p-value for the null hypothesis of their arising from the same distribution is $< 2.2e - 16$. In fact, the smallest value of `meanRewardFirm0` exceeds the largest value of `meanRewardFirm1`.

10.5.5 Number Effects

For the most part, we have focused so far on exploring and establishing the robustness of PROBE AND ADJUST and on presenting results in the case of oligopolies having two firms. In this and the following section we assume the robustness of the basic model and turn our attention to new issues. What happens when the number of firms increases beyond two? The Cournot model from standard economic theory teaches that the Cournot equilibrium will change with increasing numbers of firms, moving asymptotically toward the competitive solution. The theory predicts, however, that the *outcome* reached will continue to be the Cournot solution. That is, the standard theory asserts that there is no *number effect*; the Cournot solution will be the outcome regardless of the number of firms in the market. Considerable experimental work with human subjects, nicely summarized and extended by Huck et al.

meanRewardFirm0	meanRewardFirm1	runningAverageBid	runningAverageBidSD
Min. :8273	Min. :7911	Min. :122.8	Min. :1.831
1st Qu.:8863	1st Qu.:8727	1st Qu.:131.8	1st Qu.:2.830
Median :8960	Median :8823	Median :133.1	Median :3.245
Mean :8979	Mean :8807	Mean :133.0	Mean :3.317
3rd Qu.:9081	3rd Qu.:8906	3rd Qu.:134.3	3rd Qu.:3.670
Max. :9663	Max. :9266	Max. :143.3	Max. :7.650

TABLE 10.12: Summary data, full factorial own vs. market-own experiment. Firm0 uses Own Returns and Firm1 uses Market Returns, Constrained by Own Returns.

meanRewardFirm0	meanRewardFirm1	runningAverageBid	runningAverageBidSD
Min. : 9723	Min. : 9739	Min. : 99.42	Min. :1.740
1st Qu.: 9910	1st Qu.: 9912	1st Qu.:100.05	1st Qu.:2.437
Median : 9992	Median : 9994	Median :100.15	Median :2.672
Mean : 9992	Mean : 9994	Mean :100.16	Mean :2.644
3rd Qu.:10072	3rd Qu.:10074	3rd Qu.:100.26	3rd Qu.:3.014
Max. :10246	Max. :10260	Max. :101.00	Max. :3.568

TABLE 10.13: Summary data, full factorial, market-own vs. market-own experiment. Both Firm0 and Firm1 use Market Returns, Constrained by Own Returns.

meanRewardFirm0	meanRewardFirm1	runningAverageBid	runningAverageBidSD
Min. :10073	Min. :9773	Min. : 99.86	Min. :2.557
1st Qu.:10130	1st Qu.:9819	1st Qu.:100.07	1st Qu.:2.639
Median :10148	Median :9838	Median :100.16	Median :2.683
Mean :10147	Mean :9839	Mean :100.16	Mean :2.678
3rd Qu.:10168	3rd Qu.:9856	3rd Qu.:100.27	3rd Qu.:2.717
Max. :10212	Max. :9913	Max. :100.44	Max. :2.795

TABLE 10.14: Summary data, extreme market-own vs. market-own experiment. Both Firm0 and Firm1 use Market Returns, Constrained by Own Returns. epochLengthFirm0 = 24, epochLengthFirm1 = 36, deltaFirm0 = 2.2, deltaFirm1 = 3.8, epsilonFirm0 = 0.4, epsilonFirm1 = 1.0

[158], does find, to the contrary, number effects in repeated play by human subjects with fixed counter-players. The title of [158] summarizes the experimental findings: "Two are few and four are many." That is, with four or more firms the Cournot quantity is reached or exceeded, and with two firms there is often evidence of collusion, with Q, the total production quantity, reduced in the direction of the monopoly level. What happens under PROBE AND ADJUST? Recall the results reported in Table 10.1 for PROBE AND ADJUST when all players use the Own Returns policy: there are no number effects. In order to facilitate comparison, we now report results with `InitPIntercept` = 100, `InitDSlope` = 1.0, and `unitCostFirm0` = `unitCostFirm1` = 1.0, which duplicates the demand function and cost structure used in the experiments of Huck et al. (Note that in terms of expression (10.1), a = `InitPIntercept` = 100; *slope* = `InitDSlope` = 1.0.) The main outcome statistic used in [158, page 439] is what they call r, the ratio of the (mean of the) total production quantity, Q, to the Cournot solution quantity, Q^C (or Q^N in their notation).

$$r = \frac{\overline{Q}}{Q^C(n,k)} \tag{10.20}$$

(We assume with [158] that all firms have the same costs, k; cf. expression 10.18.) Thus, r values less than 1 indicate a degree of collusion. In `oligopolyPutQuantity.nlogo`, r is renamed `ratioOfferedCournot` internally and `ratioOC` for display on the user interface panel. Let

$$r_C^M(n,k) = \frac{Q^M(k)}{Q^C(n,k)} \tag{10.21}$$

Then, for the inverse demand function of [158] we have:

n = number of firms	$Q^C(n,1)$	$Q^M(k)$	$r_C^M(n,1)$
2	66.00	49.50	0.750
3	74.25	49.50	0.667
4	79.20	49.50	0.625
5	82.50	49.50	0.600
10	90.00	49.50	0.550

We note that r may be a misleading indicator for our special purposes, since the significance of its value varies with n. A measure that adjusts for the number of players is

$$r' = \frac{Q^C(n,k) - Q}{Q^C(n,k) - Q^M(k)} = \frac{Q^C - Q}{Q^C - Q^M} \tag{10.22}$$

Values near 1 indicate very high collusion (with quantities near the monopoly level), while values near 0 would indicate lack of collusion (and quantities near the Cournot level). We will proceed with the discussion in terms of both r and r'.

n	$Q^C(n,1)$	$\overline{Q}$	$\overline{r}$	$\overline{r'}$
2	66.00	60.44 (7.05)	0.91†	0.34
3	74.25	72.59 (4.53)	0.98	0.07
4	79.20	80.67 (4.85)	1.02	−0.05
5	82.50	88.43 (8.80)	1.07	−0.18

TABLE 10.15: Summary of Huck et al.'s experimental data. n = number of suppliers in the market. $\overline{Q}$ = average total quantity offered. (Standard deviations in parentheses.) Averages are over episodes 17–25. † As reported in Huck et al. We note that $60.44/66.0 = 0.9157575\ldots$. For computing r', $Q^M = 49.5$.

Huck et al. performed a meta-analysis on the prior human experiments that investigate number effects in Cournot markets with fixed counter-players [158]. They found, in aggregate, a modest number effect. While collusion may occur with two players, it is reduced or disappears with increasing numbers of players. To this Huck et al. added their own experimental data, which we summarize in Table 10.15. Subjects offer quantities in 25 rounds of play (episodes). Allowing for some learning, we use the Huck et al. data for rounds 17–25. (Huck et al. also report data for rounds 1–25. The results are not materially different.) Points arising on the Huck et al. data:

1. $\overline{r}$ increases uniformly with n. Since at $Q = Q^C(n,k)$, $r = 1$, only the result for $n = 2$ indicates collusion. Huck et al. find the increase statistically significant. Note also that for $n = 5$ the $\overline{r}$ and $\overline{Q}$ values suggest that the subjects systematically offered more than the Cournot amount.

2. $\overline{r'}$ (not reported by Huck et al.) decreases uniformly with increasing n and is in apparent broad agreement with $\overline{r}$.

3. The standard deviation of $\overline{Q}$ increases uniformly across $n = 3, 4, 5$. It is, however, comparatively high for $n = 2$. This suggests (only) that subjects may have used somewhat different decision procedures for $n = 2$ and $n \neq 2$, and that when $n = 2$ the subjects may have been somewhat more exploratory, perhaps sensing the possibility of collusion.

Table 10.16 summarizes number-effect results obtained with Oligopoly-PutQuantity.nlogo. We remind the reader that the interpretation of the standard deviations in Table 10.16 is different from that in Table 10.15. In the latter case, the given standard deviation is the usual standard deviation of Q. In Table 10.16, individual Q values represent the mean of the total quantities offered for episodes 5601–6600 and $\overline{Q}$ is the mean (over 30 replications) of these Q values. $\overline{Q}$ is, then, a mean of means. The standard deviation values given are the means of the standard deviations of the Q values.

line no.	N	policies	$\overline{Q}$	$\overline{r}$	$\overline{r'}$
1	2	all Own Returns	66.12 (3.24)	1.00	-0.01
2	3	all Own Returns	74.38 (3.86)	1.00	-0.00
3	4	all Own Returns	79.45 (4.35)	1.00	-0.00
4	5	all Own Returns	82.45 (4.72)	1.00	0.00
5	10	all Own Returns	90.90 (6.42)	1.01	-0.02
6	2	all MR-COR	49.39 (2.70)	0.75	0.98
7	3	all MR-COR	49.71 (3.39)	0.67	0.95
8	4	all MR-COR	50.14 (4.01)	0.63	0.93
9	5	all MR-COR	50.58 (4.71)	0.61	0.91
10	10	all MR-COR	52.66 (6.93)	0.59	0.83
11	2	1MR-COR :: 1Mixture 50:50	59.52 (3.25)	0.90	0.38
12	3	1MR-COR :: 2Mixture 50:50	66.37 (3.82)	0.89	0.31
13	4	1MR-COR :: 3Mixture 50:50	70.97 (4.35)	0.90	0.26
14	5	1MR-COR :: 4Mixture 50:50	74.48 (4.81)	0.90	0.23
15	10	1MR-COR :: 9Mixture 50:50	84.20 (6.47)	0.94	0.13
16	2	all 30:70 Mixture	62.57 (3.34)	0.95	0.20
17	2	all 50:50 Mixture	59.35 (3.08)	0.90	0.39
18	3	all 50:50 Mixture	66.09 (3.75)	0.89	0.32
19	4	all 50:50 Mixture	70.92 (4.33)	0.90	0.27
20	5	all 50:50 Mixture	74.46 (4.74)	0.90	0.23
21	10	all 50:50 Mixture	84.28 (6.52)	0.94	0.13

TABLE 10.16: Summary of PROBE AND ADJUST data on number effects. (Standard deviations in parentheses.) Averages are over 30 replications.

Points arising on the Table 10.16 data:

1. When all players use the Own Returns policy (lines 1–5) neither $\overline{r}$ nor $\overline{r'}$ shows any evidence of a number effect, further verifying that this policy replicates the Cournot equilibrium.

   When all players use the Market Returns, Constrained by Own Returns (MR-COR) policy (lines 6–10), both $\overline{r}$ and $\overline{r'}$ are uniformly decreasing in n. The high value of $\overline{r'}$ when $n = 2$ indicates a high degree of collusion. (Note that $Q^C(2,1) = 66.0$, while $Q^M = 49.50$ and $\overline{Q} = 49.39$ with $n = 2$.) As n increases, $\overline{r'}$ also decreases, but even at $n = 10$ the average quantity on offer is displaced 83% of the distance away from the Cournot quantity and toward the monopoly quantity, Q^M.

   When all players use the Mixture of Market and Own Returns, in a 50:50 combination (lines 17–21), $\overline{r}$ appears to be constant for $n = 2, 3, 4, 5$, but higher for $n = 10$. $\overline{r'}$, however, is uniformly decreasing, indicating a number effect and reduced collusion. Essentially the same result obtains if Firm0 uses MR-COR and any other firms in the market use Mixture of Market and Own Returns, in a 50:50 combination (lines 11–15).

Interestingly, the relative advantage or disadvantage of the MR-COR policy is also a function of n. At low values, MR-COR is disadvantaged, but at $n = 5$ and higher it earns more than the typical firm using Mixture of Market and Own Returns, in a 50:50 combination. See the following table:

Policy Case	Firm0: MR-COR meanRewardFirm0	Firm1: Mixture 50:50 meanRewardFirm1
1MR-COR :: 1Mixture 50:50	1146.0	1193.0
1MR-COR :: 2Mixture 50:50	708.0	717.3
1MR-COR :: 3Mixture 50:50	491.4	498.6
1MR-COR :: 4Mixture 50:50	365.3	359.4
1MR-COR :: 9Mixture 50:50	129.5	121.47

2. Line 11 contains values that are quite close to those reported by Huck et al. for $n = 2$ (see Table 10.15). The data in the two tables do not track closely as n increases. The ordering is in excellent alignment, however, suggesting a good match could be found by a simple transformation.

3. $\overline{r'}$ decreases uniformly with n (except in the all Own Returns case), while $\overline{r}$ is flat for $n = 2, 3, 4, 5$ (except for the all MR-COR case). $\overline{r}$'s flatness is an artifact, since it is sensitive to the absolute level of Q, which varies with n. $\overline{r'}$ is a better indicator.

4. The standard deviation of Q increases uniformly with n, without exception. This is hardly surprising, given that PROBE AND ADJUST is a stochastic exploration procedure, undertaken with some independence by the players. A similar explanation suggests itself for the human data. How much of that stochasticity is systematic (actually part of a learning policy as in PROBE AND ADJUST) and how much is simply error is a fascinating question for future research.

10.5.6   Differential Costs

Until now we have made the literature's standard assumption of equal (usually zero) costs for the players. What happens if the players have different costs? We investigate cases in which firms' costs are proportional to the quantities they produce. As usual, we report results from specific, but representative, runs. Here we assume that runs proceed for 6600 episodes of play. This is more than ample for the system to settle.

We revert now to our original setup with $a =$ InitPIntercept $= 400.0$ and *slope* = InitDSlope $= 2.0$. Let us assume (without loss of generality; the qualitative results are robust to this assumption) that in a duopoly Firm0 has a unit cost of 10 and Firm1 50. $Q^C(2, [10, 50])$ is then 123.33. (See discussion and formulas in §10.4.) With these costs the individual monopoly quantities are $Q_0^M = 97.5$ and $Q_1^M = 87.5$. In fact, Q settles (with some variation; SD ≈ 3.2) very near 123.33 when both firms use the Own Returns policy of play. Running for 6,600 episodes of play, in a typical run the last 1,000 episodes of play yield an average Q of 123.067, with standard deviation 2.895. Firm0, the low cost producer, obtains an average reward of 10,068.879, while Firm1 achieves only 5,498.693.

Similarly, in an experiment with 100 repetitions, using the system clock to seed the random number generator, Firm0's costs were set to 10 and Firm1's to 0. The Cournot quantity is 131.667. The mean (standard deviation) running average production across the 100 repetitions was 131.683 (1.591). Firm0 averaged a reward of 8026.41 (164.09), while Firm1 got 9305.65 (176.60).

Switching to the MR-COR policy for both players, in a typical run, with unitCostFirm0=10, unitCostFirm1=50, and referenceBidGroup=All Bids, the last 1,000 episodes of play yield an average Q of 97.638 (with standard deviation 3.047), which is very close to Firm0's monopoly quantity. Firm0, the low cost producer, obtains an average reward of 9611.584, while Firm1 achieves 7453.02. The two firms have jointly extracted more wealth from the market, but Firm0 has paid a penalty. The example is extreme, however, for it assumes that Firm1's unit costs are five times those of Firm0.

Assume now that Firm1's unit cost is 20 and Firm0's remains at 10, still leaving a substantial cost advantage to Firm0. (Firm1's monopoly Q, Q_1^M, is now 95.0, close to Firm0's of 97.5. The Cournot quantity is $Q^C(2, [10, 20]) = 128.33$.) Let both firms use the Own Returns policy of play. In a typical run, using 6,600 episodes of play, the last 1,000 episodes of play yield an average Q of 129.552, with standard deviation 3.308. Firm0, the low cost producer, obtains an average reward of 8,937.33, while Firm1 achieves 7,386.882. Switching to MR-COR for both players, the last 1,000 episodes of play yield an average Q of 98.102 (with standard deviation 2.702), which is very close to Firm0's monopoly quantity. Firm0, the low cost producer, obtains an average reward of 9640.072, while Firm1 achieves 8873.914. Switching to Own Returns for Firm0, MR-COR for Firm1, in a typical run the last 1,000 episodes of play yield an average Q of 131.358 (with standard deviation 3.215). This is very far from Firm0's monopoly quantity but close to the Cournot quantity. Firm0, still the low cost producer, obtains an average reward of 8431.236, while Firm1 achieves 7617.516.

In sum even with a 2:1 unit cost advantage for one player, the policy pair (MR-COR,MR-COR) Pareto dominates (Own Returns, MR-COR), (MR-COR, Own Returns), and (Own Returns, Own Returns).

Again, we emphasize that these results are typical. They hold up under a broad range of settings and are quite robust to changes in the random number stream.

10.6   Discussion

Recognizing that measuring success by market returns only is not a viable policy because the player is too easily exploited, we compare the firm-only (Own Returns policy) and the firm-and-market (MR-COR policy) measures of success in the following payoff table for a 2×2 game in strategic form. Table 10.17 presents the results approximately, but in strategic form.

	firm-only	firm-and-market
firm-only	(8,900, 8,900)	(9,000, 8,800)
firm-and-market	(8,800, 9,000)	(10,000, 10,000)

TABLE 10.17: Payoffs from the strategies as measured by average returns after settlement; both firms have 0 costs.

Notice that the relationships among the payoffs across the choice of objective functions is a Stag Hunt game.[3] The payoffs in the Stag Hunt game look like the payoffs in Prisoner's Dilemma, except for some key differences. The players are each better off if both cooperate (play MR-COR, or hunt stag in the case of the Stag Hunt) versus neither cooperate (play Own Returns, or hunt hare) as in Prisoner's Dilemma. If one player cooperates and the other does not, both players do worse than if both cooperate. In Prisoner's Dilemma the noncooperative player is better off if the other player cooperates. The hitch, here and in the Stag Hunt generally, is that players may be led to mutual hunting of hare by considerations of risk. The experimental literature on the Stag Hunt (e.g., [23]; see [54] for reviews of many experiments) has tended to find that in repeated play subjects are led to the risk-dominant strategy of hunting hare. These findings, however, are based on repeated play by *varying* counter-players, in distinction to *fixed* counter-players. Our concern is with the latter, with markets containing a fixed number of participants, who produce quantities over time and who have to learn to live with one another. Framing the problem as one of *learning in policy space* (Which policy should I use in setting production? Should it be Own Returns or Market Returns or Market Returns and Own Returns?), it is clear that learning to cooperate, learning to hunt stag, is amply achievable under simple learning regimes [176, 177]. Nor is it implausible to think that real players would figure out and implement their collective interests. In short, if we were to allow the players to select the policy as well as search for the maximum return with the given policy, we would expect that the players would choose the firm-and-market strategy, especially if we bring discounting into this repeated game among fixed players. The firm and market policy works only if all players either accept their proportionate

---

[3]See [296, 297] for discussion of other circumstances in which what is seemingly a Prisoner's Dilemma is transformed into a Stag Hunt game.

share or are willing to accept shares that add up to less than 1. For example, if each player believes it should have 5% above the average production, the firms in the simulation will fight for market share, driving themselves into losses rather than profits. Thus, we do not propose that this policy is actually used or should be used by firms. However, it points to the potential of a more nuanced policy that brings in industry interests and not just firm interests into capacity and production strategies. The policies could be explored using the bargaining frameworks examined by Dawid and Dermietzel [72] and Carpenter [56].

10.7 Conclusions

In this chapter we have developed a model—PROBE AND ADJUST—of an agent that explores its environment and uses that exploration to improve its performance by adjusting a set of continuous parameters. This behavior is an abstraction of typical managerial decision making and is consistent with the notions of continual improvement and of a satisficing player that learns and improves. We emphasize that the required knowledge and computational capabilities for PROBE AND ADJUST are quite credibly available to real agents. We have shown in simulations that this model of an agent reproduces the classic Cournot results in oligopoly theory, under certain assumptions (e.g., use of the Own Returns policy). We have also shown that this model can explain the emergence of tacit collusion (e.g., when players all use MR-COR as their policy of play). Thus, we have a starting point for exploring alternatives to the decision model embedded in classical economic theory and have a more realistic starting point for looking at issues such as market power. Firms operate in complex environments and there are many competing interests within and without the firm. Agents can be given objectives that are more realistic in that organizations both compete with other organizations in some dimensions and cooperate with these same organizations in others (within the bounds of the law) as described in Brandenburger and Nalebuff [41]. We see that it is possible to engage in tacit collusion by taking into account the interest of the industry as well as the firm while not engaging in explicit price fixing. Thus, being a good corporate citizen can pay. By testing alternative objectives for firms it is possible to represent the richer relationships that managers have to deal with and observe the consequences in the marketplace. This paper opens up several avenues of research. First, players should be able to choose among success measures (e.g., own returns, market returns) to see which ones emerge as most effective for enhancing firm profitability. By doing this, it is possible to see what measures of success emerge in the context of repeated play. Second, the agency problem between the firms' owners and managers can be placed in a larger context, to see how those choices impact the industry as well as the firm.

10.8   For Exploration

1. Comment on the following statement:

   > If firms can collude tacitly (and legally) to achieve extraordinary returns, and if failing to collude in this way leads to substantially poorer returns, then executive compensation should not be based on relative superiority of performance, but rather on actions that management takes that are effective in creating a profitable industry.

   Agree? Disagree? Why? Does the statement need to be qualified?

2. Comment on the following statement:

   > Tacit collusion such as that created by PROBE AND ADJUST, or even extraordinary profits attributable to Cournot competition, cannot be a very important factor in economic organization, because whenever there are extraordinary profits being made new entrants to the industry will appear and cause the extraordinary profits to disappear.

   Agree? Disagree? Why? Does the statement need to be qualified?

10.9   Concluding Notes

This chapter is an edited version of [178]. Permission from Springer to republish this material is gratefully acknowledged.

Chapter 11

Oligopoly: Bertrand Competition

11.1 Price Competition

The Bertrand model models competition on the basis of price in an oligopoly. Each firm privately proposes a unit price for the goods on order and the lowest-price bid wins all of the business. What will the market price be? If there are very many firms in the market, it is easy to believe that the competitive price will prevail. The Bertrand model teaches us that this is also true in even the smallest oligopoly, consisting of just two firms. Here is a standard summary of the reasoning from a standard microeconomics text.

> If firm 1 really believes that firm 2 will charge a price $\hat{p}$ that is greater than the marginal cost, it will always pay firm 1 to cut its price to $\hat{p} - \varepsilon$. But firm 2 can reason the same way! Thus any price higher than marginal cost cannot be an equilibrium; the only equilibrium is the competitive equilibrium. [322, page 488]

In a different text (still standard) the same author makes the following remarks on the Bertrand result.

> It may seem somewhat paradoxical that we get price equal to marginal cost in a two-firm industry. Part of the problem is that the Bertrand game is a one-shot game: players choose their prices and then the game ends. This is typically not a standard practice in real-life markets.

> One way to think about the Bertrand model is that it is a model of competitive bidding. Each firm submits a sealed bid stating the price at which it will serve all customers; the bids are opened and the lowest bidder gets the customers. Viewed this way, the Bertrand result is not so paradoxical. [321, 293]

Whether the one-shot result is paradoxical or not, we are interested in the iterated game, which is not addressed by the Bertrand model. The iterated game of price competition in an oligopoly is the subject of this chapter.

11.2   Probing and Adjusting on Price in an Oligopoly

The OligopolyBidPrice.nlogo model, available as usual on the book's Web site, models iterated play of Bertrand competition by players learning with PROBE AND ADJUST. With two or more players present, if all players update with Own Returns all players indeed compete away their profits and the price goes to the marginal cost of the high-cost producer (or rather just slightly below it). See Figure 11.1.

When both firms have identical costs (zero in the example) and both update with MR-COR, they quickly learn to set the price at the monopoly price and share the market stochastically. See Figure 11.2. This produces high returns for both firms.

When the firms have different costs they will have different monopoly prices. That is, a monopolist firm with one cost structure will have a different monopoly price than a different firm with different costs. Figure 11.3 shows what happens when two firms have different costs and both update their PROBE AND ADJUST learning on price with MR-COR. The market price goes to the monopoly price of the lower-cost firm.

With more firms in the market the firms have increasing difficulty coordinating on the monopoly price. Figure 11.4 displays the tentativeness and degradation of the coordination with five firms under the default settings. With six or more firms present they revert to a competitive market, even with everyone updating with MR-COR. The problem for these firms is an epistemic one: they fail to detect the signal of cooperation in the noise of stochastic bidding with PROBE AND ADJUST. Cooperation is improved, allowing the monopoly price to be reached with more firms present, if the firms increase their epoch lengths, thereby collecting more data before they update their base prices, or if the MR-COR firms are more patient in sticking with Market Returns and allowing a greater disparity between their own returns and the average of the industry.

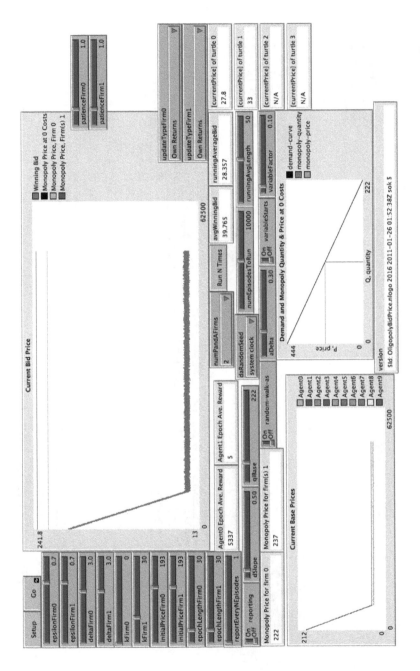

FIGURE 11.1: Iterated Bertrand competition; both players update with Own Returns; the firms have different costs.

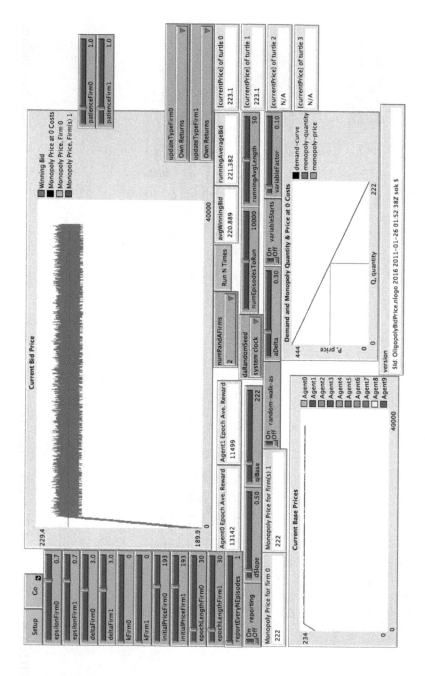

FIGURE 11.2: Iterated Bertrand competition; both players update with MR-COR.

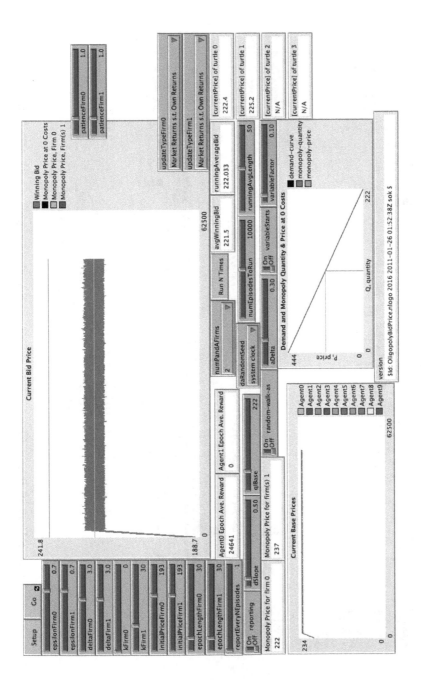

FIGURE 11.3: Iterated Bertrand competition; both players update with MR-COR; the firms have different costs.

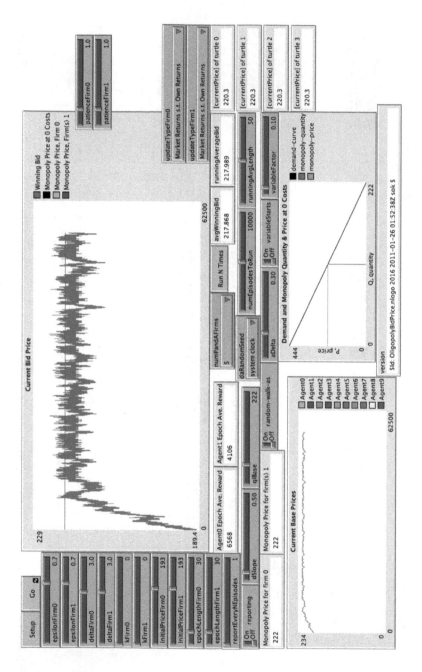

FIGURE 11.4: Iterated Bertrand competition; five players update with MR-COR.

11.3 For Exploration

1. Sticking with the default settings of OligopolyBidPrice.nlogo, how much do epoch lengths have to be increased to support eight firms coordinating on the monopoly price?

2. Sticking with the default settings of OligopolyBidPrice.nlogo, how much do the patience values have to be increased to support eight firms coordinating on the monopoly price?

3. If you increase both epoch lengths and patience values, how does the number of firms increase that can coordinate on the monopoly price?

4. Do different cost levels among the firms affect their ability to coordinate at a monopoly price? Explain.

11.4 Concluding Notes

As noted in Chapter 10 in the discussion of Cournot competition, there is a marked differentiation between the economics literature on price competition (no oligopoly problems) and the business or management literature (which emphasizes the importance of avoiding price competition). Both [252] and [231] are good representatives of the management literature on pricing. Friedman's *Oligopoly Theory* [111] is a nice introduction from the economics side.

Chapter 12

Supply Curve Bidding

12.1 Introduction and Setup

All of our market models, indeed most of our models, are "stylized" (highly approximate, justified by consensus, informal observations), some more so than others. The model in this chapter is a bit less so. Here then is the setup, still stylized but nudging closer to reality.

In an American "restructured" electric power market private entrepreneurs may build power plants, connect them to the power grid, and offer to sell electricity over the grid.[1] There is, in addition, no requirement that the entrepreneurs actually own or operate electricity generating plants. They are free to buy electricity from real producers and resell it at auction. This potentially—in fact inherently—chaotic arrangement has to be managed by some entity. That entity is commonly known as the *system operator*. More officially, the names *Regional Transmission Organization* (RTO) and *Independent System Operator* (ISO) are used for special kinds of system operators, recognized and governed by the federal government (see http://www.ferc.gov/industries/electric/indus-act/rto.asp; the relevant agency is FERC, the Federal Energy Regulatory Commission). Signifying only convenience, I will use the term ISO generically for the system operator. The connotation of "independent" is important: the system operator must be a neutral, competent third party or else the market will not function properly. It might even function disastrously.

The ISO estimates demand and conducts two kinds of auctions. In the auction for the *day-ahead market* the ISO solicits bids for electric power delivered for an hour at a time. Bids are priced in dollars per megawatt hour ($/MWh). Approximately 24 hours before the power is required the ISO holds the auction and bids are submitted. Typically, a bidder will own more than one plant and each plant will have a different cost structure. For example, the bidder might own a coal plant and a gas turbine plant, with the coal plant having a lower cost of production per MWh. We will assume that bids are in

---

[1] Often referred to as a "deregulated" power market in the mainstream media. Promoted by advocates of market-based institutions, these markets in fact are, of necessity, governed by a much larger volume of federal regulations than the "regulated" markets they replaced. For this reason, the literature refers to them as restructured [302], and so shall we.

the form of a step function. Our example producer might bid $X/MWh for a number of megawatts that hour up to the capacity of the coal plant, and $Y/MWh for a number of megawatts up to the capacity of the gas turbine plant.

With several producers bidding in the market, there will typically be a small multiple of several discrete steps bid, against which the market price must be found. Someone has to intervene to make the market work, since real-time full discovery of the price-demand intersection is simply not physically practicable. What happens is that the ISO accepts the bids and "sums" them up, much as we assumed the market did with its invisible hand in §8.1 (Competitive Story 1), creating thereby a discrete, sharply stepped supply curve, much like that in Figure 8.1 on page 163.

The ISO then uses its estimate of the demand during that hour for the next day to find the intersection of the (estimate) demand curve and the (summed, bid) supply curve, and thus to set the market price. Based on this calculation, the ISO announces the price and how much supply is being requested from each supplier. Based on these quantities, the bidders receive payments and are obligated to produce accordingly the next day. Notice two things:

1. All bidders receive the same price. The single price discovered by the ISO prevails.

2. The demand curve might not actually intersect with the supply curve, since it may cross in a step, as shown in Figure 12.1. When this happens, market price and quantity are set as shown in the figure.

12.2   Computational Model

Our computational model is embodied in the NetLogo program Supply-CurveBidding.nlogo, which may be accessed at the book's Web site. We distinguish two kinds of parameters in our model: environment parameters and agent parameters. Tables 12.1 and 12.2 list the default values for these parameters. We call these the Base Case Parameterization. Settings for specific runs can be identified as deviations from these default values.

The core structure of the computational model may be seen by reviewing the parameters and what they do, along with the program's user interface, Figure 12.2. We begin with the environment parameters. Demand is an exogenously given linear function, with price as a function of quantity. The demand curve is specified by the quantity intercept (the quantity demanded at 0 price, `QIntercept`) and the (downward) slope of the line (`DSlope`). The demand curve is assumed to have the form

$$Price = QIntercept - (dSlope \times Quantity) \qquad (12.1)$$

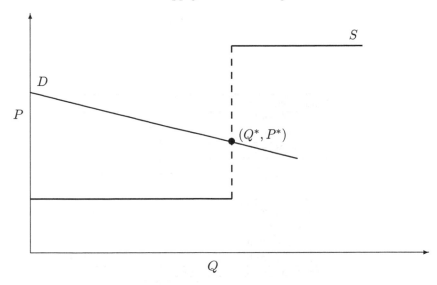

FIGURE 12.1: Market clearing quantity and price (Q^*, P^*) for demand curve D intersecting the step function supply curve S at a step ("on the vertical").

with *QIntercept* and *dSlope* set by the user on the Interface tab (see Figure 12.3 and the user interface widgets in the northeast corner of the display).

EpisodesPerRun is the number of rounds of play in a single run under the settings. NumberOfReplications is the number of runs undertaken with the same parameter settings. RunningAverageLength is the number of periods of data collected to compute the various running average statistics. When Logging? is set to true, run data are written to the file runsOutput.txt in CSV (comma separated values) format with headers for easy loading into spreadsheets and statistical programs such as R.

QIntercept (Quantity intercept of demand function)	200
DSlope (Negative of the slope of the demand function)	1.0
EpisodesPerRun (Number of episodes in a run)	50,000
RunningAverageLength (Running average length)	2,000
RandomNumberSeed (Random number seeded with)	System Clock
NumberOfReplications (Number of runs)	30
Logging? (Are the results to be logged?)	True

TABLE 12.1: Default settings of the environment parameters, Base Case Parameterization.

NumberOfFirms	2
Cost1Firm1	4
Cost1Firm2	4
MaxQ1Firm1	20
MaxQ1Firm2	20
Cost2Firm1	15
Cost2Firm2	15
MaxQ2Firm1	100
MaxQ2Firm2	100
EpochLengthFirm1	100
EpochLengthFirm2	100
InitialBasePrice1Firm1	16
InitialBasePrice1Firm2	16
InitialBasePrice2Firm1	150
InitialBasePrice2Firm2	150
InitialBasePrice3Firm1	150
InitialBasePrice3Firm2	150
QuantityBreak1Firm1	0.5
QuantityBreak1Firm2	0.5
QuantityBreak2Firm1	0.9
QuantityBreak2Firm2	0.9
UpdateTypeFirm1	Own Returns
UpdateTypeFirm2	Own Returns
Patience	1.05
QuantityDeltaFirm1	0.01
QuantityDeltaFirm2	0.01
QuantityEpsilonFirm1	0.01
QuantityEpsilonFirm2	0.01
PriceDeltaFirm1	0.2
PriceDeltaFirm2	0.2
PriceEpsilonFirm1	0.2
PriceEpsilonFirm2	0.2

TABLE 12.2: Default settings of the agent parameters, Base Case Parameterization.

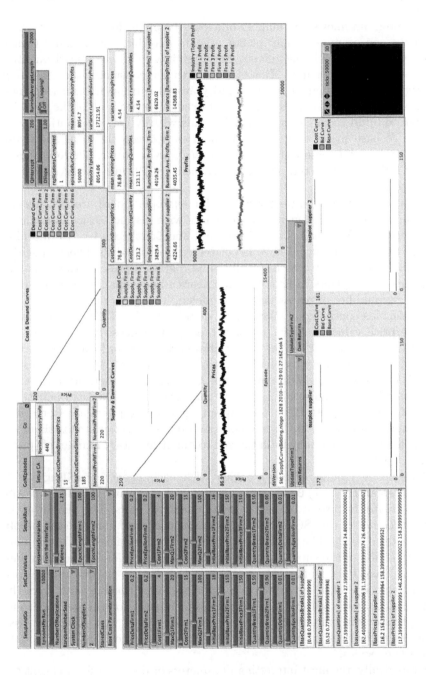

FIGURE 12.2: Base Case Parameterization, representative run (both suppliers update with Own Returns).

The agent parameters in Table 12.2 and the program interface in Figure 12.2 may best be understood in the context of a narrative of how the model works.

The program allows any number of agents (supplier firms), each of which may have one or more supply plants. Agents may bid one or more (price, quantity) pairs, committing thereby to supplying up to the stated quantity at the stated price per unit. This is subject to the constraint that the total quantity bid must equal the total quantity available from the agent's plants. Additionally, bids are constrained to be at or above cost for all of the capacity required to fulfill them. An agent making bids with N (price, quantity) pairs will use N base prices and $N - 1$ boundaries between them for specifying quantities. We call these latter values *quantity break points*. There is thus a total of $2N - 1$ parameters which are learned via PROBE AND ADJUST for each supplier.

In our Base Case Parameterization, for example, there are two agents, firm 1 and firm 2. Each agent has two plants with distinct cost and quantity characteristics. For firm 1, `Cost1Firm1` $= 4$ and `Cost2Firm1` $= 15$, meaning that the two plants for firm 1 have associated costs of 4 and 15. Their capacities are 20 and 100, respectively. Internally, the program represents the costs and capacities of the plants in lists (`costsofplants` and `capacitiesofplants`), so that any number of plants may be modeled. The agents in the Base Case each make bids consisting of three (price, quantity) pairs. These are generated by three price values and two quantity breakpoint values. In the Base Case (Table 12.2 and Figure 12.2) the three initial base prices for firm 1 are 16, 150, and 150. The initial quantity breakpoints are at 0.5 and 0.9. If these values are bid as is (without perturbation by PROBE AND ADJUST), then, since the total capacity is 120, the bid would consist of the three price/quantity) pairs: (16, 60), (150, 48), and (150, 12). Internally, an agent's base prices are stored in a list, `BasePrices`, as are its base quantities breaks, `BaseQuantitiesBreaks`. Consequently, an agent may be initialized to bid an arbitrary number of (price, quantity) pairs.

An episode (round of play) begins with each agent consulting its N `BasePrices` and $(N - 1)$ `BaseQuantitiesBreaks` and using PROBE AND AD-JUST to create its current bid as N (price, quantity) pairs. The market agent (playing the role of the ISO) sums the bids from each firm to create the supply curve for the episode. The market agent then clears the market by determining the intersection of the supply and demand curves, and thereby the prevailing price and quantities. Bidders are rewarded according to the prevailing price and the quantity taken by the market. Upon receiving their rewards from the market, each agent collects information and prepares for the next episode. The `EpochLength` parameter for each agent determines how often the base quantities and prices are updated. In the Base Case, the firms both update every 100 episodes.

Strategically, the most interesting parameters or decisions for an agent are its update type (see `UpdateTypeFirm1`). The two most important update types are Own Returns and Market Returns Constrained by Own Returns (MR-COR, pronounced "Mister Core" after "Mister Go," MRGO, the Mississippi River Gulf Outlet). An Own Returns player updates the adjustable quantities in its bids (i.e., the values in `BasePrices` and `BaseQuantitiesBreaks`) with PROBE AND ADJUST, using its profits (its own returns). That is, an Own Returns supplier keeps track of its returns from its bids on each of its $(2N-1)$ bid parameters every episode. Acting on this performance measure is a form of non-cooperative behavior where the decisions are based on the outcomes of a learning process rather than an explicit optimization. The return in a episode is coded high or low for each of the $(2N-1)$ bid parameters, depending upon whether the bid value of the parameter is above or below its base value for the episode. Learning and adjustment occurs for each price and quantity on the offer curve separately.

A Market Returns player instead updates based on maximizing the industry's returns (net of costs). This is an altruistic strategy and obviously is exploitable, but it does lead to superior returns for the industry. The Market Returns Constrained by Own Return (MR-COR) update type uses Market Returns to update its adjustable quantities provided its profits are not too much below the average profits of the industry. Here is the relevant code fragment.

```
(((MyMean myEpisodeProfits) * Patience) >=
        (MyMean industryEpisodeProfits) / count suppliers)
```

If it evaluates to true, a MR-COR player will update altruistically, based on Market Returns; if it evaluates to false, a MR-COR player will update based on Own Returns. (Note: In the NetLogo code we use the term supplier for the breed (type) of agent that represents supplier firms. `count suppliers` is simply the number of firms in the market, 2 for the Base Case.) That is, if on average the player's profits (returns net of costs) are not much below the industry's profits in the (player's) epoch just concluded, then the player updates on the Market Returns principle; otherwise the player updates on the basis of the Own Returns principle. This update criterion is a form of cooperative behavior with punishments for a lack of cooperation based on past experience and without formal and illegal communication.

12.3 Examples

12.3.1 Monopolist

In the Monopolist scenario, chosen with the InstantiateScenarios chooser, we determine the demand curve by setting QIntercept to 200 and dSlope to 1.0. Our monopolist is firm (supplier) 1. We use the Interface tab to set EpochLengthFirm1 to 50 and UpdateTypeFirm1 to Own Return. The remainder of our agent's parameters are set in the program code, in the SetupAgents procedure.[2] Our monopolist in this scenario has two plants, each with a capacity of 110. For validation purposes we assume that their costs are 0. The supplier is required to bid prices for its entire capacity, which is here 220. Our supplier will offer two prices throughout. Initially the supplier offers two prices, 10 and 15, divides the capacity at a BidBreak of 0.5. That is, $0.5 \times 220 = 110$ units are offered at a cost of 10 (talers) per unit and $(1 - 0.5) \times 220 = 110$ units are offered at a cost of 15.

Our monopolist uses PROBE AND ADJUST to learn three quantities: the two prices and the BidBreak value, that is the quantities that are associated with the two prices. The learning is simultaneous and independent. For each of the three quantities, at the end of the agent's epoch the agent adjusts according to how things have gone when that quantity was probed, regardless of the probed values of the other quantities.

We know from Chapter 9 that the monopoly quantity for this scenario is 100, as is the monopoly price, yielding an industry profit of 10,000. If our program is correct and if the supplier agent is able to learn effectively in this learning regime, then the supplier should reliably find the monopoly outcome. It does. See Figure 12.3.

To run the example, select Monopolist from the Instantiate Scenarios chooser on the user interface, click the Setup CA button, and then click the Go button. Click the Go button again to stop the run. See the program's documentation for further information on how to run it. (The documentation is contained in the Information tab of the program file, SupplyCurveBidding.nlogo, and on the Web page when it is accessed as an applet.)

---

[2]Beginning at if (InstantiateScenarios = "Monopolist").

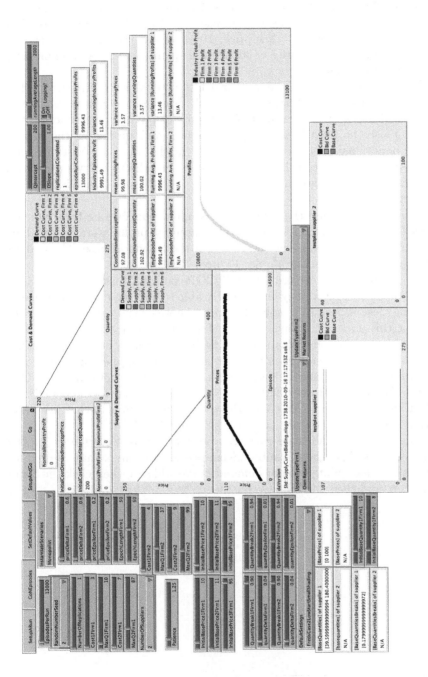

FIGURE 12.3: A run of the Monopolist scenario, in SupplyCurveBidding.nlogo.

12.3.2   One and Two Firms, Zero Cost

The One Firm, Zero Costs scenario[3] is like the Monopolist scenario, except that the agent's two plants have capacities of 55 each, instead of 110. The scenario starts them out at a low price, 10 each, and they pretty much stay there, resulting in a stable price of 90, a stable quantity of 90, and an industry episode profit of 9900.

The Two Firms, Zero Costs scenario lets us explore alternative update policies for PROBE AND ADJUST. Now each firm has two plants, both with capacities of 55, so there is a total of 220 units of capacity available. Letting supplier 1 use the Own Returns update policy, and supplier 2 the Market Returns policy makes supplier 2 an altruist in this market. We can expect that supplier 2 will graciously raises its prices until the industry's profits are maximized and its own are snuffed out. As we see in Figure 12.4, this is exactly what happens.

When both firms use the Own Returns update policy, one firm typically is forced out of the market in part. The industry profit averages at about 7000, very high, but well below the 10,000 of the monopoly case. The realized market price averages about 45. See Figures 12.5 and 12.6.

And when both firms use the MR-COR policy, Market Returns Constrained by Own Return, they are typically able to find the monopoly price, or something close to it, and to split the profits more or less equally. See Figure 12.7. But this does not always happen. They may be unable to coordinate, and one goes to Own Returns upon being forced in part out of the market. In other words, the result is as in the case of both suppliers updating with "Own Returns." Compare the results with the case when both firms use Own Returns. The firm that does better in that situation does about as well as it would if both firms used MR-COR. However, which firm gets that privileged position is random. The firms are wiser to adopt MR-COR.

(Note: To undertake these runs, choose Two Firms, Zero Costs (or One Firm, Zero Costs) on the InstantiateScenarios chooser, make any desired adjustments to the update type or demand curve widgets, then click the Setup CA button, and then click the Go button.)

---

[3]Defined beginning at if (InstantiateScenarios = "One Firm, Zero Costs") in the SetupAgents procedure.

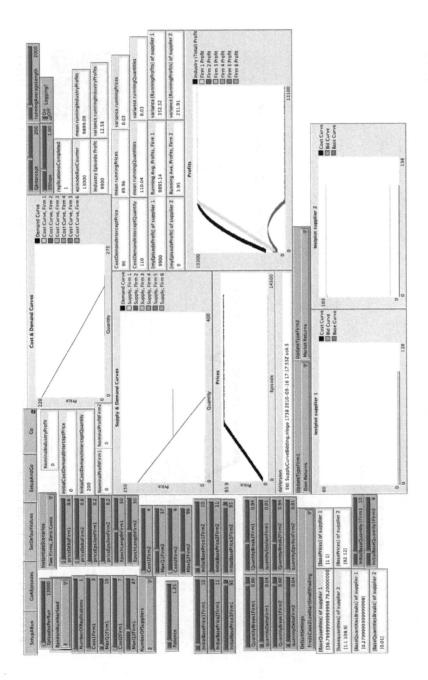

FIGURE 12.4: A run of the "Two Firms, Zero Cost" scenario, in SupplyCurveBidding.nlogo, with one altruist.

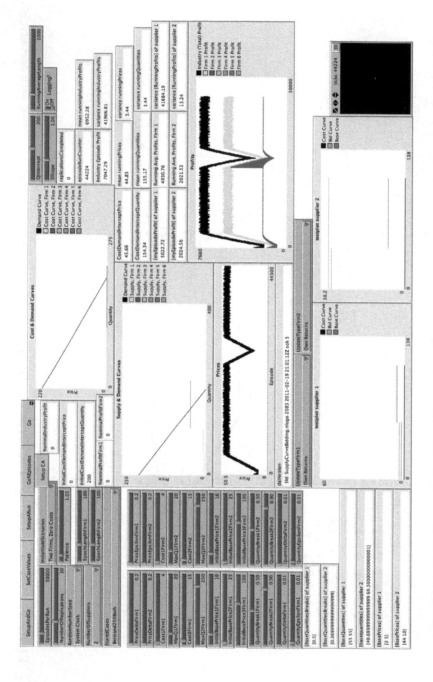

FIGURE 12.5: A run of the "Two Firms, Zero Cost" scenario, both firms update with Own Returns.

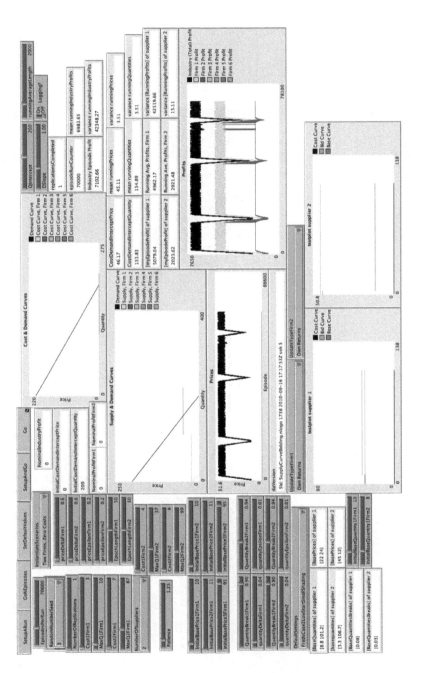

FIGURE 12.6: A run of the "Two Firms, Zero Cost" scenario, in SupplyCurveBidding.nlogo, with both selfish.

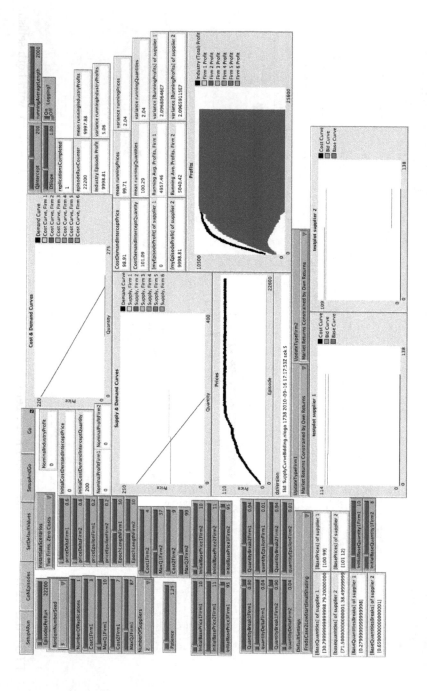

FIGURE 12.7: A run of the "Two Firms, Zero Cost" scenario, in SupplyCurveBidding.nlogo, with both MR-COR.

12.3.3 Six Firms

We consider now scenarios in which the value of `QIntercept` has been changed to 300 from its default value of 200 and the costs of all plants have been set to 0 (Tables 12.1, page 253, and 12.2, page 254). If we endow a monopolist in this situation with sufficient plant capacity (300 or more) and the monopolist uses at least two bid steps, then the monopolist will reliably (in the model, using PROBE AND ADJUST) maximize its return with a price of about 150 and an episode profit of about 22,500.

With six firms in the market,[4] all using PROBE AND ADJUST and updating with Own Returns, they do compete away their profits. We get what looks like something close to a competitive market. See Figure 12.8 on page 266. Instead of a price of about 150, achieved by the monopolist, this industry averages about 2.34 and gets episode profits of about 700 (see the figure for results from a typical run).

Figure 12.9 on page 267 shows results from a typical run resulting from just one change in the scenario: all the firms update with MR-COR. Learning is tortuous and noisy, but early on the firms are individually and collectively doing much better than in the case of everyone using Own Returns. With enough learning they indeed approximate, as an industry, the monopolist, with each firm sharing in the benefits approximately equally.

12.4 For Exploration

1. Verify for the Monopolist scenario that the update policies of Market Returns and Market Returns Constrained by Own Returns produce the same results as does the Own Returns policy. Why should this be the case?

2. Regarding, "We know from Chapter 9 that the monopoly quantity for this scenario is 100, as is the monopoly price, yielding an industry profit of 10,000" on page 258, show the calculations that support this claim.

3. Explore other scenarios (several are already built in to SupplyCurveBidding.nlogo and you can add others). In particular, examine the effects of cost. What do you find?

4. Comment on the following statement, "One of the things a procedural model, such as PROBE AND ADJUST, can do is to discover equilibria in complex, analytically intractable models."

---

[4]Choose NFirmsSimple under the InstantiateScenarios, then set NumberOfSuppliers to 6, click the Setup CA button, then the Go button. As usual, click Go again to stop.

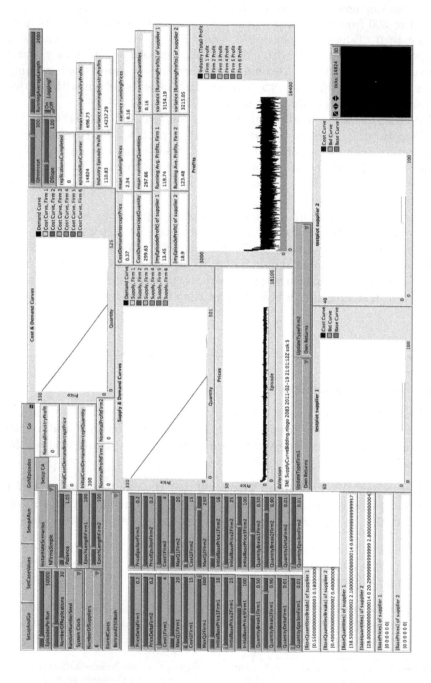

FIGURE 12.8: NFirmsSimple scenario, with six firms, all updating with Own Returns.

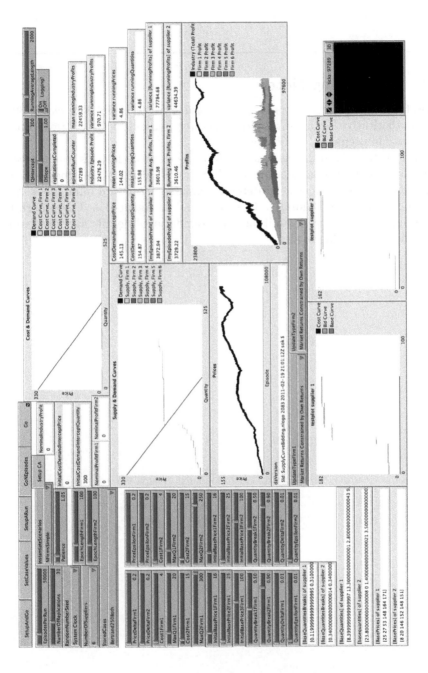

FIGURE 12.9: NFirmsSimple scenario, with six firms, all updating with MR-COR.

5. Figures 12.5 and 12.6 (pages 262 and 12.6) show "dips" in the "Prices" and "Profits" plots. Why does this happen?

12.5   Concluding Notes

Stoft's *Power System Economics: Designing Markets for Electricity* [302] is an authoritative resource on electric power markets. I have drawn on it extensively in preparing this chapter.

SupplyCurveBidding.nlogo and much of the material for this chapter were undertaken in close collaboration with Frederic H. Murphy, for whose extensive contributions I am most grateful. Remaining errors are mine alone. See our paper, "Strategic Bidding of Offer Curves: An Agent-Based Approach to Exploring Supply Curve Equilibria," for foundational details and discussion of ramifications of this model.

Chapter 13

Two-Sided Matching

13.1 Introduction: Matching in Centralized Markets

In the usual case, markets are *distributed*, with buyers and sellers mostly on their own in finding each other and in negotiating terms of trade. Distributed markets may fail in one way or another, however. A common response is to create a *centralized* market organized by a third party whose responsibility it is to set the conditions of trade, for example the price, based on the bids and asks from the buyers and sellers. Many electricity markets are organized in this way. Deregulated electricity producers, for example, offer supply schedules to a third party, often called the *independent systems operator* or ISO, who aggregates the supply schedules, observes the market demand, and sets the price of electricity (for a given period of time).

Quite a number of labor markets are similarly centralized, most famously, markets in which physicians are matched to hospitals for internships [262]. Roughly speaking, the individual doctors submit their rankings of hospitals, the hospitals submit their rankings of doctors, and a third party organization undertakes to match doctors with hospitals. This is an example of a *two-sided matching* problem, which problems are the subject of this chapter.

In a two-sided matching problem, we are given two sets ('sides') of individuals and asked to form pairs consisting of one member from each set. Standard examples dealt with widely in practice include pairing men with women, workers with employers, students with schools and so on.

A presumption in matching problems (as distinguished from assignment problems, which are treated in operations research and employ non-strategic decision making) is that both sides consist of agents who have interests of their own and capacities to act on them. Consequently, matches are ordinarily evaluated in terms of *stability*. Matching problems are inherently strategic, or game-theoretic, and stability is the accepted equilibrium concept. A match is said to be stable if there is no pair of matched couples in it containing individuals who would prefer to be matched to each other but are not. (See below for details.) The thought is that if the couple here is unstable with regard to the couple next door, divorce and remarriage will (or at least may) ensue. Requiring matches to be stable in the first place will prevent breakup and reformation among pairs and its attendant costs.

The point of departure for this chapter is the observation that two-sided matches can be evaluated—and for many applications should be evaluated—according to several objectives, particularly stability, equity, and social welfare. For present purposes, by stability we mean the count of unstable pairs of matched couples in a match. This should be minimized and at 0 the match is stable.[1] By equity we mean the sum of the absolute differences in the preference scores of each matched pair. We will be scoring preference on a ranking 1 to n scale (1 = most preferred, n = least preferred), so this too should be minimized. Finally, by social welfare we mean the sum of the agent scores in the match. Again, since scoring is from low to high, this quantity should also be minimized. (To illustrate, if agents i and j are paired in a match, and i's preference score for j is 5 and j's preference score for i is 3, then the matched pair's contribution to social welfare is the sum, 8, and their contribution to equity is the absolute value of the difference of the scores, $|5 - 3| = 2$.)

Given that we would consider designing or even centralizing a matching market (as is widely done in practice), the question arises of how best to provide the market operators and users with match options that map the Pareto frontier (as well as possible) in these three objectives. In what follows we explore two rather different algorithmic approaches to this: an agent-based model that simulates a distributed market and an evolutionary computation approach. We compare them with each other and with what can be produced by the standard approach, the deferred acceptance algorithm of Gale and Shapley [116]. The next section presents essential background.

13.2   Background

Again, now with more detail, in a two-sided matching problem we are given two sets (sides) of agents, X and Y, and are asked to find a match, μ, consisting of a decision (in or out?) for each pair $(x, y), x \in X, y \in Y$. It is helpful to view a match as represented by a matrix, $\mathbf{M}$, of size $|X| \times |Y|$, based upon arbitrary orderings of X and Y. The element $m_{i,j}$ of $\mathbf{M}$ equals 1, if $x_i \in X$ is matched with $y_j \in Y$; otherwise the element is 0. Thus the element $m_{i,j}$ of $\mathbf{M}$ represents the *pair* (x_i, y_j). Matchings pair up agents from X and Y.

Particular matching problems come with particular requirements on μ (or $\mathbf{M}$) as well as X and Y. For example in the *simple marriage matching* problem (the focus of this chapter because it is prototypical for two-sided matching problems), we require that $|X| = |Y| = n$; the number of men equals the

---

[1]If the count of unstable pairs is 1, there is no guarantee that if the two pairs rematch by exchanging partners the resulting match would be stable. In fact, it could have a higher count of unstable pairs [182].

number of women. We further require of any (valid) match that each man (or member of X (Y)) be paired (or matched) with exactly one woman (or member of Y (X)), and vice versa. In terms of $\mathbf{M}$, this means that there is one 1 in each row and one 1 in each column. $\mathbf{M}$ is thus a permutation matrix and the number of possible valid matches is $n!$. In *admissions* matching problems, which are used to model, for example, interns applying to hospitals and students applying to schools, one side of the problem, say X, is much larger than the other. There are more doctors than hospitals, more students than schools. Unlike conventional marriage problems, however, one side will have *quotas* larger than 1. Each doctor and each student will have a quota of 1, but each hospital and each school may have a much larger quota and admit many doctors or students. Thus, in a valid match for an admissions problem, each agent on one side (X or Y) is paired one or more agents on the counterpart side, not to exceed the agent's quota. With students as X, and schools as Y, $\mathbf{M}$ will have one 1 each row, and each column will have a number of 1s not to exceed the quota of the corresponding school.

Many other variations are possible and are met in practice for two-sided matching problems. As Roth notes [262], two-sided matching models are often natural for representing markets, in which agents need to be paired up. Men and women want to find partners, workers want to find employment and employers want to find workers, and so on. Moreover, many of these markets, decentralized or free markets, experience failure and unsatisfactory performance in practice. They experience unraveling, e.g., offers to students are made earlier and earlier; congestion, e.g., offerers have insufficient time to make new offers after candidates have rejected previous offers; and participants engage in disruptive strategic behavior, so that behaving straightforwardly with regard to one's preferences becomes risky, e.g., in scheduling offers and responding to them [35, 262, 264]. In consequence, there is a large and growing number of applications of two-sided matching in which decentralized markets have been replaced by centralized ones, in which a coordinating agency undertakes periodic matching between two sides of a specific market. ([262] lists over 40 labor markets, mostly in medical fields; schools in New York and Boston are using centralized markets to match students to schools; see also [35].)

How do and how should centralized market agencies produce matches? In practice, some form of, variation on, the deferred acceptance algorithm of Gale and Shapley [116] is used to find a *stable* match, which is then used. A match is *unstable* if there is a pair of matched pairs—(x_i, y_j) and (x_k, y_l)—such that x_i prefers to be matched with y_l over being matched with y_j and y_l prefers to be matched with x_i to being matched with x_k. Stable matches are the ones that are not unstable.

The deferred acceptance algorithm (DAA) was first published in a paper by Gale and Shapley [116], although the procedure was discovered and used independently before. Because the algorithm is easily understood and readily available in published works we will, in the interest of space, not repeat it here, except to present it in pseudocode, Figure 13.1 (after [116]). Instead, we

1. Assume: $|X| = |Y| = n$

2. Each $x \in X$ ranks each $y \in Y$, and each $y \in Y$ ranks each $x \in X$.

3. *Matched* ⟵ ∅, *Unmatched* ⟵ ∅.

4. For each y, *string.y* ⟵ [].

5. Each $x \in X$ proposes to its most-preferred y, appending x to *string.y*.

6. Each y with *length(string.y)* > 1 (i.e., with more than one proposal), retains in the string its most preferred member of the string, and removes the rest, adding them to *Unmatched.*

7. Each x remaining on some string is added to *Matched.*

8. Do while *Unmatched* ≠ ∅:

   (a) *Matched* ⟵ ∅, *Unmatched* ⟵ ∅.

   (b) Each $x \in$ *Unmatched* proposes to its most-preferred y, among the Ys that have not already rejected x, appending x to *string.y*.

   (c) Each y with *length(string.y)* > 1 (i.e., with more than one proposal), retains in the string its own most preferred member of the string, and removes the rest, adding them to *Unmatched.*

   (d) Each x remaining on some string is added to *Matched.*

9. Stop. Each x is matched to a distinct y, which has x as the sole member of its string.

FIGURE 13.1: Pseudocode for the deferred acceptance algorithm (DAA) for the simple marriage matching problem, Xs proposing to Ys.

will describe its key properties as we see them for present purposes. First, as proved by Gale and Shapley, under the special assumptions they made (e.g., preference ranking by agents, etc., which for the sake of discussion we retain), the stable marriage problem and the admissions problem (see above) have stable matches and the DAA will find one and will find one quickly $(O(n^2))$. Second, the DAA is asymmetric. One side proposes, the other disposes. Focusing now on the marriage problem, if the men propose, they obtain a stable match that is male optimal in the sense that no man in this match strictly prefers (does better in) any other stable match. Conversely, the match is female pessimal in the sense that no woman is worse off in any other stable match. And vice versa if the women propose [116, 137, 182].

Although here we consider it only in the context of the marriage problem, this asymmetry is a general characteristic of the DAA in its various forms. It occasions the important question of whether better matches exist and can be found. To this end, we will want to look at stable matches that may be preferable to the matches found by the DAA. As announced above, we want to examine both social welfare and equity. Further, it is natural to raise the question of multiple objectives in the context of "nearly stable" matches, by which we mean matches with relatively few unstable pairs. Decision makers, including agents participating in a centralized market, may quite reasonably want to exchange some stability for improvements in, say, social welfare or equity. We note that in many cases it may be practically difficult, or made practically difficult by the operator of the centralized market, for members of matched couples to undertake swaps, regardless of their preferences.

These issues could be neatly resolved by, for any given problem, finding all of the stable solutions and comparing them with respect to equity, social welfare, and whatever other measures of performance are relevant. Predictably, however, this is an intractable problem. Irving and Leather [159] have shown that the maximum number of stable matches for the simple marriage matching problem grows exponentially in n (see also [137, 139]). Further, they provide a lower bound on the maximum by problem size. Remarkably, for a problem as small as $n = 32$, the lower bound is $104, 310, 534, 400$ [159]. Further, they establish that the problem of determining the number stable matches is #P-complete. These are, of course, extreme-case results, but very little is known about average cases. So we are left to rely upon heuristics, and we shall for the remainder of this chapter.

13.3 Matching with Agents

We developed an agent-based model, called SimpleMarriageMatching.nlogo , that simulates a distributed market for the simple marriage matching problem. At initialization, each agent is given a preference ranking of the agents in the counterpart (opposite "gender"), and the n "men" (members of Y) are randomly paired with the n "women" (members of X) to create a valid match for the simple marriage matching problem. The program then maintains a valid match throughout its execution. (The results we report here assume the "collective" swapping regime, which we now describe.) In the main loop of the program, agents are put into a random order, then each agent in turn examines the agents of the counterpart set (the women, if the agent is a man, the men if the agent is a woman). If the agent finds a matched pair with the property that the agent prefers the counterpart member of the pair (the woman, if the agent is a man; the man if the agent is a woman) *and* the agent's counterpart member prefers the agent to its own current match, then matching of the two pairs may be swapped. Starting with (x_i, y_j) and

(x_k, y_l) we get (x_i, y_l) and (x_k, y_j). The agent identifies all potential swaps in the counterpart set, that is matched couples such that the agent prefers its counterpart to its present match *and* the counterpart prefers the agent to its current match *and* the preferences are positive net of the transaction cost. The agent picks the most attractive of these from its point of view, the swap is made, and the agent's turn is over. (If there are no potential swaps, this also terminates the agent's turn.) Note: the swapping does *not* depend on the preference of the agent's mate for the mate of the counterpart pair, or vice versa. Further, whatever swapping that is done is on the basis of preference net of transaction cost, which is measured in rank units. Thus, for example, if the transaction cost is 2, and x_i is the focal agent, then a swap only occurs if x_i's preference for y_l is more than 2 ranks superior to x_i's preference for its current mate, y_j; and similarly for y_l's preferences.

Table 13.1 summarizes results from an experiment in which $n = 20$ and the transaction cost was set to 0. There were 100 runs in which the agents were initialized with a random preference regime. For each of these runs, there were 100 replications, with the preference regimes constant, but the initial matches randomly varying and the order of swap consideration randomly changing. Table 13.2 summarizes results from a similar experiment, with transaction cost set to 1. In both experiments (Tables 13.1 and 13.2), every replication of every run produced a stable match. Neglecting transaction costs (NTC) the median number of unstable pairs in a replication is 3 (Table 13.2), rather small, and the statistics for social welfare, equity, and agent scores are very similar to those in Table 13.1, with no transaction cost. A small amount of instability has not led to significant changes in social welfare, agent welfare, or equity. From a random start, in which the various measures of performance show poorly, these myopic and greedy agents manage to improve their positions, and that of their entire society, quite nicely. (We remind the reader that the search space here is huge: $20! \approx 2.4 \times 10^{18}$.)

At $n = 40$, Tables 13.3 and 13.4 tell a story similar to that for $n = 20$. Not every replication, however, leads to a stable match when transaction cost is 0. What stops the search in a replication is the number of swaps (changing partners) reaching `MaximumSwaps` $= 100,000$. With a transaction cost of 2, Table 13.4, the swap count is again low and the measures of performance again similar to those with transaction cost 0. Notice that again, the random starts yield poor performance measures, but the agents individually and collectively achieve much improvement.

Finally, when $n = 100$ and transaction cost is 0, the runs do not terminate in acceptable time, so no results are available. The issue of scaling is one we will return to below. Table 13.5 shows results for when transaction cost is 5. Now each replication results in a stable match, although neglecting transaction costs no replication found a stable match and the median number of unstable pairs has risen to about $\frac{n}{3}$. Even so, the agents make remarkable improvements in the various measures of performance, from their random starts.

	1st Qu	Median	3rd Qu
Init. # unstable pairs	73	83	93
Final # unstable pairs	0	0	0
InitialSocialWelfareSum	395	420	444
Final SocialWelfareSum	161	171	183
Initial Equity	119	133	148
Final Equity	61	69	80
SwapCount	42	55	79
InitialSumXScores	192	210	228
Final SumXScores	74	85	97
InitialSumYScores	192	210	227
Final SumYScores	75	85	97
Number of Runs	100		
Number of Replications	100		
TransactionCost	0		
SwappingRegime	Collective		

TABLE 13.1: 20×20, transaction cost = 0.

13.4 Evolutionary Matching

We also developed a genetic algorithm-based system, called Stable Matching GA, that samples the solution space of the simple marriage matching problem. Solutions are represented as permutations of $1 \ldots n$. After experimentation we settled on a mutation rate of 0.4 per solution, a two-point crossover rate of 0.6, a population size of 50, runs of 2,000 generations, and 100 repetitions or trials per problem. Mutation was effected by randomly choosing alleles at two loci and swapping them, thereby maintaining a valid solution, as a permutation. We used order crossover, OX, as our two point crossover operator ([225, page 217], [129, page 174]).

The fitness of a given solution is its number of unstable pairs. The lower the fitness value, the fitter a solution is and the higher chance it has for participating in the reproduction process. To a given population, one generation of solutions goes through three processes: fitness evaluation, selection (we used tournament-2), and reproduction (with mutation and crossover as described above). The evolution stops after a specified number of generations are reached.

We ran the GA seeking to minimize the number of unstable pairs in the match solutions. We report here with regard to the *stable* matches found by the GA for these 25 random 20×20 problems, and we compare the GA's solutions with the DAA solutions for these problems. The results may be summarized as in Table 13.6.

Here, in Table 13.6, the column labeled Case is for the 25 random 20×20

	1st Qu	Median	3rd Qu
Init. # unstable pairs	59	69	78
Final # unstable pairs	0	0	0
Init. # unstbl pairs NTC	73	83	93
Fin. # unstbl pairs NTC	2	3	4
InitialSocialWelfareSum	395	420	445
Final SocialWelfareSum	165	175	186
Initial Equity	118	132	147
Final Equity	61	70	78
SwapCount	26	31	38
InitialSumXScores	192	210	228
Final SumXScores	77	88	100
InitialSumYScores	192	210	227
Final SumYScores	76	86	98
Number of Runs	100		
Number of Replications	100		
TransactionCost	1		
SwappingRegime	Collective		

TABLE 13.2: 20×20, transaction cost = 1.

random problem instances; "D(DorE) 1G.S. Soln" means the count of strictly dominant (better than the DAA (Gale-Shapley) solutions on the two dimensions of fairness and social welfare) or (in parentheses) the count of weakly dominant (at least as good as the DAA solution)—for at least 1 of the solutions (and there may be only one). "D(DorE) 2G.S.Soln" means the count of strictly dominant (better than the DAA on the two dimensions of fairness and social welfare) or (in parentheses) the count of weakly dominant (at least as good as the DAA solutions)—for both DAA solutions (although there may be only one). So the form X(Y)/Z means X solutions strictly dominating, Y solutions weakly dominating, and Z solutions found by the GA overall. (All of these solutions are stable solutions.) Finally, "Found G.S.Soln/#GS" with the form X/Y means of the Y DAA solutions (from Gale-Shapley), X of them were found by the GA.

To summarize, Table 13.6 shows:

- In 18 of 24 cases, the GA found at least one stable solution that strictly dominates both of the GS/DAA solutions. (In case 20, there is only 1 GS/DAA solution.)
- Excluding case 20 with only 1 GS/DAA solution, in 24 of 24 cases the GA found one or more stable solution that strictly dominates one of the two GS/DSS solutions.

Our study thus shows promising results on using a GA to search for favorable matching schemes. In terms of searching for stable match schemes,

	1st Qu	Median	3rd Qu
Init. # unstable pairs	314	340	367
Final # unstable pairs†	0	0	0
InitialSocialWelfareSum	1570	1638	1709
Final SocialWelfareSum	476	499	529
Initial Equity	492	532	572
Final Equity	200	224	245
SwapCount	1610	5731	22389
InitialSumXScores	769	819	869
Final SumXScores	217	246	280
InitialSumYScores	772	820	869
Final SumYScores	224	253	293
Number of Runs	100		
Number of Replications	100		
TransactionCost	0		
SwappingRegime	Collective		

TABLE 13.3: 40×40, transaction cost = 0 (†Mean=3, Max=124).

the GAs effectively found either strictly better or at least equally good solutions comparing to the Deferred Acceptance algorithm results with regard to fairness and social welfare. When stability is not the only objective, GAs provide many other dominant solutions. Depending on how different objectives are weighted, sometimes a minimum number of unstable pairs may be a cheap price to pay for the improvement on other objectives (for example, higher individual satisfaction or greater assignment fairness). The key point of our finding is about the GA's capability for providing decision makers with information about the otherwise unseen alternatives.

13.5 Agent Model Comparison

For comparison purposes we report on runs from the agent model on the same 25 20×20 randomly generated problems on which we ran the GA. It is a bit tricky to make the computational efforts of the two programs commensurate. Standardly in GA work comparisons are made on the basis of the number of fitness evaluations, but in the agent model we do not have a fitness function to evaluate. In the two models, however, there is a common basis for comparison: the number of calculations of whether two matched pairs are mutually unstable. In the GA, this is how fitness is evaluated. The fitness of a solution is the number of (mutually) unstable matched pairs it has. To calculate this number requires $\frac{n(n-1)}{2}$ comparisons of two matched pairs. Thus,

	1st Qu	Median	3rd Qu
Init. # unstable pairs	258	284	311
Final # unstable pairs	0	0	0
Init. # unstable pairs NTC	314	341	369
Final # unstable pairs NTC	7	9	11
InitialSocialWelfareSum	1569	1641	1711
Final SocialWelfareSum	483	505	529
Initial Equity	493	533	574
Final Equity	198	218	239
SwapCount	89	105	127
InitialSumXScores	771	820	869
Final SumXScores	226	251	279
InitialSumYScores	772	821	870
Final SumYScores	225	251	278
Number of Runs	100		
Number of Replications	100		
TransactionCost	2		
SwappingRegime	Collective		

TABLE 13.4: 40×40, transaction cost = 2.

the number of comparisons undertaken for a single case (there are 25 in the present context) is $\frac{n(n-1)}{2} \times K \times G \times R$, where K is the population size, G is the number of generations per replication, and R is the number of replications per case.

For the agent model, in each round (tick) during a replication each of the $2n$ agents undertakes a comparison with the n agents of opposite gender, so we have $2n^2$ comparisons per round/tick. Since the number of rounds/ticks varies by replication, we estimated the number of comparisons empirically at about 3500 per replication in the 20×20 cases. In consequence we set the number of replications per case to be 1000, putting us roughly on parity with the GA effort.

Table 13.7 shows the results. The column labeled "D1GS/TC" has the form X/Y, where X is the number of stable solutions found that strictly dominate one of the GS/DAA solutions and Y is the total number of stable solutions found. Similarly, in the column labeled "D2GS/TC" with form X/Y, X is the number of stable solutions found that strictly dominate both GS/DAA solutions and Y is the number of stable solutions found. Comparing Tables 13.7 and 13.6 we find that the agent model and the GA perform about equally as measured by the numbers of dominating solutions found.

	1st Qu	Median	3rd Qu
Init. # unstable pairs	1709	1808	1908
Fin. # unstable pairs	0	0	0
Init. # unstbl pairs NTC	2061	2165	2270
Fin. # unstbl pairs NTC	28	32	37
InitialSocialWelfareSum	9830	10104	10376
Final SocialWelfareSum	1984	2048	2113
Initial Equity	3173	3329	3492
Final Equity	878	934	994
SwapCount	655	1049	1901
InitialSumXScores	4859	5050	5242
Final SumXScores	941	1020	1106
InitialSumYScores	4858	5055	5249
Final SumYScores	939	1018	1103
Number of Runs	100		
Number of Replications	100		
TransactionCost	5		
SwappingRegime	Collective		

TABLE 13.5: 100×100, transaction cost = 5.

13.6 Nearly Stable Matches

A stable match is one in which there are no pairs of matched couples that are mutually unstable. A nearly stable match is one in which there are few pairs of matched couples that are mutually unstable. In a "one-away" match there is only one pair of matched couples that is mutually unstable. We emphasize that in such a one-away match there is no guarantee that swapping the unstable pair of couples will produce a stable match. The new couples resulting from the swap may be unstable with many other couples and unraveling may well be possible [182].

Our GA was able to find typically very many one-away solutions and very many of these were Pareto superior to the GS/DAA solutions. See Table 13.8 and Figure 13.3 for a graphical presentation.

13.7 Related Work

Two-sided matching problems, and the stable marriage problem in particular, have received exploratory investigation as dual objective problems in [19],

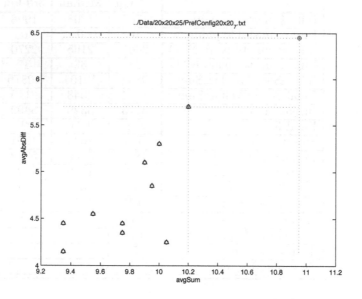

FIGURE 13.2: Plot of alternate stable solutions found, test case 7. GA in ∘, agent model in △, GS/DAA with ∗ (filled in).

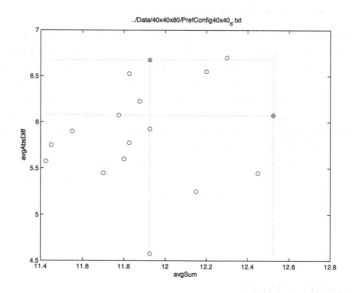

FIGURE 13.3: One-away solutions compared to GS/DSS solutions, 40x40, case 6; see Table 13.9. GS/DAA with ∗ (filled in).

Case Case Case	D(DorE) 1G.S.Soln / TC	D(DorE) 2G.S.Soln / TC	Found G.S.Soln / #GS
1	9(10) / 10	4(5) / 10	2/2
2	6(8) / 8	5(5) / 8	2/2
3	4 (5)/ 6	0(1) / 6	2/2
4	7(8) / 8	3(4) / 8	2/2
5	3(5) / 6	3(3) / 6	2/2
6	4(5) / 6	1(2) / 6	2/2
7	10(11) / 11	9(10) / 11	2/2
8	5(6) / 6	1(2) / 6	2/2
9	4(5) / 6	0(2) / 6	2/2
10	6(7) / 7	2(4) / 7	2/2
11	9(11) / 12	2(4) / 12	2/2
12	4(5) / 6	1(3) / 6	2/2
13	8(9) / 9	4(5) / 9	2/2
14	1(3) / 4	0(1) / 4	2/2
15	4(5) / 6	1(2) / 6	2/2
16	4(5) / 5	1(2) / 5	2/2
17	5 (7)/ 7	3(3) / 7	2/2
18	10(11) / 11	2(4) / 11	2/ 2
19	4(6) / 6	4(4) / 6	2/2
20	0(1) / 1	0(1) / 1	1/1
21	8(10) / 10	7(8) / 10	2/2
22	4(5) / 5	2(3) / 5	2/2
23	1(3) / 3	0(1) / 3	2/2
24	2(3) / 6	0(1) / 6	2/2
25	5(6) / 6	1(2) / 6	2/2

TABLE 13.6: 20x20x25 TPXOver strongly dominating solution counts (m04x06p50t100g2000).

[115], and [323]. Aldershof and Carducci [5] report optimistically on application of a genetic algorithm to two-sided matching problems, but the problems they examine are smaller than the 20×20 problems discussed here, so they do not address scaling issues. In [232] a genetic algorithm is used to find gender-unbiased solutions for the stable marriage problem. [69] has useful and suggestive findings pertaining to multiobjective evolutionary algorithms generally, as do [65] and [75]. [324] presents an innovative use of ant colony optimization for the stable marriage problem. We believe that population-based metaheuristics generally, whether or not they involve evolutionary computation, are very promising for two-sided matching problems, seen as multiobjective problems.

Case	D1GS/TC	D2GS/TC
1	8 / 8	3 / 8
2	6 / 8	5 / 8
3	4 / 6	0 / 6
4	7 / 7	3 / 7
5	3 / 6	3 / 6
6	4 / 6	1 / 6
7	10 / 10	9 / 10
8	8 / 8	3 / 8
9	4 / 6	0 / 6
10	6 / 7	2 / 7
11	9 / 12	2 / 12
12	4 / 6	1 / 6
13	8 / 8	4 / 8
14	1 / 4	0 / 4
15	4 / 6	1 / 6
16	4 / 5	1 / 5
17	5 / 6	3 / 6
18	10 / 10	2 / 10
19	4 / 6	4 / 6
20	0 / 1	0 / 1
21	8 / 9	7 / 9
22	4 / 4	2 / 4
23	1 / 3	0 / 3
24	2 / 6	0 / 6
25	5 / 6	1 / 6

TABLE 13.7: Agent 20x20x25 SD.

13.8   Summary and Discussion

We find and are reporting in compressed form the following:

1. Typically in simple marriage matching problems there are stable matches that are Pareto superior to the deferred acceptance matches, in regard to equity and social welfare (as we have characterized them).

2. With roughly equal computational effort, both the agent-based model and the genetic algorithm we built find similar numbers and quantities of stable solutions for the simple marriage matching problems we examined.

3. There are typically very many "near-stable" matches that are superior to the deferred acceptance (stable) matches on equity and/or social welfare (or both). We found these solutions with the genetic algorithm.

Test Case	D1GS/TC	D2GS/TC
1	121 / 128	31 / 128
2	100 / 143	88 / 143
3	31 / 140	1 / 140
4	78 / 97	49 / 97
5	49 / 96	39 / 96
6	28 / 83	7 / 83
7	113 / 124	75 / 124
8	53 / 67	15 / 67
9	61 / 94	9 / 94
10	78 / 110	14 / 110
11	75 / 154	20 / 154
12	10 / 97	7 / 97
13	82 / 106	18 / 106
14	27 / 65	12 / 65
15	73 / 134	7 / 134
16	56 / 87	14 / 87
17	72 / 87	56 / 87
18	52 / 89	11 / 89
19	43 / 84	27 / 84
20	1 / 38	1 / 38
21	124 / 148	109 / 148
22	64 / 82	35 / 82
23	11 / 61	1 / 61
24	20 / 96	8 / 96
25	76 / 107	8 / 107

TABLE 13.8: 20x20x25 TPXOver, 0.8, mutation, 0.3, one-away strongly dominating solution counts.

Test Case	D1GS/TC	D2GS/TC
1	6 / 8	4 / 8
2	19 / 20	8 / 20
6	13 / 16	6 / 16

TABLE 13.9: 40x40 Case1,2,6 GA TPXOver strongly dominating optimal solution counts.

4. Scale is an important issue. As the size of the problem exceeds 50 or so, the agent-based model becomes generally unable to find any stable solution with zero transaction costs with fewer than millions of swaps (match alterations). The GA generally continues to perform well, see Table 13.9, but this needs more extensive testing.

5. Adding transaction costs to the agent model generally results in better finding of stability, but even costs in the range of $\frac{n}{10}$ are overwhelmed when n is 100 or more. The agents simply do not find stable solutions even in this attenuated sense. The agents do, however, generally improve overall social welfare and equity scores as a side effect of their swapping.

We see both a practical, applied upshot of these findings and a theoretical one. On the applied side:

1. The case is now quite strong for insisting that two-sided matching problems be viewed as multiobjective, and that policy should look well beyond the GS/DAA-style solutions. Alternate solutions that are Pareto superior do (often) exist and may be found by heuristic methods such as those on display here.

2. The necessity of stability, or equilibrium in a matching, can be questioned, at least in many applications. When the number of unstable pairs is small but non-zero, there will in many cases be no realistic means for the pairs to find each other and unravel the matching. We agree with Kreps's more general comment that

> Unless a given game has a self-evident way to play, self-evident to the participants, the notion of a Nash equilibrium has no particular claim upon our attention. [185, page 31]

Matching applications should reconsider their requirements for (exact, full) stability, especially when, as we have seen in an example, there may be very many nearly stable matches with superior equity and social welfare properties.

3. Our agent-based simulation of a distributed market raises the possibility that centralized markets might become superior to distributed markets by the tactic of simulating them under various conditions, realizable or not, and allocating results based on the simulations. This is an intriguing idea for future consideration.

On the theoretical side, there are two main points arising. First, the agent model is an epistemically very generous model of how real agents cope. That even this model will fail to achieve equilibrium under non-drastic scaling up raises the question concretely of whether real markets can find equilibrium.

Second, looking forward we note that matching problems in general and the simple marriage matching problem in particular may have application broader than heretofore conceived. Consider a model of a market with n buyers of a good such as a house and n sellers of the good. Assume that for each possible pair there is a negotiated price at which they would transact the sale of the house in question. This provides a ranking of the buyers for each seller, since sellers we presume only care about price. On the buyer side, however, price is only one factor among many in determining the value of the property. So buyers have preference rankings on the houses (sellers) that are influenced by, but not determined by, the negotiated prices. At equilibrium, all the houses are sold, and a stable matching is achieved. What we have learned from the agent and evolutionary models described here is that even with the rather heroic epistemic powers of our agents, they will not be able to attain stable matches in any reasonably large market context. Nor is there much assurance that on any one run of the market good equity or social welfare outcomes will be achieved. The upshot is (1) centralized markets, of the sort in which deferred acceptance algorithms are commonly run, may be more widely desirable and (2) heuristic alternatives to deferred acceptance may well be able to offer practical improvements and complements to deferred acceptance in these centralized markets.

13.9   For Exploration

1. Comment on the requirement for stability in matching in a centralized market. When will this be a genuine requirement? Examples? When is it likely not to be genuinely necessary? Examples?

2. Review one or more examples of centralized markets. ([262] has a good list of those using some form of the deferred acceptance algorithm.) How are they working? Are they successful? Why or why not? Is stability of matching guaranteed? How to they score on fairness and resource efficiency (social welfare)?

3. Undertake a study on the lines of the one described in this chapter, but for admissions problems.

4. Develop an agent-based model that more accurately represents the "dating and mating" process, perhaps by including "hook-up culture" in which many very tentative hook-ups (non-date dates) occur without leading to lasting intimate relationships. (NetLogo would be a good choice for the programming environment.) Is this perhaps a mechanism that facilitates finding good matches? Why or why not?

13.10   Concluding Notes

This chapter is an edited version of [173]. Permission to republish this material is gratefully acknowledged to the Association for Computing Machinery, Inc.

Chapter 14

IDS Games

14.1 Introduction

Imagine you live in a large condominium and there is a one-time opportunity to invest in fire-protection infrastructure. Specifically, for a price, you can have a sprinkler system installed in your apartment. If the sprinkler system is installed and a fire ever starts in your apartment (from whatever cause), you may assume that the system will reliably extinguish the fire and that any damage will be minimal. Your neighbors are also facing the same one-time decision. The complication in all of this is that even if you install a sprinkler system, if your neighbor does not do likewise *and* happens to experience a fire, the collateral damage to your condo (from smoke, etc.) will be the same as if you had not installed the sprinkler system and had had the fire yourself. Under what conditions would you elect to install the sprinkler system?

This is an example of what is called an *Interdependent Security* (IDS) game [189]. What is characteristic of these strategic situations is the possibility that a player may be harmed ("contaminated") through inaction by other players in investing in security, even if the player itself invests in security. When other players do not invest in risk reduction, the possibility of contamination (from realized risk events they experience) reduces the incentive for a player to invest for itself. Heal and Kunreuther [145] discuss the wide scope applicability of these games. Examples include: airline security, industrial accidents (e.g., chemical plants, nuclear power plants), protecting buildings against terrorist attacks, protecting against risky behavior by other members of an organization (e.g., rogue traders, risk-seeking business units), and protecting networked computers from security violators.

For starters we can model the IDS story as a 2×2 one-shot game in strategic form. See Table 14.1.

(Expected payoffs are given with Row on top and Column underneath.) The interpretation is this. Y is the wealth position of each player before the game is played. c is the cost of investing in the security or protection device. p is the probability that a security event happens. It may or may not do damage, depending on what the players have chosen. If damage is done, then the loss is L. We see in Table 14.1 that if both players invest, then they each have a resulting wealth position of $Y - c$, no matter what, regardless of whether

	Invest (I)	Not Invest ($\neg I$)
Invest (I)	$Y - c$	$Y - c - q \cdot L$
	$Y - c$	$Y - p \cdot L$
Not Invest ($\neg I$)	$Y - p \cdot L$	$Y - [p \cdot L + (1 - p) \cdot q \cdot L]$
	$Y - c - q \cdot L$	$Y - [p \cdot L + (1 - p) \cdot q \cdot L]$

TABLE 14.1: Basic IDS game: Expected outcomes associated with investing or not investing in security.

there has been a loss event. L is the resulting loss to the agent (here, Row or Column) that has not invested against it. So we see in Table 14.1 that if Row invests and Column does not, then Column's expected loss is $p \cdot L$, the probability of a security event times the loss to Column if it materializes. Column's net expected wealth position is then $Y - p \cdot L$.

The quantity q is the probability of contagion, that is the probability of "contaminating" your neighbor if you do not invest in security. Clearly we require that $q \leq p$. If $(I, \neg I)$, that is if Row invests and Column does not, then the expected loss to Row is $q \cdot L$, the probability of the security event happening times the probability of contamination, given that the event has happened (i.e., q) times the loss amount, L. Row's net expected wealth position is then its wealth position after investing, minus its expected loss: $(Y - c) - (q \cdot L)$. The stories are the same, but with the roles reversed, if $(\neg I, I)$.

Finally, if neither player invests $(\neg I, \neg I)$, there are two sources of expected loss for each player. First, a loss event may occur that directly affects a player. That expected loss is $p \cdot L$. Second, if the direct loss event does not occur, an indirect loss event might occur. We are assuming that the probability of this event is independent of the probability of a direct loss. Thus, the probability of a direct loss not occurring and a contaminating loss occurring is just the product of these two quantities: $(1 - p) \cdot q$ and the associated expected loss is $(1 - p) \cdot q \cdot L$. Summing everything up we get the entries shown for $(\neg I, \neg I)$ in Table 14.1. What we neglect is the expected loss from two security events occurring in the same period: a direct event to a player and an indirect event, that is, a direct event occurring to the player's counterpart.

Depending on the parameter values, particular IDS games will *in expectation* have payoffs that are characteristic of familiar 2×2 games, notably including Prisoner's Dilemma and Stag Hunt. What makes IDS games distinctive and especially interesting is that their payoffs are stochastic (except in the (I, I) case, when both players invest in security). An investing row player playing against a non-investing column player (see Table 14.1) will almost certainly *not* receive a payoff of $Y - c - q \cdot L$. The row player will actually receive a payoff of $Y - c$ with probability $(1 - q)$ or a payoff of $(Y - c - L)$ with probability q. In short, the row player is facing a lottery: $[(Y - c), 1 - q;$

$(Y - c - L), q]$. And so on for $(\neg I, I)$ and $(\neg I, \neg I)$. IDS games are games with lottery payoffs.

It is usual to represent games as having fixed and certain payoffs, rather than lottery payoffs. Standard solution techniques require this. Indeed, a main benefit of representing payoffs as utilities is that the decision maker is, in theory, indifferent between the expected value of a lottery whose outcomes are given in utilities and the lottery itself. If so, then any lottery outcome can be reduced to a certain outcome, its expected value. A similar situation often obtains when game models are applied in biology and the payoffs are given in fitnesses, usually defined as expectations of offspring. Roughgarden [265] presents several nice examples. The main point for our purposes is that if payoffs are denominated in utilities or in expectations of offspring, then it is plausible to collapse lotteries (whose outcomes are in expectations of offspring) into expected values.

And if not, perhaps not. Behavioral experiments (e.g., [190, 335]) have indicated systematic differences in subjects' behavior between IDS games (with lottery payoffs) and their deterministic analogs (with lotteries replaced by their expected values). Setting the payoffs so that the games are Prisoner's Dilemmas (stochastic or deterministic), Kunreuther et al. find generally less cooperation in the stochastic IDS case than in the deterministic version of the game [190].

Even if we stick with convention in game theory and give payoff end-values in terms of utilities, the fact that payoffs are realized through lotteries presents new strategic considerations. A player, for example, may choose to act in one way after experiencing the consequences of a security event and act in another way when the security event does not occur. So, let us see what happens when various strategies for iterated play of IDS games confront one another.

14.2 Comparing Strategies in Iterated Play

Our focus as usual is on the behavior of strategies, here in iterated play between two players. The basic IDS game (Table 14.1) thus becomes the stage game in the supergame we investigate. Reminiscent of Axelrod's Prisoner's Dilemma tournaments [15, 18], 17 plausible IDS strategies were identified, based on observations of subjects' behaviors in experiments and on brainstorming. Tournament software was developed in the form of a NetLogo program, IDS-2x2-Tournaments.nlogo. See Figure 14.1 on page 290.

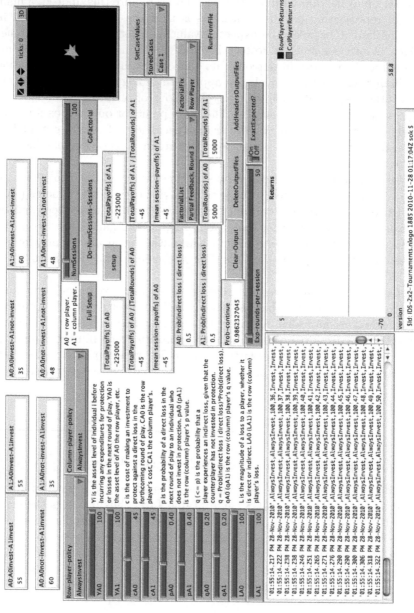

FIGURE 14.1: IDS-2x2-Tournaments.nlogo, AlwaysInvest plays itself with the Case 1 settings.

14.2.1   Description of the Strategies

IDS-2x2-Tournaments.nlogo comes with 17 built-in strategies for playing IDS games. A strategy is referred to in the program as a `policy-of-play`. The purpose of this subsection is to describe these strategies. The casual reader may want to skim lightly over the section, or simply use it for reference later. IDS-2x2-Tournaments.nlogo is designed to be extensible. Essentially any IDS strategy can easily be added to the existing framework. The following detailed description of the 17 built-in strategies, including their code in NetLogo will be useful to anyone wishing to add new strategies to the tournament.

1. PROB(I)=0.7

   In each round of play, invest in risk mitigation with probability 0.7.

   ```
   if (policy-of-play = "Prob(I)=0.7")
      [if-else (random-float 1 <= 0.7)
       [set next-move "Invest"]
       [set next-move "NotInvest"]
      ] ; end of if-else (random 1 <= 0.7
   ```

2. PROB(I)=0.2

   In each round of play, invest in risk mitigation with probability 0.2.

   ```
   if (policy-of-play = "Prob(I)=0.2")
      [if-else (random-float 1 <= 0.2)
       [set next-move "Invest"]
       [set next-move "NotInvest"]
      ] ; end of if-else (random 1 <= 0.2
   ```

3. ALWAYSINVEST

   In each round of play, invest in risk mitigation.

   ```
   if (policy-of-play = "AlwaysInvest")
      [set next-move "Invest"]
   ```

4. NEVERINVEST

   Never invest in risk mitigation.

   ```
   if (policy-of-play = "NeverInvest")
      [set next-move "NotInvest"]
   ```

5. STRICKTITFORTAT

A player using STRICKTITFORTAT invests on the first round and after that invests only if the counterpart invested in the previous round.

Note: This strategy only applies to the full feedback situation.

```
if (policy-of-play = "StrictTitForTat")
  [if-else (HisLastMove = "NotInvest")
  [set next-move "NotInvest"]
    [set next-move "Invest"]
  ]
```

6. FFTITFORTATNSTICKY

Under full feedback (the player knows whether the counterpart has invested in security during the previous rounds of play), the player plays a tempered form of TIT FOR TAT. The player cooperates under the counterpart defects, then defects and continues to defect until the counterpart has cooperated $N = 3$ times in a row.

```
if( policy-of-play = "FFTitForTatNSticky")
   [ let N 3
   let DaCount 0
   ; The HisMoves list is appended at the end.
   let DaMoves reverse HisMoves
   if (length DaMoves = 0)
     [set DaCount 0]
   if (length DaMoves > 0 and length DaMoves < N)
     [foreach n-values length DaMoves [?] [
       if (item ? DaMoves = "NotInvest") [
       set DaCount DaCount + 1]
     ]
     ]
   if (length DaMoves >= N)
     [foreach n-values N [?] [
       if (item ? DaMoves = "NotInvest") [
       set DaCount DaCount + 1]
     ]
     ]
   if-else (DaCount > 0)
       [set next-move "NotInvest"]
       [set next-move "Invest"]

   ]
```

7. FFGRIMTRIGGER

Under full feedback, a player with the FFGRIMTRIGGER policy cooperates (plays "Invest") until the counterpart defects (plays "NotInvest"), and then plays "NotInvest" ever after, regardless of what the counterpart subsequently does.

```
if (policy-of-play = "FFGrimTrigger")
  [if-else (HisLastMove = "NotInvest")
    [set next-move "NotInvest"]
    [if-else (MyLastMove = "NotInvest")
      [set next-move "NotInvest"]
      [set next-move "Invest"]
    ]
  ]
```

8. INVESTAFTERLOSS

A player using INVESTAFTERLOSS does *not* invest on the first round, and continues not to invest, except in rounds immediately following a loss, whether direct or indirect.

```
if( policy-of-play = "InvestAfterLoss")
  [ if-else( prev-direct-loss-realized = 1 or
              prev-indirect-loss-realized = 1 )
    [set next-move "Invest"]
    [set next-move "NotInvest"]
  ]
```

9. INVESTNAFTERLOSS

A player using INVESTNAFTERLOSS does *not* invest on the first round, and continues not to invest, except for the N rounds immediately following a loss, whether direct or indirect. N is set to 3.

```
if( policy-of-play = "InvestNAfterLoss")
  [ if( prev-direct-loss-realized = 1 or
        prev-indirect-loss-realized = 1 )
    [set invest-count 3] ; invest for 3 rounds after loss
  if-else( invest-count >= 1)
    [set next-move "Invest"
     set invest-count (invest-count - 1)
    ]
    [set next-move "NotInvest"]
  ]
```

10. DONTINVESTAFTERLOSS

A player using DONTINVESTAFTERLOSS invests, unless it has suffered a loss (direct or indirect) in the previous round, in which case it does not invest in mitigation.

```
if( policy-of-play = "DontInvestAfterLoss")
  [ if-else( prev-direct-loss-realized = 1 or
                prev-indirect-loss-realized = 1 )
    [set next-move "NotInvest"]
    [set next-move "Invest"]
  ]
```

11. PFTITFORTATPLUSLOSSINVEST

PF = partial feedback. Under partial feedback, players do not know what their counterparts do, whether they invest in protection or not, unless they receive an indirect loss.

A player using PFTITFORTATPLUSLOSSINVEST invests except in the round after it receives an indirect loss.

```
if-else (first MyIndirectLosses < 0)
    [set next-move "NotInvest"]
    [set next-move "Invest"]
```

12. PFTITFORTATPLUSLOSSNOTINVEST

The same as PFTITFORTATPLUSLOSSINVEST, except do not invest after any loss.

```
if-else (first MyDirectLosses < 0 or
              first MyIndirectLosses < 0)
    [set next-move "NotInvest"]
    [set next-move "Invest"]
```

13. FF2TITSFOR1TAT

Assuming full feedback, FF. Defect twice in response to a single defection; otherwise cooperate.

```
if( policy-of-play ="FF2TitsFor1Tat" )
    [; If the oponent didn't invest in the last round
     ; then we always play tit for two rounds.
     ; Hence play tit once in this round and set the
     ; counter so that tit is also played in
```

```
;the next round.
  if-else( HisLastMove = "NotInvest" )
  [ set next-move "NotInvest"
    set tit_count 1
  ]
  [ if-else( tit_count > 0 )
    [ set next-move "NotInvest"
    set tit_count (tit_count - 1) ]
    [ set next-move "Invest"] ;there is an investment
    ; only if the other player invested or his play
    ; was not known AND tit_counter is 0.
  ]
]
```

14. FF1TitFor2Tats

Assuming full feedback, FF. Defect once in response to two consecutive defections; otherwise cooperate.

```
if( policy-of-play ="FF1TitFor2Tats" )
  [ ; Check if the other player has defaulted twice and
    ; only then play tat.
    ; Update info from the last round.
    if( HisLastMove = "NotInvest" )
    [ set tat_count (tat_count + 1)]
    if-else( tat_count = 2 )
    [ set next-move "NotInvest"
      set tat_count 0
    ]
    [ set next-move "Invest" ]
  ]
```

15. PFGrimTrigger

Cooperate until an indirect loss is realized and defect (play "NotInvest") forever after.

```
if( policy-of-play = "PFGrimTrigger")
  [if-else (first MyIndirectLosses < 0)
    [set next-move "NotInvest"]
    [if-else (MyLastMove = "NotInvest")
      [set next-move MyLastMove]
      [set next-move "Invest"]
    ]
  ]
```

16. PFTitForTatNSticky

Assuming partial feedback, PF. Begin by cooperating (playing "Invest")
and continue to do so until receiving an indirect loss. After receiving an
indirect loss, play "NotInvest" until the counterpart has played "Invest"
for $N = 3$ rounds in a row.

```
if( policy-of-play = "PFTitForTatNSticky")
   [ let NN 3
   let DaLosses 0
   if (length MyIndirectLosses = 0)
     [set DaLosses 0]
   if (length MyIndirectLosses > 0 and
                   length MyIndirectLosses < NN)
     [set DaLosses sum MyIndirectLosses]
   if (length MyIndirectLosses >= NN)
     [foreach n-values NN [?] [
       set DaLosses (DaLosses + item ? MyIndirectLosses)
     ]
     ]
   if-else (DaLosses < 0)
       [set next-move "NotInvest"]
       [set next-move "Invest"]
   ]
```

17. M1FictiousPlay

Attending to the counterpart's and your previous moves, attempt to
estimate the counterpart's next move and play your best response to it.

```
if( policy-of-play ="M1FictiousPlay" )
[ ; In this policy we maintain the action that the
  : opponent took
  ; in reaction to our play in the last round. It is
  ; assumed that this policy works in a FF case.

  ; Update the event in the last play
  if-else( MyMoveBeforeLast = "Invest")
  [ if-else( HisLastMove = "Invest" )
    [set times-invest-invest
               (times-invest-invest + 1)]
    [set times-invest-notinvest
               (times-invest-notinvest + 1)]
  ]
  [if-else( HisLastMove = "Invest" )
```

```
    [set times-notinvest-invest
              (times-notinvest-invest + 1)]
  [ set times-notinvest-notinvest
              (times-notinvest-notinvest + 1)]
 ]

 let invest 0

 ; Check what I played in the last round, and given
 ; that, what the counterpart is most likely to play.
 if-else( MyLastMove = "Invest" )
 [ if-else( times-invest-invest >
                  times-invest-notinvest )
   ; Means the counterpart is going to invest with
   ; higher probability than if I invested
   ; in the last round.
   [set invest 1]
   [set invest 0]
 ]
 [ if-else( times-notinvest-invest >
              times-notinvest-notinvest )
   [set invest 1]
   [set invest 0]
 ]
 ; Compute optimal policy given that the
 ; counterpart is going to Invest or to NotInvest.
 let payoff-invest (-1 * cvalue - qvalue *
                  LValue * (1 - invest))
 let payoff-notinvest
 ( -1 * pValue * LValue - (1 - pValue) * qvalue *
     LValue * (1 - invest) )
 if-else( payoff-invest > payoff-notinvest )
 [ set next-move "Invest" ]
 [ set next-move "NotInvest" ]

if (MyMoveBeforeLast = "Null")
[ if-else (random 2 = 1)
  [ set next-move "Invest" ]
  [ set next-move "NotInvest" ]
]
]
```

14.2.2   Setup for the Experiments

IDS-2x2-Tournaments.nlogo is quite flexible and easily accommodates arbitrary parameter settings. Nevertheless, we shall focus here on reporting and discussing experiments with just one such setting, Case 1, whose values are given in Figure 14.2, page 299. I encourage the reader to experiment with alternative settings.

With these parameter values, the expected losses in 2×2 form constitute a Prisoner's Dilemma. See Table 14.2. We distinguish play under two conditions, as noted above. Under full feedback each player has available to it the moves—"Invest" or "NotInvest"—made by the counterpart in the previous rounds of play. §14.3 explores the performance of strategies appropriate for the full feedback condition. Under partial feedback, a player only knows what the counterpart did in previous rounds by inference: if the player suffers an indirect loss, this implies that the counterpart did not invest; otherwise the player has no knowledge of what the counterpart did. §14.4 explores strategies appropriate for this condition.

	Invest (I)	Not Invest ($\neg I$)
Invest (I)	−45, −45	−65, −40
Not Invest ($\neg I$)	−40, −65	−52, −52

TABLE 14.2: Basic IDS game: Expected losses associated with investing or not investing in security, Case 1.

14.3   Full Feedback

Under the condition of full feedback (each player knows whether its counterpart has invested or not in previous rounds of play in the session), the following policies (among the 17 overall) apply.

1. PROB(I)=0.7

2. PROB(I)=0.2

3. ALWAYSINVEST

4. STRICTTITFORTAT

5. NEVERINVEST

6. INVESTAFTERLOSS

7. INVESTNAFTERLOSS

```
if (StoredCases = "Case 1")
 [; Expected number of rounds of play per session:
  set Exp-Rounds-Per-Session 50
  ; Number of sessions to be played:
  set NumSessions 100
  ; Row player's (A0's) policy/strategy:
  set Row-Player-Policy "StrictTitForTat"
  ; Column player's (A1's) policy/strategy:
  set Column-Player-Policy "StrictTitForTat"
  set YA0 100 ; Yi is the assets level of individual i before
  set YA1 100 ; incurring any expenditures for protection or
            ; losses in the next round of play.
            ; YA0 is the asset level of A0 the row player, etc.
  set cA0 45  ; c is the cost of making
            ; an investment to protect
  set cA1 45  ; against a direct loss in the forthcoming
            ; round of play. cA0 is the row player's cost,
            ; cA1 the column player's.
  set pA0 0.4 ; p is the probability of a direct loss
  set pA1 0.4 ; in the next round of play to an individual
            ; who does not invest in protection. pA0 (pA1)
            ;  is the Row (Column) player's p value.
  set qA0 0.2 ; q (<= p) is the probability that a
  set qA1 0.2 ; player experiences an indirect loss,
            ; given that the counterplayer has not invested
            ; in protection. q =
            ; Prob(indirect loss | direct loss)*
            ; Prob(direct loss). qA0 (qA1) is
            ; the Row (Column) player's q value.
  set LA0 100 ; L is the magnitude of a loss to a player,
  set LA1 100 ; whether it is direct or indirect. LA0 (LA1)
            ;  is the Row (Column) player's loss.
 ]
```

FIGURE 14.2: IDS-2x2-Tournaments.nlogo: Case 1 default settings.

8. DontInvestAfterLoss

9. FF1TitFor2Tats

10. FF2TitsFor1Tat

11. M1FictiousPlay

12. FFGrimTrigger

13. FFTitForTatNSticky

(See §14.2.1 for descriptions of these strategies.)

As usual, our first interest is in discovering good strategies for play, here in the iterated IDS game. Also as usual, we pursue MR-P, the method of rationality, pragmatic. As a reminder here is a summary.

1. Assemble a tractable consideration set, $\mathcal{U}$, of alternatives.

   For full feedback, the case at hand, we will work with the 13 strategies listed above, which are implemented in IDS-2x2-Tournaments.nlogo.

2. Design a discrimination test procedure for the strategies in the consideration set $\mathcal{U}$.

   We shall conduct a series of tournaments, selecting the better performers in each round for survival.

3. Apply the test procedure and arrive at the set of contenders, $\mathcal{C}$, for actual use.

   A description, along with discussion of additional considerations, follows immediately.

Table 14.3 summarizes the results of a tournament among all 13 members of our consideration set. Points arising:

1. There is considerable variation evident in the returns to the various policies, as they play one another.

2. The variation in returns is generally skewed to the down side, sometimes quite dramatically so.

3. There are ultimately two sources of variation in the model: losses caused directly by loss events, and losses caused by contagion.

4. InvestAfterLoss and InvestNAfterLoss, which seem to approximate the behavior of many subjects, are not especially attractive policies. Their performance differs little from AlwaysInvest.

5. StrictTitForTat and FF2TitsFor1Tat appear to be robust, generally well-performing strategies.

Strategy	1st Qu.	Median	Mean	3rd Qu.
P(I)=0.7	−45.0	−45.0	−52.68	−45.0
P(I)=0.2	−100.0	−45.0	−49.39	0.0
ALWAYSINVEST	−45.0	−45.0	−51.36	−45.0
STRICTTITFORTAT	−45.0	−45.0	−48.02	−45.0
NEVERINVEST	−100.0	0.0	−48.11	0.0
INVESTAFTERLOSS	−100.0	−45.0	−51.67	0.0
INVESTNAFTERLOSS	−45.0	−45.0	−52.54	−45.0
DONTINVESTAFTERLOSS	−100.0	−45.0	−51.57	0.0
FF2TITSFOR1TAT	−45.0	−45.0	−47.28	−45.0
FF1TITFOR2TATS	−45.0	−45.0	−49.76	−45.0
M1FICTITIOUSPLAY	−100.0	0.0	−47.84	0.0
FFGRIMTRIGGER	−45.0	45.0	−46.66	0.0
FFTITFORTATNSTICKY	−45.0	45.0	−46.97	−45.00

TABLE 14.3: Round 1 FF tournament: Mean results to Strategy, playing the other strategies and itself, 100 sessions, fixed termination, 50 rounds exactly.

6. DONTINVESTAFTERLOSS messes up STRICTTITFORTAT, but not FF2TITSFOR1TAT or FF1TITFOR2TATS.

Selecting the 6 strategies whose mean performance is better than −50.0 and conducting a round 2 tournament among them, produced the results summarized in Table 14.4.

Strategy	1st Qu.	Median	Mean	3rd Qu.
STRICTTITFORTAT	−45.0	−45.0	−47.39	−45.0
NEVERINVEST	−100.0	0.0	−52.45	0.0
FF2TITSFOR1TAT	−45.0	−45.0	−47.26	−45.0
M1FICTITIOUSPLAY	−100.0	−100.0	−51.96	0.0
FFGRIMTRIGGER	−45.0	45.0	−47.21	0.0
FFTITFORTATNSTICKY	−45.0	45.0	−47.44	−45.00

TABLE 14.4: Round 2 FF tournament (strategies better than −49.0 in round 1): Mean results to Strategy, playing the other strategies and itself, 100 sessions, fixed termination, 50 rounds exactly.

Points arising:

1. The relative performances of the strategies in this group differs from their performances in the first round of competition, with the full consideration set in play. The performance of a strategy in a tournament depends in general on the strategic constitution of the entire tournament. If we would conduct a replicator dynamics experiment we would find in general that the performances of these strategies are frequency dependent. What the other strategies are matters as do their frequencies in the population.

2. NEVERINVEST and M1FICTITIOUSPLAY are clearly the worst-performing strategies in this group.

3. If we remove NEVERINVEST and M1FICTITIOUSPLAY, and then conduct a tournament with the remaining four strategies they would all always cooperate with each other, always playing "Invest," and achieving -45.0 at every round.

4. Taking these four strategies as provisionally constituting our collection of contenders, $\mathcal{C}$, it is immediate that each of the strategies does well when played against itself and when played against every other member of $\mathcal{C}$.

5. Further, none of the four strategies in $\mathcal{C}$ is exploitable by any of the considered strategies. In fact, NEVERINVEST and M1FICTITIOUSPLAY are the most exploitative strategies in the original consideration set and they fare comparatively poorly against the members of $\mathcal{C}$.

We may conclude, always provisionally, that given the consideration set we started with, both players would be rationally justified in the MR-P sense in picking any of the four strategies in $\mathcal{C}$: STRICTTITFORTAT, FF2TITSFOR1TAT, FFGRIMTRIGGER, or FFTITFORTATNSTICKY.

14.4   Partial Feedback

Under the condition of partial feedback a player does not know whether its counterpart has invested or not in previous rounds of play in the session, unless the player suffers contagion from its counterpart. The following policies (among the 17 overall) apply.

1. PROB(I)=0.7

2. PROB(I)=0.2

3. ALWAYSINVEST

4. NEVERINVEST

5. INVESTAFTERLOSS

6. INVESTNAFTERLOSS

7. DONTINVESTAFTERLOSS

8. PFTITFORTATPLUSLOSSINVEST

9. PFTITFORTATPLUSLOSSNOTINVEST

10. PFGRIMTRIGGER

11. PFTITFORTATPLUSNSTICKY

Strategy	1st Qu.	Median	Mean	3rd Qu.
P(I)=0.7	−45.00	−45.00	−50.4	−45.00
P(I)=0.2	−100.00	−45.0	−46.69	0.00
ALWAYSINVEST	−45.0	−45.0	−50.85	−45.0
NEVERINVEST	−100.0	0.0	−45.60	0.0
INVESTAFTERLOSS	−100.0	−45.0	−49.61	0.0
INVESTNAFTERLOSS	−45.0	−45.0	−50.84	−45.0
DONTINVESTAFTERLOSS	−100.0	−45.0	−48.82	0.0
PFTITFORTATPLUSLOSSINVEST	−45.00	−45.00	−50.22	−45.00
PFNOTINVEST	−45.00	−45.00	−49.75	−45.00
PFGRIMTRIGGER	−45.00	−45.00	−46.87	−45.00
PFTITFORTATNSTICKY	−45.00	−45.00	−49.42	−45.00

TABLE 14.5: Round 1 PF tournament: Mean results to Strategy, playing the other strategies and itself, 100 sessions, fixed termination, 50 rounds exactly. PFNOTINVEST = PFTITFORTATPLUSLOSSNOTINVEST.

Table 14.5 shows the results of a tournament among these 11 strategies constituting our initial consideration set. Taking strategies scoring better than −50.0, Table 14.6 shows the results of the round 2 tournament.

Strategy	1st Qu.	Median	Mean	3rd Qu.
P(I)=0.2	−100.00	−45.0	−49.38	0.00
NEVERINVEST	−100.0	0.00	−47.62	0.00
INVESTAFTERLOSS	−100.0	−45.0	−49.61	0.00
DONTINVESTAFTERLOSS	−100.0	−45.0	−51.81	0.00
PFNOTINVEST	−45.00	−45.00	−51.84	−45.00
PFGRIMTRIGGER	−45.00	−45.00	−47.90	0.00
PFTITFORTATNSTICKY	−45.00	−45.00	−50.9	−45.00

TABLE 14.6: Round 2 PF tournament: Mean results to Strategy, playing the other strategies and itself, 100 sessions, fixed termination, 50 rounds exactly. PFNOTINVEST = PFTITFORTATPLUSLOSSNOTINVEST.

Drawing the line at −51.0 and conducting a third round tournament, we get the results shown in Table 14.7.

Strategy	1st Qu.	Median	Mean	3rd Qu.
P(I)=0.2	−100.00	−45.00	−50.47	0.00
NEVERINVEST	−100.00	0.00	−47.96	0.00
INVESTAFTERLOSS	−100.00	−45.00	−53.30	−45.00
PFGRIMTRIGGER	−100.00	−45.00	−48.35	0.00
PFTITFORTATNSTICKY	−45.00	−45.00	−51.78	−45.00

TABLE 14.7: Round 3 PF tournament: Mean results to Strategy, playing the other strategies and itself, 100 sessions, fixed termination, 50 rounds exactly.

Now, P(I)=0.2 and INVESTAFTERLOSS are distinctly inferior strategies, as is PFTITFORTATNSTICKY. We are left with just two contenders, NEVER-INVEST and PFGRIMTRIGGER. Table 14.8 shows what happens when they play each other: NEVERINVEST beats PFGRIMTRIGGER by a wide margin.

Strategy	1st Qu.	Median	Mean	3rd Qu.
PFGRIMTRIGGER	−100.00	−100.00	−54.34	0.00
NEVERINVEST	−100.00	−100.00	−50.8	0.00

TABLE 14.8: Round 4.1 PF tournament: Mean results to Strategy, playing the other strategy, 100 sessions, fixed termination, 50 rounds exactly

Are we left, then, with only NEVERINVEST as a contender? Consider what happens when NEVERINVEST plays itself. In one experiment (50 rounds per session, 100 sessions), the two NEVERINVEST players scored −51.78 and −52.84 for an average of 52.31. PFGRIMTRIGGER and for that matter PFTIT-FORTATNSTICKY played against themselves (and each other) always gets −45.0. Representing PFGRIMTRIGGER versus NEVERINVEST in strategic form gives us:

	PFGRIMTRIGGER	NEVERINVEST
PFGRIMTRIGGER	−45.00, −45.00	−54.34, −50.80
NEVERINVEST	−50.80, −54.34	−52.31, −52.31

Reducing this to rank representation gives us:

	PFGRIMTRIGGER	NEVERINVEST
PFGRIMTRIGGER	4, 4	1, 3
NEVERINVEST	3, 1	2, 2

We see that this situation is rather like a Stag Hunt game. Is it less favorable to cooperation because the "temptation" to defect, −50.80 is better than the "penalty" for mutual defection, −52.31?

It is interesting to compare PFTITFORTATNSTICKY. Table 14.9 shows what happens when they play each other: NEVERINVEST beats PFGRIMTRIGGER by a very wide margin.

Strategy	1st Qu.	Median	Mean	3rd Qu.
PFTITFORTATNSTICKY	−100.00	−45.00	−60.2	−45.00
NEVERINVEST	−100.00	0.00	−45.14	-0.00

TABLE 14.9: Round 4.2 PF tournament: Mean results to Strategy, playing the other strategy, 100 sessions, fixed termination, 50 rounds exactly.

Representing PFTITFORTATNSTICKY versus NEVERINVEST in strategic form gives us:

	PFTITFORTATNSTICKY	NEVERINVEST
PFTITFORTATNSTICKY	−45.00, −45.00	−60.20, −45.14
NEVERINVEST	−45.14, −60.20	−52.31, −52.31

Treating −45.0 as morally equal to −45.14 and reducing this to rank representation gives us:

	PFTITFORTATNSTICKY	NEVERINVEST
PFTITFORTATNSTICKY	3, 3	1, 3
NEVERINVEST	3, 1	2, 2

In the partial feedback case (with these strategies and the Case 1 setup) NEVERINVEST is the safest strategy on the downside, but it foregoes even the possibility of benefit from mutual cooperation. When, if ever, would we enjoin a pragmatically rational agent to play PFGRIMTRIGGER? PFTITFORTATNSTICKY?

14.5 Discussion

IDS situations occur naturally. They present themselves to us as contexts of strategic interaction, as games in the wild. So far, however, we have confined ourselves to discussing IDS situations as represented by models: 2×2 games in strategic form played iteratively. This is appropriate and has been useful, but the confinement is, well, ... confining. Let us foray a little into the wild. To that end, I want to put before us a hypothetical example that serves to convey something of the larger, socially important issues attending IDS situations.

Consider a neighborhood consisting mostly of detached houses, with twins and small apartment buildings mixed in. The structures are situated generally close to one another, the lot sizes tend to be small, but there is variation in this regard. A pleasing profusion of mature trees graces the scene. Thundershowers, and with them lightning, appear with a known regularity. The probability is known that in any given month lightning will strike a building in the neighborhood. If the building is protected with a lightning rod system, there will be no damage. If the building is unprotected, it will catch fire and cause extensive damage. Further, there is a good chance that the fire will spread to neighboring buildings and damage them significantly as well. The probabilities of contagion and the levels of damage resulting are imperfectly understood and vary by lot size, the placement of the trees, and other factors. It is known, however, that such damage will generally occur and be substantial. While residents may invest to protect their buildings from lightning, they have essentially no options, beyond those already embodied in the local fire department, for limiting damage from any fire that results from lightning.

Let us further assume, for the sake of the example, that the cost of protecting a building from lightning is substantially less than the expected cost to the building and associated property. In consequence, the property owners have simply purchased lightning insurance. Even quite ample policies cost much less than the insurance premiums. When lightning strikes and a building is destroyed, the owner simply collects the insurance and rebuilds. Neighbors damaged by the fire spreading from the primary event—the lightning strike on a near-by building—also collect insurance from their fire insurance policies. Prudent smallholders insure themselves against direct hits from lightning and against collateral damage from fire when lightning strikes a neighbor. Nothing can prevent lightning from striking, but when it does the damage is made good and the neighborhood restored. What's not to like?

One more thing: the cost to each property owner of both lightning insurance and fire insurance for lightning-based collateral damage is substantially more than the cost of putting up lightning rods. If everyone installs a lightning rod system, no one will need either of the insurance policies and everyone will be better off. At present we have an insurance regime: everyone resorts to insurance for recovering damages from lightning and for recovering damages from fire resulting from lightning striking another property. If we changed to a protection regime, everyone would have lightning rods and so be protected from all consequences of a lightning strike. And the insurance arrangement is more expensive than the protection arrangement.

What, if anything, can and should be done to move the neighborhood from its current (all puns intended) insurance regime for lightning to the socially cheaper protection regime? We have here an IDS problem. It is a social dilemma with lottery payoffs, complicated by the fact that the high-consequence events (lightning strikes) have low probability in any given year. There are many positions one might take on this question of what to do. Here in outline are a few.

We have a status quo. The first position is to advocate it. But what if not everyone has insurance? Should we chalk this up to individual freedom? What if not everyone can afford insurance? Should we suggest they move to a less expensive neighborhood? No doubt there will be many who would resist any suggestion favoring collective action and would advocate complete *laissez-faire*. What about collective action—say through government funded by taxes—to ascertain the risks, to estimate them precisely, and disseminate the results so that individuals could decide for themselves? Do we think this job can be adequately handled by competition among insurance providers? What about NGOs?

The comment I would make is that the lower the benefit-cost ratio of a universal protection regime, the more attractive this first position is. In terms of a Prisoner's Dilemma game, if the reward for mutual cooperation is not much larger than the penalty for mutual defection, if the R/P ratio is small (or P/R in the case of losses), then we do not have much of a dilemma. Conversely, as the relative benefit of cooperation compared to defection increases, the first position becomes less tenable. Instead of lightning striking a single house, think of a nuclear reactor accident in a densely populated region with many nuclear reactors. Think of other examples. Can we find principles that would help us decide when a line has been crossed and this first position is not credible?

Universal defection in the Prisoner's Dilemma, and social dilemmas in general can be avoided if the players can make an enforceable agreement to cooperate. The second position, then, is to advocate that the owners in the neighborhood find a way to agree to cooperate. To make any agreement enforceable the participants need to invoke a credible enforcement mechanism. When will social norms be enough? Robert Ellickson's field study of cattle ranchers in Shasta County, California, *Order without Law: How Neighbors Settle Disputes* [88], documents intriguing possibilities for relying on social norms. What about contracts freely entered into among neighbors, enforceable in court? What about a law passed by a state or local government? What if the neighborhood is privately governed by a condominium association? Can we expect the association to vote for and enforce investment in a protection regime? In any of these regimes, there are important questions arising regarding how enforcement is to work. How much punishment is needed? How much is desirable? (See [335] for experiments with punishment in the context of IDS games.)

A third position, or suggestion, is to try persuasion in one form or another. Without requiring any more or less explicit agreement, try to establish norms of cooperation with respect to investing in protection. What about celebrity endorsements? These are widely practiced in consumer goods advertising and are known to work there [1]. Can the approach be made to work for IDS situations, for moving them to protection regimes? Who might the celebrities be? Who is going to persuade them to act and on what basis? Will anything like this work internationally? Proponents of collective action to reduce greenhouse gases have hoped to persuade large economic blocs (e.g., the European Union, the United States) to act vigorously and "become virtuous" on greenhouse gas emissions, as a way of persuading other countries to follow suit. The implicit stick has been the potential treat of taxing imports from polluting countries.

14.6   For Exploration

1. Continue the discussion begun in the previous section, §14.5. Elaborate and discuss in more depth any of the three positions sketched there.

2. Continue the discussion begun in the previous section, §14.5. Are there plausible positions distinctly different from the three described in §14.5? Explore—describe and discuss—them.

3. If we take into account the risks associated with the various strategies, how would this affect our choice of strategies? In particular, consider choice under strong risk aversion versus under strong risk seeking. Prospect theory [165] suggests that people are often risk averse on gains but risk seeking on losses. For IDS games with losses, as in Case 1, do you think it is sensible or wise to choose a strategy based on risk seeking? Why or why not?

4. How could we change Case 1 so that payoffs were on balance gains, rather than (as is the case for Case 1) losses? How would this affect your choice of strategy in playing the game? Why?

5. "Depending on the parameter values, particular IDS games will *in expectation* have payoffs that are characteristic of familiar 2×2 games, notably Prisoner's Dilemma and Stag Hunt." What characterizes parameter settings that lead to Prisoner's Dilemma? Stag Hunt?

6. Heal and Kunreuther [145] discuss a number of risk management strategies for IDS situations, such as: collecting more information and making it available; targeting influential players in an IDS context and given them incentive to invest in protection, thereby encouraging others to follow; and mandating insurance. Discuss and critically assess these and other relevant risk management strategies for IDS situations.

7. Define an interesting IDS game setup differing substantially from Case 1. Conduct a series of experiments to investigate how various strategies perform, and report your results. Use the example in this chapter as a model.

8. Can you think of interesting IDS strategies besides the ones described in this chapter? What are they and why are they interesting? Implement them in IDS-2x2-Tournaments.nlogo and investigate their performance.

14.7   Concluding Notes

The originating papers on IDS games are [145, 189, 190]. See also the useful [335]. The work reported in this chapter was undertaken collaboratively with Min Gong, Howard Kunreuther, Yevgeniy Vorobeychik, and Erte Xiao. Thanks to all.

See the works of Christina Bicchieri, e.g., [29], Jon Elster, e.g., [90, 91], and Richard Sennett, e.g., [284], for insights on and discussion of social norms.

Chapter 15

Organizational Ambidexterity

15.1 Introduction

Investments in new product development (NPD) are strategic in the broad sense of being important and consequential. They are also strategic in our technical, game-theoretic sense: the rewards received by a firm for its NPD decisions are in part determined by NPD decisions made by other firms. Our concern in this chapter is, as always in this book, focused on CSIs, contexts of strategic interaction, as conceived of in the theory of games.

The possibility of developing substantive models for "gaming business strategies" [336, page 342] has been in the air for some time and has been explicitly envisioned as a future development by Zenobia et al. [336] in a recent and thorough review of artificial markets. The aim of this chapter is to investigate just such "gaming [of] business strategies" in the context of new product development, and specifically on the subject of organizational ambidexterity with regard to new product development.

Organizational ambidexterity may be defined roughly as the simultaneous pursuit by organizations of exploitation and exploration activities in product and service offerings. A vibrant literature on it has become established in the field of management studies. In a recent comprehensive literature review, however, Raisch and Birkinshaw [251] conclude that despite the increasing number of studies on organizational ambidexterity, the concept is still in the process of developing into a new research paradigm in organization theory. They go so far as to say that "Despite the rapidly expanding number of studies referring to organizational ambidexterity, empirical tests of the ambidexterity–performance relationship remain scarce" and "...the empirical evidence of the organizational ambidexterity–performance relationship remains limited and mixed" [251, page 393].

In fact, the bulk of the extant literature on organizational ambidexterity relies on thoughtful, experience-based conceptualizing and on case studies with convenience samples. There is much room—and need—for additional information on this important subject, including further reflections, new case studies, data, and modeling.

This chapter introduces and discusses an agent-based model (ABM [237]), AmbidexterityStrategyExplorer.nlogo, that captures many of the important aspects and tradeoffs, as identified in the literature, of the exploration–exploitation dilemma faced by firms in the context of organizational ambidexterity. It is this fundamental dilemma that the organizational ambidexterity literature addresses. Our model is hardly complete, but it is public, it is extensible, and it can be used to gain insight into the general domain and into specific hypotheses. Because it is a model, its assumptions are explicit and accessible. By virtue of being a computational model, the assumptions are readily modified. The model, or rather its implementation in NetLogo as AmbidexterityStrategyExplorer.nlogo, is available on the book's Web site. Documentation of the model is contained in the NetLogo application and may be viewed online or downloaded.

We turn now to a review of the essential background for understanding current thinking on organizational ambidexterity.

15.2   The Basic Story

A number of key points, taken together, constitute the core of the received views, or at least theses, regarding organizational ambidexterity. In the interest of brevity, we cite only a few of the many references. The review by Raisch and Birkinshaw [251] is at present the most comprehensive source on organizational ambidexterity.[1] The following points capture the essentials of the extant thought on organizational ambidexterity.

1. Product innovations come in various forms, which may be classified according to how radical, or not, the innovations are from current practice [251].

   Some innovations involve modest improvements over current practice. These are commonly said to be *incremental*. Other innovations, said to be *radical*, consist of very substantial departures from existing products. More radical innovations are normally thought to be harder and costlier to discover and implement than more incremental innovations. Incremental innovations are commonly associated with exploitation-weighted learning, while radical innovations are normally associated with exploration-weighted learning.

   Researchers in organizational ambidexterity often identify three distinct kinds of innovation. The following passage is representative.

   > Great firms compete over time by actively shaping innovation streams. These streams include incremental innovation

---

[1]See also the special issue of *Organization Science* (July–August 2009), which is devoted to organizational ambidexterity.

(e.g., thinner mechanical watches), architectural innovation (e.g., continuous aim gunfire, the ITE [inside the ear] hearing aid, or SMH's Swatch watch), and discontinuous innovation (e.g., Seiko's quartz movement substituting for mechanical movements). By actively managing streams of innovation, firms take advantage of fundamentally new markets for existing technology and proactively introduce substitute products, which, even as they cannibalize existing products, create new markets and competitive rules. [316, page 160]

Both architectural innovations and discontinuous innovations, when they succeed, are usually classified as radical. It is recognized that they involve substantial departures from current thinking, including modes of production and strategic positioning.

2. Products, and ways of producing them, are generally thought to have life cycles.

An innovation appears and after some turmoil becomes established in an industry. There follows a comparatively static period of incremental improvement and refinement. Eventually, a radical innovation of some kind or other appears, disrupts the industry, and results in the displacement of the previous innovation. And then the process repeats itself.

Putting the point hyperbolically,

> ...the only difference between high-technology industries and low is the length of time between the emergence of a dominant design and the subsequent product substitution. [316, page 164]

More straightforwardly,

> To understand how managers can cope with the contradictory requirements of these innovations, and shape evolving streams of innovation, we need to understand technology cycles— patterns of technological change where early product variation ...is followed by the emergence of an industry standard ...and a period of incremental technological change punctuated, in turn, by a subsequent technological breakthrough ...[316, page 160]

3. The timing, appearance, constitution, and other key attributes of product innovations cannot be precisely determined.

This point is largely presumed in the extant literature, which emphasizes the managerial and organizational consequences of the above points and which applies as well when managers may be fairly certain that a development effort will lead to discovery of an implementable radical product innovation.

4. These conditions, above, present organizations with a form of the exploration-exploitation dilemma.

In seeking innovations of any kind, organizations are attempting to adapt to their environments, and

> A central concern of studies of adaptive processes is the relation between the exploration of new possibilities and the exploitation of old certainties .... Exploration includes things captured by terms such as search, variation, risk taking, experimentation, play, flexibility, discovery, innovation. Exploitation includes such things as refinement, choice, production, efficiency, selection, implementation, execution. Adaptive systems that engage in exploration to the exclusion of exploitation are likely to find that they suffer the costs of experimentation without gaining many of its benefits. They exhibit too many undeveloped new ideas and too little distinctive competence. Conversely, systems that engage in exploitation to the exclusion of exploration are likely to find themselves trapped in suboptimal stable equilibria. As a result, maintaining an appropriate balance between exploration and exploitation is a primary factor in system survival and prosperity. [209, page 71]

5. Different kinds of innovations require different kinds of management and organizational arrangements.

The exploitation-exploration dilemma in this context presents both investment dilemmas (How much and when to invest in incremental versus radical innovation?) and managerial dilemmas (Given investment in different kinds of innovations, how best to manage it organizationally?). The organizational ambidexterity literature has been largely, but hardly exclusively, concerned with the latter question.

> Because the power, resources, and traditions of organizations are usually anchored in the more traditional units, these units usually try to ignore, trample, or otherwise kill the entrepreneurial units. Thus, the management team must not only protect and legitimize the entrepreneurial units, but also keep them physically, culturally, and structurally separate from the rest of the organization.
>
> ...
>
> The challenge for managements and their teams, then, is to create co-existing highly differentiated *and* highly integrated organizations. [316, page 171]

And indeed

> Shifts in innovation streams can be executed only through discontinuous organizational change. Managers can attempt to rewrite their industry's rules only if they are willing to rewrite their organization's rules. [316, page 176]

The organizational ambidexterity literature has been less focused on the problems of discovering innovations and very much focused on the problem of *managing* the discovery and implementation ("execution") of the various kinds of product innovations. The current point is central to this literature: different kinds of innovation (incremental versus radical, for example) require different organizational and managerial approaches.

6. In an ambidextrous organization both exploitation-weighted learning (for more incremental innovations) and exploration-weighted learning (for more radical innovations) are supported [251, page 380].

This is a definitional point. The underlying idea is that different goals have different best organizational and managerial arrangements, and that an ambidextrous organization explicitly recognizes the need to pursue different—"inconsistent"— goals. One way of doing so is by fostering separate organizational structures, norms, and cultures for the different main goals it wishes to achieve.

> Organizations can sustain their competitive advantage by operating in multiple modes simultaneously—managing for short-term efficiency by emphasizing stability and control, as well as for long-term innovation by taking risks and learning by doing. Organizations that operate in this way may be thought of as ambidextrous—hosting, multiple, internally inconsistent architectures, competencies, and cultures, with built-in capabilities for efficiency, consistency, and reliability on the one hand and experimentation, improvisation, and luck on the other. [316, page 167]

7. Ambidexterity is (at least in many cases) superior to any homogeneous organizational arrangement.

This, in the context of all of the above, is the *ambidexterity hypothesis* or the *ambidexterity premise* [251, page 392]. The thought is that success requires continual innovation of all kinds—

> The route to sustained competitive advantage is not through succeeding at either incremental, architectural or discontinuous innovation, but through producing streams of innovation. [316, page 165]

—and that the best way to manage multiple forms of innovation is through a heterogeneous organizational arrangement (ambidexterity), rather than any homogeneous formula:

> The source of sustained competitive advantage, and a way
> to avoid being trapped by success, is through building and
> leading ambidextrous organizations. [316, page 215]

It is this approach that best assures the long-term survival and prospering of the organization.

> Through the diversity inherent in ambidextrous organizations,
> managers develop options from which they can shape innovation streams. Organizations can, then, renew themselves
> through proactive strategic reorientations coupled to bets on
> dominant designs, architectural innovation, and/or product
> substitution. [316, page 178f]

Points arising:

A. This account of the basic story is intendedly brief and in consequence it
   leaves out much, for example:

   > The literature [on organizational ambidexterity] has focused
   > on three broad approaches that enable ambidexterity within
   > an organization: structural solutions that allow two activities
   > to be carried out in different organizational units, contextual
   > solutions that allow two activities to be pursued within the
   > same unit, and leadership-based solutions that make the top
   > management team responsible for reconciling and responding
   > to the tensions between the two activities. [251, page 389]

   For useful overviews of the literature see especially [28] and [251].

B. This account, in accordance with much of the literature, takes the perspective of the managers of the firm, not the investors, at least explicitly. The emphasis is on what can be done to insure "the long-term growth and survival" of the organization. Ambidexterity—or the prudent and effective managing two or more learning efforts involving different exploitation-exploration tradeoffs—is thought to promote the survivability of the organization. How can then ambidexterity be most effectively accomplished? How is it in fact accomplished, or attempted?

   These are certainly legitimate and very interesting questions. Surely the organizational ambidexterity literature is on to something important and has made significant progress. These core questions, however, leave unaddressed the question of under what conditions investment ought to be made in ambidexterity. From the point of view of the stockholders, the principals, it may well be best seek a higher-risk strategy and accept that the firm is more likely to fail. The point is touched upon, very lightly and only indirectly, in the review by Raisch and Birkinshaw:

> Thornhill and White ... found that firms with a one-sided fo-
> cus on either cost leadership or differentiation outperformed
> firms with mixed strategies in terms of short-term operating
> margin. They also found that despite pure strategies' observ-
> able performance benefits, the vast majority of firms occupy
> strategic space's middle ground. The explanation was found
> in a follow-up study ...: Although pure players are more prof-
> itable, they also have higher risks and higher exit rates. The
> authors conclude that a middle position may be a rational
> choice that reflects firms' preference for growth and survival
> rather than short-term profit maximization. [251, page 400]

Rational perhaps, but in whose interests? The principals or their agents the managers? Raisch and Birkinshaw do note explicitly that "evidence of the linkage between organizational ambidexterity and performance remains weak" [251, page 399f]. Herein lies opportunity for further research.

C. Also barely considered in the literature, even noting the alternatives presented in point A above, is the possibility of acquiring the fruits of exploratory learning by purchasing them in the open market. Raisch and Birkinshaw, again, touch lightly upon the issue in their review. The context is that of linking the organizational ambidexterity literature to other literatures that seem promising.

> ... external development through acquisitions was found to
> stimulate the exploration of new capabilities ... as well as to
> hard exploitation by diverting attention and resources away
> from internal growth and innovation .... [251, page 398]

Are there effective ways by which an organization may pursue explo-ration through a policy of aggressive environmental scanning and acqui-sition? What are the pluses and minuses of such an approach? Has it been tried and has it worked? Herein, too, lie many opportunities for further research.

D. Neither this basic account, nor it would appear the larger literature, focuses on decision making by the organization with regard to the exploitation-exploration tradeoff. Both the basic account, above, and the reference literature focus instead on the important and core prob-lem of how to *manage* for ambidexterity (see point A above).

E. The basic account, above, is bereft of support from formal modeling. The reference literature is thin on modeling, with some notable excep-tions, which we discuss in the sequel. The comprehensive recent review by Raisch and Birkinshaw [251] does not discuss any formal (analytic or computational) modeling pertaining to organizational ambidexter-ity. This is quite remarkable because a computational model is at the core of one of the founding papers of the (sub)field of organizational ambidexterity, March's 1991 paper, "Exploration and Exploitation in Organizational Learning" [209], which Raisch and Birkinshaw describe as a "landmark article that has frequently been cited as the catalyst for the current interest in the concept" of organizational ambidexterity (page 376).

Briefly for the present, Tay and Lusch [306] present a complex agent-based model, based on fuzzy logic and evolutionary computing, for markets in which firms may pursue various mixtures of exploration and exploitation. Important for the organizational design literature but less directly relevant to our present concerns are several notable papers using NK models, for example, [289], which we also discuss below.

Organizational ambidexterity is avowedly an "emerging" theory [251, page 403]. Bringing to bear formal models holds the promise of affording significant enrichment, especially as the models are connected to other research findings and are made to address gaps in our current understanding.

With the foregoing as context, we offer, beginning in the next section, an "entry level" model for organizational decision making for exploration-exploitation tradeoffs.

15.3   Description of the Agent Model

The purpose of the AmbidexterityStrategyExplorer.nlogo model is to support the investigation of exploration-exploitation tradeoffs in the context of organizational ambidexterity and investments in incremental and radical innovation. The model represents various policies of investment in organizational learning and affords their evaluation in a competitive market. A key aspect of the model is that returns from investments—whether in incremental or in radical innovation—are stochastic as, in consequence, are the rewards from committing to the various policies.

Our aim is to build and explore at first a very simple model of ambidexterity decisions and subsequently to articulate it in light of unfolding theory and experience. We report here on this first model. The salient aspects of the model, with comments, are as follows. Our aim in this chapter is to present a model and a simulation tool to the reader, and indeed to the pertinent research community.

In AmbidexterityStrategyExplorer.nlogo,

1. The industry consists of two or more supplier firms, the number of firms

being set by the user. Each firm produces and sells product 0 to the market.

2. Time unfolds discretely. Each tick of the clock is called an *episode* of activity.

3. There are **NumberOfCustomers** customers in the market for product 0. Each customer has an inherent demand for product 0, which it seeks to satisfy during each *episode* of activity. The inherent demand for product 0, common to all customers, is governed by the parameter **InitialDemand0**, which itself is set by a slider on the Interface tab. The default value is 5. Similarly, each customer has an inherent demand for product 1, which it also seeks to satisfy during each episode. It is governed by the parameter **InitialDemand1**, which itself is set by a slider on the Interface tab and has a default value of 5.

4. The products of each vendor have a price and a quality index, either of which may change over time. Vendors may compete on quality, price, and reduction of cost of manufacture.

5. Each customer focuses on one of the two producers as its primary vendor, using an "epsilon-greedy" choice policy [304]. During any given episode a customer seeks with probability $(1 - \varepsilon)$ to fulfill its requirements for product 0 from its focal supplier. If the focal supplier has sufficient quantity on hand, the transaction is made; otherwise the customer attempts to purchase product 0 from the other supplier. Similarly, with probably ε the customer first attempts to purchase its supply of product 0 from the non-focal vendor. After each transaction each customer records the supplier(s) and product qualities it experienced with each vendor.

6. Each supplier has a unit cost of manufacture and a price for its product 0. Profits (and losses) are accumulated after transactions occur.

7. Each customer reconsiders its focal supplier after a number of episodes have occurred, called the **epochLength** for the customer. At the end of its epoch, a customer will probabilistically refocus on a supplier based on which of the suppliers has provided it higher quality on average. (See [178] for a full discussion of a related learning model, using epochs. We use a parameterized Boltzman distribution to determine the relevant probabilities. See §B.4 and [304] on Softmax.)

8. Each customer also has a latent demand for product 1, which is not offered on the market at the inception of the run. If any vendor comes to offer product 1 the market proceeds in a manner structurally identical to that for product 0.

9. Vendors also organize their time (episodes, ticks of the clock) into epochs. At the end of its epoch a vendor considers whether to invest in

R&D and if so, whether to invest in incremental innovation or in radical innovation. That is, whether to invest in cost or quality improvements in product 0 or in discovering an unknown new product (which we know as product 1). Each investment has a cost, a probability of success, and an incubation period (in number of episodes or ticks). The vendor must pay for the R&D from accumulated profits when the decision is made. Accumulated profits cannot go negative.

10. There is a complexity cost incurred when investing in both an incremental innovation and a radical innovation at the same time. This is in addition to the base cost of each investment. Multiple investments of the same type are permitted without incurring the complexity cost. (The program variable `AmbidexterityCostMultiplier` by default is set at 1.5.)

11. At the end of an investment's incubation period success or failure is probabilistically realized. In the case of success with an incremental innovation the result is an improvement in the quality index of product 0 and/or a reduction in the cost of manufacture. In the case of success with a radical innovation the result is that the vendor may sell product 1 and realize profits, which by default are set quite high.

12. To prepare (initialize) a run (with two firms):

    (a) Key customer parameters are set by default and may be changed by sliders on the Interface tab:

    ```
    set NumberOfCustomers 120
    set InitialProbOfProducer0Focus 0.5
    set CustomerEpochLengthLow 20
    set CustomerEpochLengthHigh 50
    ```

    The number of customers is by default 120, half of whom focus initially on producer 0, the others on producer 1.

    (b) Each customer has a characteristic `epochLenth`, set during initialization. Epoch lengths are uniformly drawn for each customer:
    ```
    epochLength (random (CustomerEpochLengthHigh -
    CustomerEpochLengthLow)) +
    CustomerEpochLengthLow
    ```
    By default the range is (20, 50).

    (c) The two producers are initialized as described below.

13. In each episode (using the `Go` procedure):

    - Both producers replenish their supplies of products 0 and 1, according to the values of their `product0production` and `product1production` variables.

In the default setting case "Base Case: No Investment," these values are initialized from the parameters set by:

```
set Producer0Product0InitialProduction 1000
set Producer1Product0InitialProduction  1000
set P0P1IP 0 ; Producer 0 Product 1 Initial Production
set P1P1IP 0 ; Producer 1 Product 1 Initial Production
```

Then each customer:

(a) Resets its demand for products 0 and 1 to `InitialDemand0` and `InitialDemand1`.

(b) Determines its `firstSupplier` and `secondSupplier` for the episode, using greedy–epsilon: with probability $1 - \varepsilon$ the customer uses its `focalSupplier` as its `firstSupplier` and the other supplier as its `secondSupplier`, and with probability ε, vice versa.

(c) Attempts to fill its demand for product 0 from its `firstSupplier` and then (if the supplier is exhausted) from its `secondSupplier`.

In doing so, when it acquires a unit of supply, the customer updates two accumulators: the count of purchased product from the supplier, `purchasedProduct0Producer0` and the total value received from the supplier during the epoch, `value0Producer0`.

(d) Attempts to fill its demand for product 1 from its `firstSupplier` and then (if the supplier is exhausted) from its `secondSupplier`.

14. To postpare an episode (that is, to undertake processing after completion of each episode):

(a) Producers realize any gains from trade, decide whether to make any investments in development projects, and conclude any projects under way that have completed during the episode.

(b) Customers reset **demand0** and **demand1** for the next episode. Further, if the episode concludes an epoch for the customer, the customer reconsiders its focal supplier, based on quantity and price performance for product 0.

15.4 Explorations with the Model

Our aim in this section is to discuss the behavior of our model implementation, AmbidexterityStrategyExplorer.nlogo, with regard to key strategies the firms have available to them; to use these results in revisiting the hypotheses from the literature, which we have described above; and to discuss the robustness of the model. The model implementation—see Figure 15.1 for a view of the interface with the default settings displayed—is available at the book's Web site, as noted above.

In undertaking this discussion we focus on two scenarios, specified by settings of the model's parameters. The first scenario, called the *default setting* or *scenario*, has the parameter values shown in Figure 15.1. The second scenario, called the *high competition scenario*, alters some of these parameter values, as will be discussed in the sequel.

15.4.1   Strategies

We have identified, and implemented in AmbidexterityStrategyExplorer.nlogo, several basic strategies of interest that firms may employ with regard to ambidexterity and the exploration-exploitation dilemma. They are:

A. NO INVESTMENT

No investment in either exploitation or exploration. Firms simply do business and accumulate retained earnings. Neither firm invests in R&D. This is called the NO INVESTMENT strategy in AmbidexterityStrategyExplorer.nlogo. It, like all of the available strategies, may be selected using the StrategyFirm0 and StrategyFirm1 choosers on the Interface tab.

B. MAX INCREMENTAL INVESTMENT LIMITED, value

Invest available funds exploitatively, in incremental improvements in product 0, with the locus of improvement on value. This is implemented by using the MAX INCREMENTAL INVESTMENT LIMITED strategy in AmbidexterityStrategyExplorer.nlogo and setting LocusOfIncrementalImprovementFirmX (0 or 1) to Value.

Firms following the MAX INCREMENTAL INVESTMENT LIMITED strategy invest available R&D funds in incremental projects, constrained by budget (the cost of the incremental project times RetainedEarningsMultiplier must be less than the current value of RetainedEarnings for the firm), by total number of incremental projects undertaken (IncrementalLimit), and by a limit on the number of projects under way simultaneously (MaxSimultaneousProjects). When an incremental project is successful, the proceeds are invested in improving the value score of the firm's product 0 (by the amount IncrementalProjectValueIncrement).

C. MAX INCREMENTAL INVESTMENT LIMITED, cost

Invest available funds exploitatively, in incremental improvements in product 0, with the locus of improvement on cost. This is implemented by using the MAX INCREMENTAL INVESTMENT LIMITED strat-

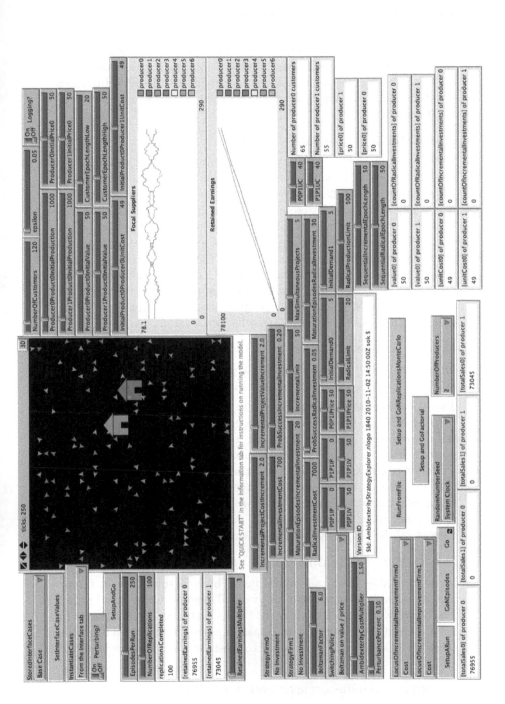

FIGURE 15.1: AmbidexterityStrategyExplorer.nlogo with parameter settings for the default scenario.

egy in AmbidexterityStrategyExplorer.nlogo and setting LocusOfIncrementalImprovementFirmX (0 or 1) to Cost.

This strategy is identical to B, except that when an incremental project is successful, the proceeds are invested in improving the cost score of the firm's product 0 (by the amount IncrementalProjectCostIncrement). The firm's unit cost for product 0 is reduced by this amount; customers see the same value.

D. MAX RADICAL INVESTMENT LIMITED

Invest available funds exploratively, searching for a new product (called product 1) having a latent demand. This is implemented by using the MAX RADICAL INVESTMENT LIMITED in AmbidexterityStrategyExplorer.nlogo. Strategy D resembles B (and C), except that all available R&D funds are directed at radical investments. As in the case of B and C, the firm is limited by budget (the cost of the radical project times RetainedEarningsMultiplier must be less than the current value of RetainedEarnings for the firm), by a limit of 1 on the number of successful radical projects, and by the total number of radical projects undertaken (RadicalLimit). When a radical project is successful the firm begins to offer product 1 to the market and sells to the latent demand.

E. MAX INCREMENTAL LIMITED & RADICAL INVESTMENT LIMITED, value

Invest available funds ambidextrously, both in incremental improvements focused on value and on radical search for a new product.

This is MAX INCREMENTAL LIMITED & RADICAL INVESTMENT LIMITED in AmbidexterityStrategyExplorer.nlogo, with LocusOfIncrementalImprovementFirmX (0 or 1) set to Value. Successful incremental projects lead to improvements in product 0 value, as in strategy B. The conditions for this strategy are the combination of B and D, except that the budget constraint for a prospective project is its cost times RetainedEarningsMultiplier (as before) times AmbidexterityCostMultiplier. Also, when an investment is made, RetainedEarnings is decremented by the project's cost times AmbidexterityCostMultiplier.

F. MAX INCREMENTAL LIMITED & RADICAL INVESTMENT LIMITED, cost

Invest available funds ambidextrously, both in incremental improvements focused on cost and on radical search for a new product. This is MAX INCREMENTAL LIMITED & RADICAL INVESTMENT LIMITED in AmbidexterityStrategyExplorer.nlogo, with LocusOfIncrementalImprovementFirmX (0 or 1) set to Cost, and is otherwise identical to E.

G. SEQUENTIAL AMBIDEXTERITY, cost, I=50, R=50

Invest available funds ambidextrously, beginning with incremental investments for I=50 periods (episodes), followed by radical investments for R=50 periods, then repeating until the run is completed. During the incremental periods, the firm behaves as a C player; during the radical periods, as an F player. There is no extra cost for following this kind of ambidexterity. Unlike E and F, the AmbidexterityCostMultiplier is not applied.

15.4.2 Results

Given these seven distinct strategies and two players there are 7×7 combinations, but 21 of these are duplicates because of symmetry, leaving 28 distinct combinations of interest. Working with the default scenario of the model (Figure 15.1), we conducted computational experiments for the 28 distinct cases of interest. The results are summarized in Tables 15.1 and 15.2. We draw the reader's attention to the following key points.

1. Notice in Tables 15.1 and 15.2 the large variability associated with strategy D (pure exploration). Its means are often very different from its medians. Exploration is inherently risky. While *for these parameter settings* exploration pays off on average over 100 repetitions, in any given run a firm following the D strategy is more likely not than to discover product 1 and so reap extra rewards. This is, we submit, an indicator that the default settings have captured an interesting and indeed typical situation: radical improvement is risky.

2. Comparing strategies BB and EE (both of which invest in product values) with CC with FF (their analogs that invest in cost reduction), Tables 15.1 and 15.2 reveal very poor performance (from the firm's perspective) of B compared to C and E compared to F. For these parameter settings, a firm that invests in improving product value will succeed in drawing customers away from a firm that does not do this. But when both firms invest in value the customers gain and the firms lose.

Stra-tegy Pair	Retained Earnings by Firm							
	Firm 0				Firm 1			
	1st Qu.	Median	Mean	3rd Qu.	1st Qu.	Median	Mean	3rd Qu.
AA	73,422	74,860	74,788	76,154	73,846	75,140	75,212	76,578
AB	62,299	66,820	65,686	68,122	46,178	47,480	48,614	52,001
AC	73,914	74,870	74,870	76,692	309,997	374,569	385,821	465,054
AD	73,216	75,188	75,143	76,750	16,596	19,185	159,637	178,065
AE	61,929	64,302	64,367	67,462	22,770	26,090	26,098	28,870
AF	73,049	74,475	74,694	76,421	328,256	564,962	574,461	843,266
AG	73,660	75,150	75,144	76,665	135,948	375,487	477,710	911,411
BB	36,326	40,005	39,943	43,268	35,332	38,595	38,657	42,274
BC	46,802	48,978	48,820	51,270	246,757	303,007	303,328	358,963
BD	47,156	49,602	49,122	51,171	16,018	18,995	147,668	216,480
BE	34,142	37,810	38,296	43,015	17,735	22,310	21,824	25,652
BF	46,231	48,358	48,197	50,519	193,190	456,073	474,557	742,239
BG	47,070	49,378	48,438	49,295	90,461	371,743	438,438	841,162

TABLE 15.1: Part 1: Retained earnings by firm for the default scenario, by strategy pair. Summary for 100 repetitions. Strategy pair XY = firm 0 plays X, firm 1 plays Y.

Stra-tegy Pair	Retained Earnings by Firm							
	Firm 0				Firm 1			
	1st Qu.	Median	Mean	3rd Qu.	1st Qu.	Median	Mean	3rd Qu.
CC	317,180	366,936	377,334	455,405	312,932	395,075	387,162	462,246
CD	311,402	394,545	381,793	448,292	16,474	19,028	169,936	286,329
CE	249,279	300,125	298,754	338,166	22,915	25,022	25,816	29,079
CF	313,212	383,514	385,360	465,394	292,268	576,251	574,337	842,472
CG	305,989	367,563	372,559	444,879	145,286	429,363	475,752	841,311
DD	17,014	19,335	130,363	132,308	16,541	19,372	152,537	195,444
DE	16,069	18,012	115,067	20,260	22,975	26,152	26,233	29,528
DF	17,549	19,682	135,313	159,849	291,225	601,782	599,003	870,801
DG	16,419	18,800	103,332	20,902	151,924	411,906	447,794	654,381
EE	16,904	21,660	20,956	24,871	17,482	21,140	20,684	24,859
EF	24,044	26,512	26,540	29,361	185,665	540,229	505,992	784,069
EG	23,760	26,495	26,267	28,518	77,355	357,159	409,944	765,089
FF	259,302	491,031	495,980	691,248	365,582	538,092	550,693	739,873
FG	272,362	460,057	520,885	755,181	176,293	291,650	419,337	752,812
GG	134,650	315,904	363,669	471,603	147,883	338,640	408,409	638,622

TABLE 15.2: Part 2: Retained earnings by firm for the default scenario, by strategy pair. Summary for 100 repetitions. Strategy pair XY = firm 0 plays X, firm 1 plays Y.

	A	B	C	D	E	F	G
A	*74,788*	*65,686*	*74,870*	*75,143*	*64,367*	*74,694*	*75,144*
	75,212	*48,614*	*385,821*	*159,637*	*26,098*	*574,461*	*477,710*
B	49,164	*39,943*	*48,820*	*49,122*	*38,296*	*48,197*	*48,438*
	65,136	*38,657*	*303,328*	*147,668*	*21,824*	*474,557*	*438,438*
C	359,488	289,871	*377,334*	*381,793*	*298,754*	*385,360*	*372,559*
	75,086	48,596	*387,162*	*169,936*	*25,816*	*574,337*	*475,752*
D	158,324	152,973	150,791	*130,363*	*115,067*	*135,313*	*103,332*
	74,896	48,147	359,898	*152,537*	*26,233*	*599,003*	*447,794*
E	26,007	21,970	26,972	26,109	*20,956*	*26,540*	*26,267*
	64,143	38,045	293,801	117,136	*20,684*	*505,992*	*409,944*
F	579,823	482,080	540,453	567,876	461,136	*495,980*	*520,885*
	75,377	48,783	374,464	93,695	26,626	*550,693*	*419,337*
G	452,287	399,737	439,972	405,844	361,118	309,873	*363,669*
	75,091	48,733	364,322	109,600	25,907	547,193	*408,409*

TABLE 15.3: Pairwise strategy comparison: mean retained earnings (firm 0 row, top; firm 1 column, bottom). All runs using the default settings. Results in *italics* are based on data in Tables 15.1 and 15.2.

Table 15.3 extracts data from Tables 15.1 and 15.2, and adds results from symmetric runs. We should think of Table 15.3 as presenting the (always stochastic) payoffs of the seven strategies (A–G) in strategic form, as is standard in game theory.

Examining the table tells us much about is going on (in this model with these parameter settings). AA, with both firms declining to invest in R&D, constitutes a baseline case; each firm earns about 75,000 units on average during a 250 period run. If either firm switches to strategy B (AB or BA), then *both* firms are worse off, although the A player fares better than the B player. By investing the proceeds of successful incremental R&D projects in improving the value of its product 0, a B-playing firm will draw customers from an A-playing firm but not enough to actually pay for the investment in R&D. Here the B strategy cannot be justified by simple return on investment, although it may have net strategic value for the damage it does to the counter-player. Notice (again) that if both firms play B, both are substantially worse off than if they both play A.

CC is the strategy combination of both firms making incremental investments in R&D and using the results to reduce their production costs. Because neither firm improves the value/cost ratio for product 0 from the customer's perspective, each firm retains about half of the customers and both firms do quite well compared to AA. Notice that if one firm opts for B (BC or CB) to draw business away from the other firm, the defecting firm is both comparatively and absolutely worse off.

D (investing in R&D for a radical new product) is on average quite profitable compared to the baseline AA scenario, although its value is suppressed somewhat by other radical investment strategies (D, E, F and G), which may lead to sharing the market for product 1 with the counter-player who may also discover the new product.

E and F are our simultaneous ambidextrous strategies, improving on value and cost, respectively, when incremental projects are successful. We see that E is by far the *worst* of these six strategies (for the default parameter settings). G is our sequential ambidextrous strategy. It is the analog of F, in which proceeds are invested in cost reduction rather than value improvement for product 0. (We do not find the sequential analog of E worth reporting.)

Notice that the total retained earnings from FF (= 495980 + 550693 = 1046673) is essentially identical to that of CC + DD (= 377334 + 387162 + 130363 + 152537 = 1047396) in spite of the fact that ambidexterity (in this model with these parameter settings) imposes an extra 50% cost burden on every R&D project (AmbidexterityCostMultiplier = 1.50). How is this possible? Investment in R&D is always limited by available funds. We represent this in our model by requiring that the cost of a project times the AmbidexterityCostMultiplier (if applicable) times the RetainedEarningsMultiplier (=3 in the default settings) be less than the retained earnings on hand. What the model teaches here is that ambidextrous firms may be gaining because they use incremental improvements to increase funding for radical R&D, thereby broadening and deepening their search processes and increasing the chances of finding a substantial return. Thus the model is demonstrating in a computationally rigorous way a phenomenon suggested by observations from case studies.

Returning to E, the explanation of its performance complements that of F. In the case of E, retained earnings used for incremental projects reduce the funds available for radical projects, greatly reducing the chances of a radical success because fewer projects are funded. The behavior of G, our model of a sequentially ambidextrous strategy, coheres with this basic story. G does very well, but not as well as F. The reason for this is apparent in the log files. G strategists consistently field fewer incremental R&D projects *and* fewer radical R&D projects than F strategists. What limits the Gs is time and available funding. In a nutshell, incremental projects whose successes are invested in cost reduction are very profitable. By (in this case) halving the time during which incremental projects can be undertaken, fewer are initiated, resulting in lower levels of retained earnings that can be invested in radical projects, leading to lower probabilities of making radical discoveries. Even the 50% project cost penalty in the F strategy for simultaneous ambidexterity is not overcome by this handicap in the sequential G strategy.

Stepping back and viewing Table 15.3 as a game in strategic form it is quite clear that F, ambidexterity with incremental successes invested in cost reduction, is the best response to each of the seven strategies, although its value will be suppressed somewhat by "value" strategies B and E. G is also an excellent strategy, but it has no evident advantage over F. Ambidexterity in either form wins. E, on the other hand, is a poor strategy, simply because the profits from investment in R&D are given to the customers.

These results were obtained, of course, for a single scenario and it is true that with sufficient variation of the default settings very different results can be obtained. For example, if the rewards from incremental investment are small enough, then D would become a dominant strategy. It can happen. In this regard we make the following points:

1. We believe that the default parameter settings are plausibly realistic and applicable in some cases.

2. Under mild perturbations of the default parameter settings, the model is quite robust.

3. While it would be entirely inappropriate to claim any sort of veridicality with predictions based on such subjectively assessed parameter values as are on display here, it is appropriate to use the settings for exploration seeking insight and for hypothesis generation, which is our main purpose.

4. Other scenarios or parameter settings are easily investigated. We have made the model available and accessible and we encourage the reader to explore with it. In particular, we note that the AmbidexterityStrategy-Explorer.nlogo implementation may be used in DSS fashion when data or subjective judgments are available to create credible scenarios.

5. Something the model teaches us is that outcomes and strategic values will depend upon many parameters, even in this simplified case. These parameter values are unlikely to be constant across all industries, firms, and times. Correct understanding of ambidexterity phenomena very likely will need to be conditioned upon particular ranges of parameter values. The model helps us understand which are most crucial and to "test drive" hypotheses.

15.4.3 Tournament with the Base Case Scenario

Abbrev.	Strategy	Locus of Incremental Improvement
A	No Investment	Cost
B	Max Incremental Investment	Value
C	Max Incremental Investment Limited	Cost
D	Max Radical Investment Limited	Cost
E	Max Incremental Limited and Radical Investment Limited	Value
F	Max Incremental Limited and Radical Investment Limited	Cost
G	Sequential Ambidexterity	Cost

TABLE 15.4: Table of strategy abbreviations.

	Min.	1st Qu.	Median	Mean	3rd Qu.	Max
A	64840	72578	75205	75387	78208	88020
B	27980	37015	39928	39764	42532	51710
C	182270	304266	371326	380971	458546	655072
D	14125	15820	19032	163924	283864	825005
E	14055	18574	21782	21547	24848	29910
F	22855	270437	474369	525146	780483	1387228
G	35126	94084	357079	408378	561835	1220665

TABLE 15.5: Tournament results for the Base Case scenario, all seven strategies.

15.5 Robustness Tests

15.5.1 Second Scenario

With these comments in mind, let us briefly examine a second scenario. The literature on organization ambidexterity attends, as we have seen, to degrees of competitive intensity. We can model this in AmbidexterityStrategyExplorer.nlogo by altering the rate at which customers switch focal suppliers when they encounter improved value/cost ratios. Our *high competition*

	Min.	1st Qu.	Median	Mean	3rd Qu.	Max
C	202038	301809	378318	381391	459826	653026
F	23521	298908	485031	553475	832346	1240172
G	50292	166291	409848	431156	538404	1139547

TABLE 15.6: Tournament results for the Base Case scenario, top three strategies from the all strategies tournament play a new tournament.

	Min.	1st Qu.	Median	Mean	3rd Qu.	Max
A	238659	279741	302213	302718	320074	374745
B	204114	249355	264254	267153	285059	329322
C	728164	1399597	1625969	1671527	1885266	3229406
D	86662	181927	813511	704885	1103019	1330547
E	21339	31186	250950	356606	664714	1040484
F	887104	2428558	2889973	2903179	3318421	4434431
G	499655	1238386	1766074	1753147	2202246	3152097

TABLE 15.7: Tournament results for the Base Case, Scaled 1 scenario. All seven strategies are in play, with Monte Carlo perturbing ±10%.

scenario is like the default scenario, with the following changes: CustomerEpochLengthLow=10, CustomerEpochLengthHigh=20, epsilon=0.03, BoltzmanFactor=0.6. (Note that in the default scenario the BoltzmanFactor is 6.0, so this is an order of magnitude change in that parameter value.) The upshot of these changes is that customers switch with great alacrity, but of course this only applies when strategies B and E are being played since it is only these strategies that affect what the customers see with regard to product 0.

Table 15.9, analogous to Table 15.3, contains the results. In the default scenario a non-value player playing against a value player could expect to lose more than 50% of its customers. In the high competition scenario, essentially all of the customers would be lost by about 100 episodes into the run. As a small point, notice that playing B instead of A against A now has a positive payoff. The B player benefits on balance from using its R&D investments to attract customers away from A. However, it remains the case that C is a much better strategy to play against A, taking into account only retained earnings. The larger picture is that even with this drastic change in customer behavior F remains an excellent strategy. Although some of the payoffs are substantially changed, F clearly remains the best response to all strategies, possibly excepting E. From the payoffs reported in the table, D is the best response to E, but only by a small margin, and E remains a foolish strategy, one that is dominated by F. Ambidexterity with investment in cost reduction is robust to customer stickiness.

	Min.	1st Qu.	Median	Mean	3rd Qu.	Max
C	603924	1330986	1696134	1700011	1984432	2833093
F	608919	2215361	2703093	2754812	3314945	4967593
G	337750	1321094	1637841	1653926	1999148	2970698

TABLE 15.8: Tournament results for the Base Case, Scaled 1 scenario. The top three strategies from the all strategies tournament play a new tournament, with Monte Carlo perturbing ±10%.

	A	B	C	D	E	F	G
A	74,946	28,627	75,252	75,160	25,903	74,932	75,203
	75,054	85,673	382,301	166,050	36,342	553,586	452,061
B	86,753	42,219	86,785	87,091	40,749	85,724	87,830
	27,547	40,322	47,269	53,559	22,808	61,732	88,287
C	378,753	49,717	382,915	400,250	37,618	375,670	377,287
	74,909	86,949	387,754	171,962	45,277	604,008	471,131
D	138,176	28,874	120,081	145,639	41,172	117,352	121,678
	74,944	87,226	371,785	91,251	43,803	557,386	410,998
E	50,355	18,927	47,278	36,509	16,870	41,187	35,214
	26,600	40,733	36,999	62,526	20,947	44,608	102,127
F	562,677	50,800	621,758	517,172	34,075	482,342	517,920
	75,028	87,145	375,839	148,277	41,560	513,677	406,401
G	483,775	95,466	421,578	459,400	111,577	403,138	447,312
	74,934	86,962	392,205	93,347	44,495	525,545	349,996

TABLE 15.9: Pairwise strategy comparison: mean retained earnings (firm 0 row, top; firm 1 column, bottom). All runs using the default settings modified for a "high competition" scenario: CustomerEpochLengthLow=10, CustomerEpochLengthHigh=20, epsilon=0.03, BoltzmanFactor=0.6.

15.5.2 Hypotheses

We now visit briefly the hypotheses that may be derived from the literature.

1. The ambidexterity hypotheses.

   (a) *The simultaneous pursuit of both exploitative and exploratory activities, or ambidexterity, positively affects firms' performance.*

   (b) *Firms that invest in both exploitative and exploratory activities, or ambidexterity, perform better than those only in one or the other activity.*

These hypotheses are confirmed—or perhaps better to say *instantiated*—by our model and the scenarios we investigated, but with some qualifications. Investing in incremental R&D to improve quality (or reduce price) is seen here as a very poor strategy for maximizing retained earnings.

This is because the benefits of innovation are paid for by the producer and enjoyed by the customer. Moreover, as demonstrated by strategy E, firms that fail to avoid this trap are starved of funds that might be used for radical exploration and subsequent higher profits. If a competing firm opts for B, a value-directed strategy, does this create a competitive necessity to follow suit? Not in either of our two scenarios. Even in the extreme case of the second, high competition scenario in which customers move rapidly to abandon inferior offerings, D, F, and G remain better responses than either B or E. The situation is unfortunate from the firm's perspective, but the best response remains to search for another market. Of course, these remarks do not represent findings in the real world; rather, they are credible hypotheses occasioned by the model.

2. Sequential ambidexterity.

   (a) *The sequential pursuit of both exploitative and exploratory activities, or ambidexterity, positively affects firms' performance.*

   (b) *Firms that invest in ambidextrous activities sequentially perform better than those that invest simultaneously.*

In our model and the scenarios we investigated, we find ample support (or clear instantiation) for hypothesis (2a). Strategy G is a robust strong performer. We are not able, however, to instantiate hypothesis (2b), for reasons discussed above. With these parameter settings, including a 50% cost penalty on simultaneous ambidexterity (strategy F), the opportunity cost of undertaking fewer projects because of sequential arrangements is simply too high. Of course, different parameter settings may yield different results, however robust the present scenarios are. What the model does achieve is a principled questioning of hypothesis (2b) and a means to incorporate more precise empirical findings as they become available. As estimates of project opportunity costs become well grounded they can be incorporated into revised scenarios and used to revisit the question. See also our discussion of hypothesis (3) for further insight about G, sequential ambidexterity.

3. Environmental competitiveness.

   As environmental competitiveness increases, the positive relationship between organizational ambidexterity and firm performance is strengthened.

Our model and the findings we have extracted from it suggest the need for a nuanced, even qualified, interpretation of this hypothesis. On the face of it, if the environment becomes more competitively intense profits can be expected to decline whether or not radical exploration is undertaken. Further, if results from incremental projects are used to fund exploration, and as a result of a declining competitive situation these results become less profitable, then there will be less money to support exploration. Comparing the best responses to strategy B (another surrogate for high competition) in Tables 15.3 and 15.9 we see that things are uniformly *worse* for the firm in the high competition scenario. Compared to the baseline case of AA in the high competition scenario, only G as a response to B yields improved profits for the firm. If we focus on improved profitability, then under lower competition (the default scenario) non-ambidextrous strategies can be improving, but under high competition only G, the sequentially ambidextrous strategy improves profits, although F leads to less diminishing of profits than either C or D.

Why does sequential G do better than simultaneous F as a response to B in the high competition scenario (Table 15.9)? Under the F strategy (as implemented) more costly exploratory projects have priority. They are initiated during an episode if funding is available. Incremental projects are then initiated if funding is available, but as a second priority. Given that the B player is stealing away the customers, it would be better for an ambidextrous player to undertake exploitative projects and grab the profits while there are still customers to exploit. Then, after building up funds, investment in radical projects can begin. G approximates this. We note that we ran G′ = SEQUENTIAL AMBIDEXTERITY, cost, I=70, R=175 against B, under the conditions of Table 15.9. B's averaged return (over 100 repetitions) was 87,079 while G′'s was 107,747. Sequential ambidexterity may be advisable in a dynamic environment.

15.6 Discussion

The parameters of AmbidexterityStrategyExplorer.nlogo are not calibrated to any real data. That is, the many parameters of the model have not been set by fitting them to real-world observations. In this regard, AmbidexterityStrategyExplorer.nlogo is like many other agent-based models, some of which are widely accepted as providing significant scientific value, for example Schelling's models in [278] especially his segregation model and all of the models in Epstein and Axtell's *Growing Artificial Socieities* [93]. What, then, can we say about the scientific value of the model to hand, AmbidexterityStrategyExplorer.nlogo? We note the following points.

1. As noted in the important review by Raisch and Birkinshaw [251], real-world data that are applicable to understanding organizational ambidexterity are in short supply. While obtaining new relevant data is always valuable, in the absence of such data other forms of knowledge can be very useful. Modeling is surely one such form.

2. We may think of a formal model as a device that reliably determines consequences of the assumptions that go into it. AmbidexterityStrategyExplorer.nlogo is no exception. Even very approximate assumptions may have surprising and useful consequences. We have tried to give some illustrations of this in the case of AmbidexterityStrategyExplorer.nlogo. Recalling two:

   (a) In the experiments summarized in Tables 15.1 and 15.2 we noted the high variability of the "Max Radical Investment" strategy, compared to "No Investment." Although its *mean* retained earnings are nearly double, its *median* is about one-fourth of "No Investment." While different parameterizations of the AmbidexterityStrategyExplorer.nlogo model will of course yield different results here, our parameterization is sensible and robust, and serves to draw attention to, and explain, an important aspect of exploratory product development.

   (b) AmbidexterityStrategyExplorer.nlogo provides computationally an explanation of why ambidexterity can be valuable to an organization: exploitative returns may be used to fund exploratory investigation and development. This is an example of an "emergent" property. It is a consequence of the model, but a surprising one and not explicitly designed into the model.

3. If real data are not available, subjective data can be used to "ballpark" the model, to gain insights into the phenomena, and to provide something of a "sniff test" for the entirely informal models or hypotheses in the literature. Reliance on subjective data, supplemented with robustness studies, is afforded by AmbidexterityStrategyExplorer.nlogo and other agent-based models. Moreover, this is a time-honored, well-established approach to practical decision support. In this regard, agent-based models by themselves break no new ground.

   As is also the case in traditional modeling, an important use of subjective data is for robustness studies, which may then be used to direct more careful data collection and model calibration.

4. While AmbidexterityStrategyExplorer.nlogo is not calibrated to real-world data, it is in principle *calibratable*. For example, if we take an episode to correspond to a week in the real world (this was our intention in setting the default parameters, however roughly), then it should be possible in a given case to estimate differences in quality or price for a product from two different vendors, and then to calibrate (or test and replace) the Boltzman function used to determine the speed at which customers will change focal vendors. All or nearly all of the possible calibrations would, we think, be specific to particular cases. Even so, this hardly mitigates against the calibratability of the model.

5. Perhaps the most important aspect of AmbidexterityStrategyExplorer.nlogo (and other agent-based models) is that it is both procedural and public. In virtue of being procedural, one can assess whether the procedures are credible. If they are, calibration can be pursued. If they are not, since the model is public and open to inspection, the model may be modified to improve the plausibility of any of its basic procedures. In this regard, the model is highly *decomposable*. The incremental investment procedure may, for example, be changed without having to change the customer switching procedure. All of this affords using the model to suggest testable hypotheses.

In sum, AmbidexterityStrategyExplorer.nlogo contains, we believe, credible procedural models of how firms invest in exploitative and exploratory product development projects. Given credible procedures, even without close calibration with real-world data, the model can be very useful and can support surprises and insights. If procedures are questioned or found wanting, they can incrementally be replaced with more satisfactory versions. In any event, the model is "a tool to think with" regarding organizational ambidexterity.

15.7 For Exploration

1. Discuss how the various parameters in the AmbidexterityStrategyExplorer.nlogo model might be determined. For which are firms likely to have data that could be used to estimate the parameter values? For which would subjective assessment likely be necessary? Who are the subjects that would provide this information?

2. Determine, as best you can, realistic values of the model's parameters, pooling subjective assessments as needed. Then conduct an investigation to compare the various strategies. What do you find?

3. Discuss ways in which the model might be extended or modified. Which do you think would be most interesting and why?

4. Discuss how more extensive sensitivity or post-solution analysis would be helpful for supporting decision making with this model or a model like it.

15.8   Concluding Notes

Duncan [82] first introduced the concept of organizational ambidexterity and asserted that a dual organizational structure is needed to initiate innovation and implement innovation. As a further step, March's [210] influential paper proposed exploitation and exploration as two different learning activities which should be pursued in balanced ways but, at the same time, may compete for organizationally scarce resources. Based on March's work, many scholars applied exploitation and exploration concepts into their research in different contexts such as organizational learning, technological innovation, organizational adaptation, strategic management, and organizational design (Ancona et al. [8], Atuahene-Gima [12], Benner and Tushman [28], Burgelman [51], Gupta, Smith, and Shalley [136], He and Wong [144], Katila and Ahuja [166], Levinthal and March [193], Raisch and Birkinshaw [251], Tushman and O'Reilly [314], Tushman and Smith [315]).

In spite of conflicting, or at least varying, definitions, the literature agrees that exploitation and exploration refer to very different learning activities within an organization. Exploitation includes activities associated with refinement, extension or improvement of current components, competences, and technologies; exploration includes activities including experimentation, innovation or a shift to a different technological trajectory (Benner and Tushman [28], March [210]). A combination or simultaneous pursuit of exploitation and exploration in an organization is defined as ambidexterity (Tushman and O'Reilly [314], He and Wong [144], Lubatkin et al. [202]). Studies including concepts such as "reconciling exploitation and exploration," the "simultaneity of induced and autonomous strategy processes," "synchronizing incremental and discontinuous innovation," and "balancing search and stability" belong to the same stream of literature (Raisch and Birkinshaw [251]).

Because of the complex nature of organizational ambidexterity, the linkage between organizational ambidexterity and performance remains controversial. On one hand, following March's assertion that firms run the risk of being mediocre at both exploitation and exploration resulting from the inherent challenge of doing both, some scholars propose to pursue only one direction instead of both (Barny [21], Ghemawat and Ricart i Costa [120]). Firms simultaneously pursuing both activities are likely becoming internally inconsistent so as to lead to inferior performance (Wernerfelt and Montgomery [331]). The knowledge processes of exploitation and exploration are contradictory because they tap different administrative routines and managerial behaviors. As suggested by Lubatkin, Simsek, Ling, and Veiga [202], exploitation primarily involves learning from a top-down process because managers are used to those organizational routines, while exploration involves a bottom-up process which managers are persuaded to abandon their old routines and make a change or take innovative action.

On the other hand, Tushman and OReilly [314] point out that firms capable of simultaneously pursuing exploitation and exploration are more likely to achieve superior performance than firms only emphasizing one. Levinthal and March [193] also argue that "an organization that engages exclusively in exploration will ordinarily suffer from the fact it never gains the return of its knowledge ... an organization that engages exclusively in exploitation will ordinarily suffer from obsolescence." In other words, the exclusive pursuit of exploration may end up with endless search efforts or R&D expenditure without appropriate return (Raisch and Birkinshaw [251]). In contrast, the exclusive pursuit of exploitation may enhance short-term return but may lead the organization to being incapable of adapting to a new environment (Levitt and March [195]). Firms should exploit existing competencies and explore new ones and these two activities are inseparable (Floyd and Lane [100]).

Besides these conceptual works, there are a few empirical studies regarding this topic. Knott [180] found that exploitation and exploration coexist in Toyota's product development, and concluded that these two activities are likely to be complementary. Katila and Ahuja [166] found a positive interaction between exploitation and exploration on new product development but did not test their effects on firm performance. In their empirical work, He and Wong [144] found that the interaction between explorative and exploitive innovation strategies is positively related to sales growth rate, and the relative imbalance between explorative and exploitative innovation strategies is negatively related to sales growth rate at the firm level. Gibson and Birkinshaw [121], focusing on the business unit level, found that the capacity to simultaneously achieve alignment and adaptability positively affects performance. Lubatkin et al. [202] suggest that the joint pursuit of an exploitative and exploratory orientation positively affects performance in small- and medium-sized enterprises. Other empirical work, however, did not support the ambidexterity-performance linkage. Rather, it found that temporal cycling between exploitation and exploration has a positive effect on firm performance.

Despite the rapidly increasing number of studies concerning ambidexterity, the empirical data for the ambidexterity-performance remain relatively scarce (Raisch and Birkinshaw [251]), and the causal link has neither been theoretical clear nor empirically established (Lubatkin et al. [202]).

The previous literature exploring formal models pertaining to organization ambidexterity is thin. Siggelkow and co-workers have produced some excellent related work, based on implementations of Kauffman's NK model framework (e.g., [258], [259], [288], and [289].) See also [194] for an early treatment of NK modeling in the context of organizational studies. The aims of these papers are quite distinct from those of the present work, focusing instead on such matters as comparing regimes of information flow within an organization. The work by Tay and Lusch [306] is perhaps closest in outlook to the model reported here. It offers a complementary perspective, assessing how a population of firms will behave and imputing to them fuzzy-logic-based learning.

See [306] for a different kind of ABM, one that models organizational ambidexterity, but without a strategic focus in our sense.

* * *

The material in this chapter is based on joint work with Professor Christine Chou. I am very grateful for her contributions and collaboration. See also [63].

Chapter 16

Bargaining

16.1 Introduction

There is much interest in modeling human learning in strategic contexts, both in economics [55, 94, 227, 263] and in cognitive science, e.g., [191, 257]. Previous efforts have focused on modeling learning in repeated extensive and strategic form games. Here, we extend the literature by modeling the outcomes reported in an experimental study of a characteristic function form game [163].

Our goal in this chapter is to investigate whether agents endowed with a minimal learning rule can approximate play by humans in the coalition formation game of [163]. As part of the investigation we compare two action selection rules—Greedy and Matching—described in detail below. Because the human data available are limited, our approach is not to produce entire trajectories of play, but to see if the overall behavior of our agents generally corresponds to that of humans.

Previous work [83, 84, 85, 86] used genetic programming to explore coalition games similar to those in this chapter, where agents were explicit strategies in the space of offers and players. However, such agents cannot be said to have a cognitive component, as each agent represents a single rule. The agents are *strategy-centric*. In contrast, our agents represent players of low or minimal rationality, who must make decisions on how much to offer and to whom. Our agents are *identity-centric* in that a particular agent may change its strategy while maintaining its identity. Finally, while characteristic function games have been investigated widely in terms of outcomes or solutions through both experimental and mathematical approaches (see [164, 318] for overviews), less attention has been paid to the learning processes associated with such games.

In the next section, we describe our coalition formation game, and we present the experimental study of [163]. After that, we describe our model, and present our procedural modeling results. Although we only explore three-person games with a certain characteristic function, we can hope our results generalize to games with larger numbers of players.

16.2   Coalition Formation

16.2.1   Description

In our coalition formation game with a set of n players P, any of the players can join together to form a *coalition* $S \subseteq P$. Such a coalition can attain a guaranteed payoffof $v(S)$, called the *coalition value*, where v is defined over all coalitions and is known as the *characteristic function* of the game; along with P, this defines the coalition game. While it generally is advantageous for a player to be included in some coalition,[1] it is up to each member of the coalition to secure a portion of $v(S)$ for herself or himself. Once a coalition S forms an agreement of how to split $v(S)$, then this agreement is enforced. While some coalitions may have greater guaranteed payoffs than others, individual players should be drawn to those coalitions where they can attain the greatest individual reward.

To provide an example, imagine a situation in which there are three researchers (players), A, B, and C, each of whom has some resource needed to run a study. No researcher can run the study alone (so that $v(A) = v(B) = v(C) = 0$), but any two of them can collaborate to run a study. Assume that the combined resources of A and B permit them to run the "best" study (say $v(AB) = 95$); that A and C can run the next "best" one (let $v(AC) = 90$); and B and C the worst one (let $v(BC) = 65$). Assume further that the coalition ABC does not yield any value, so that $v(ABC) = 0$ (there is a limit on the number of researchers who can be involved). Thus, the characteristic function v has been defined completely, and so the situation constitutes a coalition game.

Now researcher A must ask the question, "With whom should I propose a coalition, and how should I propose to allocate the resources assigned by v?" For example, A might be greedy and propose the split $(65, 30, 0)$, where A receives 65 (say, in units of recognition), B receives 30, and C is not included in the *winning coalition* AB (note $65 + 30 = 95 = v(AB)$). But realizing that C would be better off receiving even a small payoff, B might then propose the allocation $(0, 40, 25)$ for the coalition BC, where both B and C do better than they would under A's proposal. The bargaining might continue, with C trying to increase her payoff by proposing the allocation $(60, 0, 30)$ (excluding B from the coalition). This last proposed allocation is special because neither player in the coalition (e.g., A) can make an offer to the excluded player (B) without the excluded player (B) being able to make a counteroffer to the remaining player (C) in which they both do better than under the original

---

[1] If the condition $v(AB) > v(A) + v(B)$ holds for all players (where AB is shorthand for the coalition $\{A, B\}$), then v said to have the property of superadditivity. Such a set of relations specifies that players achieve a higher joint payoff in a coalition compared to the sum of their payoffs when acting alone, and represents comparative advantages. Superadditivity holds for all games studied here.

	Game:				
	I	II	III	IV	V
Char. Function					
$v(AB)$	95	115	95	106	118
$v(AC)$	90	90	88	86	84
$v(BC)$	65	85	81	66	50
Quota Values					
ω_A	60	60	51	63	76
ω_B	35	55	44	43	42
ω_C	30	30	37	23	8

TABLE 16.1: Characteristic function and quota solutions.

allocation; in this sense, the proposed split is *stable* [13]. For example, say A was not satisfied with the proposed allocation $(60, 0, 30)$, and decides to propose a coalition with B with the split $(61, 34, 0)$. In this case, B could offer $(0, 35, 30)$ to C, which is another stable split. Along with $(0, 0, 0)$ (where no players form a coalition) and $(60, 35, 0)$, these stable allocations form the solution concept of the "bargaining set" of [13][164]. As [163] point out, the bargaining set solution does not predict *which* of the four above allocations will emerge.

16.2.2 An Experimental Study

In their study, Kahan and Repoport [163] used human subjects in a computerized experiment designed to test behavior in situations similar to the one described above. Forty-eight undergraduate male subjects were divided into groups of 16 and participated in three separate experiments. In the first experiment, messages were public, so that all players were aware of the others' offers. Subjects had to send messages publicly in a fixed order (as opposed to being able to speak at will). In the second experiment, messages could be private, but again were sent in order. In the last experiment, messages could be private, but were sent at will. The 16 players in each experiment were broken up into four quartets, each of which played five three-person characteristic function games for four iterations. For each game, one member of the quartet would sit out as an observer – this procedure was employed to allow subjects to reflect upon the task, and to increase the validity of the assumption of independence between games. Order of play between and within games was randomized subject to the condition that no player would be observer in two consecutive rounds. Subjects were given an extended practice session. The five games are shown in Table 16.1.

16.2.3 Quota Values

The type of characteristic function game considered here is a special case known as the *quota games* [163, 164], where the conditions $v(ABC) = v(A) = v(B) = v(C) = 0$ and $v(AB), v(AC), v(BC) > 0$ hold. Such games have *quota solutions*, which are generally accepted by cooperative game theory, and are given by the following equation for player i:

$$\omega_i = \frac{1}{2} \sum I(j, k) \times v(jk) \quad \forall j, k \in P, j \neq k \qquad (16.1)$$

where $I(j, k) = 1$ if $i = j$ or $i = k$, and equals -1 otherwise (note that $\omega_i + \omega_j = v(ij)$). So for example, player A's quota simply is calculated by $\frac{1}{2}(v(AB) + v(AC) - v(BC))$. These quotas represent normative predictions; when players follow a set of weak rationality conditions (e.g., of the bargaining set [13]), they will arrive at the quota values. The quota solution is a solution concept for characteristic function form games, just as the Nash equilibrium is one for strategic form games. Table 16.1 shows the quota values for players A, B, and C in all five games.

	I	II	III	IV	V
	Experimental MRAC, Averaged across Experiment				
A	57.43	63.00	53.73	62.07	71.6
B	38.90	54.40	43.40	45.20	45.70
C	29.53	26.93	34.67	19.27	18.17
	Experimental Frequency of Coalition Structures				
A,B,C	.0208	.0000	.0000	.0000	.0000
AB,C	.5625	.5208	.3542	.7292	.8750
AC,B	.3125	.1667	.3333	.1458	.1250
BC,A	.1050	.3125	.3125	.1250	.0000

TABLE 16.2: Human data.

Kahan and Repoport [163] consider for each player *the mean reward as a member of the winning coalition* (MRAC). Because solution concepts such as the quota make predictions about how much players will get *given that* they are in a coalition, MRAC is the appropriate measure if one wishes to test the theory. [163] report that human subjects' overall deviations from the quotas are not significantly different from zero, for each of the experimental conditions; this reinforces that idea that the quota is an important theoretical notion. Table 16.2 (from [163]) displays the human data in an aggregate form.

While MRAC is the appropriate measure for comparing human performance with theoretical predictions, if one wishes simply to measure how well the subjects do in terms of wealth extracted from the game, then the mean reward (MR) is more appropriate. MRAC and MR can differ greatly; if a stubborn player s always refuses to accept any amount below his quota $+ 20$, for example, then he may be included in the winning coalition once or twice out of a hundred trials, so that his MRAC would be $\omega_s + 20$, but his MR close to zero. The difference arises because in rounds when a player is not in the winning coalition, he or she receives a payoff of zero, which the MRAC ignores. We report both MRAC and MR below.

16.3   The Model

We wish to consider the behavior of agents using a simple learning rule in the context of the five coalition formation games above, and to see if these agents achieve outcomes close to those of humans in the experimental study of [163]. The model assumes that each player updates a belief about how much payoff it can expect from every other agent. We shall refer to this value as player i's *aspiration level*, following [207], to player j at a given time t, or $A_i^j(t)$ for short. Aspiration levels are updated over time by adding a fraction of the difference between actual reward received from the environment, and the payoff level expected, as given by the equation, e.g., [207, 304]:

$$A_i^j(t) = A_i^j(t-1) + \alpha[r_i^j(t) - A_i^j(t-1)] \tag{16.2}$$

where $\alpha \in (0,1)$ represents a recency constant, and $r_i^j(t)$ is the reward received by player i from being in a coalition with player j at time t, as specified by their agreement. For all results reported here, we set $\alpha = 0.2$, a typical value in the reinforcement learning literature. In simulations not reported here, α values between 0.1 and 0.4 yielded results similar to those given here. In addition to the updating rule given in Equation (16.2), we considered two methods for agents to decide to whom to offer. They are given in Figure 16.1.[2]

In addition, each agent makes offers *at* its aspiration levels, and not below. So for example, if agent A has aspiration levels of 50 to agent B and 65 to agent C, then it would offer $v(AC) - 65$ to agent C with probability 1 under the Greedy rule. Under the Matching rule, it would offer $v(AB) - 50$ to agent B with probability $65/115$ and $v(AC) - 65$ to agent C with probability $50/115$. While [276] note that Greedy (i.e., non-probabilistic) action selection is in line with the traditional economic precept of choice as maximization over beliefs, others have found that humans and other animals use probability matching to select between actions associated with a reward, e.g., [117]. Thus, we investigate the performance of both types of selection rules.

Also, our agents are myopic subjective maximizers in their offer behavior,

---

[2]In simulations not reported here, other relatively exploitative selection methods drawn from reinforcement learning literature, such as ϵ-greedy and Softmax selection [304], produced results similar to those reported below for the Greedy agent.

Greedy:

$$P_i(j) = \begin{cases} 1 & \text{if } j = \arg\max_{p \in P} A_i^p(t) \\ 0 & \text{else} \end{cases} \qquad (16.3)$$

Matching:

$$P_i(j) \;=\; \frac{A_i^j(t)}{\sum\limits_{p \in P} A_i^p(t)} \qquad (16.4)$$

FIGURE 16.1: Selection rules: Probability of i offering to j, $P_i(j)$.

in that they make offers based on the maximum they currently "think" they can get, without considering the possible ramifications of their offer behavior on the future state of the system. Such agents represent players who vastly simplify their objective environment, collapsing the available history of offer behavior for each other player into a single real value, A_i^j [276]. Again, we wish to explore the behavior of agents representing players of minimal rationality, and to see if they can model the overall results of humans. The simple agents described in this section satisfy such a condition.

Finally, simulations were run as follows. At the start of a simulation, each agent i's initial aspiration level to j was initialized from the uniform distribution on $[0, v(ij)]$.[3] Next, for each episode, an agent was selected at random as the initial offerer. This agent made an offer to a receiving agent, according to the rules described above. The receiving agent accepted if the offer amount was greater than or equal to its current aspiration level to the offering agent. If it declined, then the receiving agent would become the new offering agent, and so on. The process continued until a receiving agent accepted an offer, which always happened within six rounds. When an episode ended, agents' aspiration levels were updated, and the next episode was started. This process continued until a maximum number of episodes was reached, whereupon the simulation would terminate.

16.4   Results

The MRAC values for the data from [163] for human subjects are presented along with results for Greedy and Matching agents (20 simulations of 1,000 episodes each) in Figure 16.2. As Figure 16.2 shows, the MRAC values of the

---

[3]We found that setting $A_i^j(0) = v(ij)/2$—that is, to half of the coalition value between i and j—did not affect MRAC or MR in expectation (it resulted in less variance).

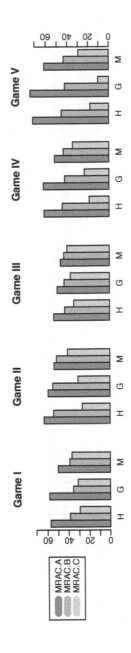

FIGURE 16.2: Mean reward as members of winning coalition for **H**uman subjects, **G**reedy agents, and **M**atching agents.

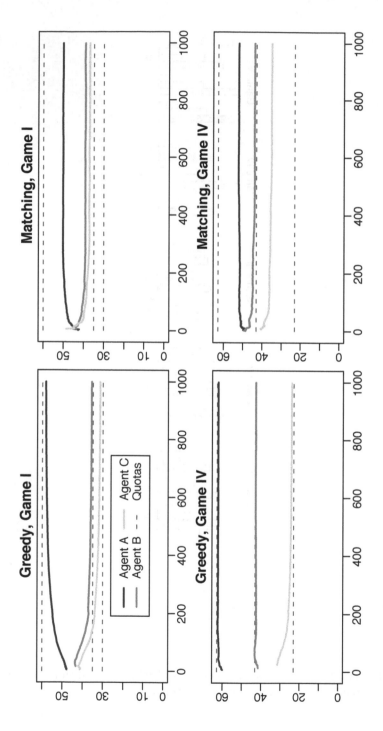

FIGURE 16.3: MRAC over time in games I and IV, for greedy and matching agents.

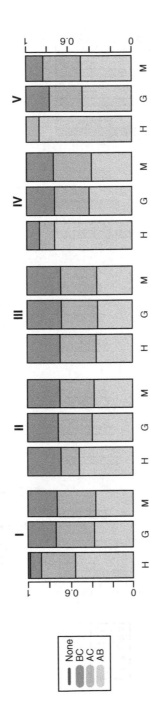

FIGURE 16.4: Frequency of coalition cormation for Human subjects, Greedy agents, and Matching agents.

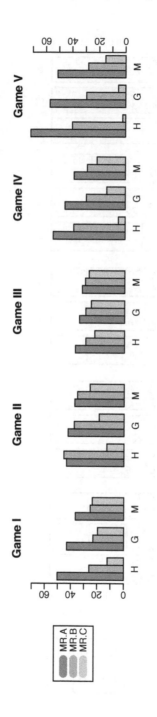

FIGURE 16.5: Mean reward: <u>H</u>uman, <u>G</u>reedy, <u>M</u>atching agents.

agents, especially for the Greedy type, closely fit those of human subjects ($r =$ .98 for the Greedy agents, and $r = .96$ for the Matching agents). Figure 16.3 shows the rate of convergence of MRAC values for Greedy and Matching agents in Games I and IV (Games II, III, and V are not displayed for space reasons). As can partially be seen from the figure, MRAC values for Greedy agents rapidly approach the quota solutions in Games II, III, and IV, more slowly in Games I and V, and never for Matching agents (this is explained later). Figure 16.4 shows frequencies of coalition structures for human subjects and for simulations. The model does not predict some important differences in these frequencies. Namely, player A is included in the winning coalition more often than in the human data than in the simulations, and player C less often (this can be seen for player A by looking at the AC and AB blocks together, and for player C by looking at the BC and AC blocks together).

As discussed previously, to gauge the overall performance of players, it is useful to consider the measure of mean reward (MR) of players. MR values for human subjects, Greedy agents, and Matching agents are shown in Figure 16.5. The MR values for both agent types fit the data of human subjects well, with Greedy agents again doing better than the Matching agents ($r = .97$ vs. $r = .95$). However, there are some noticeable departures. For example, player A has lower MR values and player C higher, for agents vs. humans. This is attributable directly to the differences in the frequencies of coalition formation (Figure 16.4), where player A is in the winning coalition less often (and player C more often) for agents vs. humans.

We would like to make two statements here about our findings. The first is that systems of relatively exploitative agents (e.g., Greedy agents) converge to neighborhoods of quota solutions, and that this outcome is robust to variations of initial aspiration levels. The second is that the difference between an agent's initial aspiration level and its quota value is related to its speed of convergence in MRAC value (the smaller the difference, the faster the convergence). This may be important because although human subjects arrived at the quota solution in only four iterations of play, their initial offers appear to have been very close to their quota values.[4] Whether this occurred as a result of practice sessions, transfer between games or from acting as the observer, or from deliberation is not clear. We merely point out that if the initial aspiration levels of the agents are close to their quota values, then their MRAC values converge almost immediately to the quotas (see, e.g., agents A and B in Game IV of Figure 16.3).

---

[4]For example, [163] report that first offers made to the winning coalition had an average deviation over games and coalitions of -2.95.

16.5   Conclusion

There is now an extensive literature in behavioral game theory that aims at describing, modeling, and explaining human behavior in games (see for example the literature reviews [54, 162]). This literature has discovered that a variety of simple learning models provide reasonably good predictions of human behavior in laboratory games. A number of authors have found that reinforcement learning models supplemented with additional heuristic rules (e.g., pertaining to the specifics of the game) achieve good predictive power. Moreover, other simple models of learning, e.g., using activation-based recall [191], have also provided good accounts of human behavior in games.

In the present work, we found that systems of agents using a simple learning rule can model human data reasonably well, and that their performance converges to the theoretical predictions of quota values in a class of three-player, coalition formation games. The MRAC values for both Greedy and Matching agents were very close to those of humans, while their MR values showed some systematic differences from the human data. Although we do not claim the learning rule presented in Equation (16.2) fully accounts for the way people play these games, we submit that it is a fruitful and practicable way of investigating strategic environments. While much remains to be done, it is intriguing that the agents' behavior so closely matches the human data, much more so than has been reported for simple reinforcement learning models in non-cooperative games.

The results of both agent and human bargaining results are interesting, but perhaps the most important questions regarding bargaining games pertain to their dynamics. When will (and when should) a player accept an offer? If a player chooses to reject, how much and to whom should he/she offer? The answers to these questions may depend not only on the coalition values and other initial parameters of the game, but also on the history of offers and round number. For example, if only two rounds are left before the game is ended by default, can player A (with the highest total coalition value) extract more reward by making an aggressive offer? or, perhaps he/she must accept to avoid forfeiting any positive reward for the game. To answer these questions, different approaches such as (learning) classifier systems are needed, for example to find bargaining rules such as (for Agent A): "If round is 4 and offeringAgent is B and amount is at least aspirationLevel$-0.1 \times v(AB)$, then accept."

In sum, the prospects for mutual benefit between the cognitive modeling community and the behavioral game theory community are bright indeed.

16.6   For Exploration

1. A process may be *described* by rule or procedure, indeed any representation, without being *governed* by it. Put otherwise, a process may be *rule-described* without being *rule-governed*. On the evidence presented in this chapter, with our aspiration level model we can *describe* human subject behavior (approximately) in certain characteristic function games with certain rules or procedures. But is it likely that the decision processes these subjects are following do map closely to the descriptive procedures? How might we go about resolving the question?

2. What are some plausible alternatives to the aspiration level model discussed above?

3. What are some real-world strategic situations in the wild that might credibly be modeled as a characteristic function game?

16.7   Concluding Notes

The work reported here was supported in part by NSF grant number SES-9709548 and funding from the University Scholars Program at the University of Pennsylvania. An earlier version of this paper was published as [58]. We thank two anonymous reviewers, Frank Lee, and Dario Salvucci for helpful comments. James D. Laing introduced me to this class of games and was the senior collaborator on our previous work with the games, in which Garett O. Dworman made strong contributions [83, 84, 85, 86]. Permission to republish this material is gratefully acknowledged to Taylor & Francis.

Part IV

Topics in Strategic Analysis

Chapter 17

Lying and Related Abuses

17.1 Introduction

Individual stakeholders communicate with one another during negotiations for the sake of reaching favorable outcomes. When their interests are not perfectly aligned, there may be (and usually is) an incentive to lie or engage in other abuses in these communications. A potential buyer usually will not want to signal accurately the value of the envisioned acquisition; nor will the owner of the good want to indicate accurately the lowest price at which she is willing to sell. Such situations do *not* necessarily lead to lying, of course. The buyer might lie and make a definite statement of falsehood ("I saw one like it yesterday for only $250"), or he might conduct the discussion entirely without lying, simply refusing to say anything very definite on the matter. Similarly, the seller might lie ("I can't let it go for less than $300"), or she might indicate unwillingness to sell at various prices, without ever saying anything explicitly that would count as a lie. And so on.

Lying, for starters (and we shall go further), might be thought of as the deliberate, intentional expression of falsehood with the goal of achieving gain at the expense of the target of the expression, the person to whom the utterance of the lie is directed. Putting it more formally, we can think of lying as being constituted by three conditions. If S says that P to H, then S is lying if and only if

1. P is false.

2. S knows that P is false.

3. S utters P with the intention of deceiving or misdirecting H.

Let us call this the *know-false-deception* account of lying. I'll also refer to it as the *standard account* of lying. We may take it provisionally as constituting an answer to the question *What is lying?* Let us call this question the *what-is-it* question.

A main claim of this chapter is that this provisional definition of lying is wanting in some ways. It shall be our goal to understand why and how this is so, mainly in the context of negotiation.

A second question naturally attends the subject of lying, especially in negotiations: *Ought one to lie?* Let us call this the *ought* question. A provisional answer is that

- One ought not to lie in negotiations.

We can call this the *ought-never* answer to the ought question for lying.

This sets us up for the chapter. Our topic is lying and matters related in negotiations. We focus for starters on two positions: (1) the standard account of lying, and (2) the ought-never thesis. As such, we are dealing centrally with two kinds of concepts: concepts for lying and concepts for what one ought to do or not to do. It will be useful, before we take up the two positions in earnest, to say a bit on the nature of concepts and particularly on concepts for the category "ought." The next two sections essay just that.

17.2   Concept Concepts

We may take concepts to be representations (think: "thoughts in the head") of categories (think: "real things in the real world").[1] So there are particular shoes, ships, and sealing waxes. There are the categories of shoes, ships, and sealing waxes, as kinds of things in the real world. And there are concepts of these categories, ways in which we represent them. The category for shoes is (more or less) just the collection of all the shoes. The conceptual question, which is our concern now, is the question of how to specify, represent, describe the category. What, in particular, distinguishes the category of shoes, that is the things that are shoes, from everything else? How do we do this? How does it work?

The classical theory of concepts, going back to thinkers in ancient Greece, is that every thing in any given category has something (or things) in common with every other thing in the category that is different from things in all other categories. This thing is often called an essence; it is that which without the thing wouldn't be what it is. Fancy talk, simple idea. The classical view is also called the *definitional* view or theory of concepts: a concept of something is a definition of it, and definitions consist of necessary and sufficient conditions. They have the form:

Something, X, is in the category, C, if and only if X has properties $P_1, P_2, \ldots P_n$.

In short, concepts are definitions of categories and definitions consist of necessary and sufficient conditions for something to be in the category. One is

---

[1]Here I follow the terminology of Gregory L. Murphy [229], who quite correctly notes that these terms do not have consistent usage in the literature. We do the best we can.

reminded of H. L. Mencken's remark that "For every complex problem there is an answer that is clear, simple, and wrong." While the definition theory of concepts has held sway from ancient times until today (it still has many followers, at least as the default theory), it is wrong. Moreover, there are better theories available and all of this matters for our present discussion.

There are two fundamental difficulties with the definitional theory of concepts.[2] First, after all these years and after much trying, there are very few concepts that have satisfactory definitions. If the theory is right, why is this? Second, the definitional theory requires a certain crispness: something either does or does not fall under the definition and hence belong to the category. Yet it seems that category boundaries are fuzzy, inclusion decisions are often fraught and subject to much disagreement. Chairs and tables are furniture. Is a telephone or TV? Is a car seat a chair and if so is it furniture? Examples abound.

The most famous and influential critique of the definitional theory of concepts appears in a short passage in Ludwig Wittgenstein's *Philosophical Investigations*, which I reproduce in part here.

> 66. Consider for example the proceedings that we call "games". I mean board-games, card-games, ball-games, Olympic games, and so on. What is common to them all?—Don't say: "There *must* be something common, or they would not be called 'games'"— but *look and see* whether there is anything common to all.—For if you look at them you will not see something that is common to *all*, but similarities, relationships, and a whole series of them at that. To repeat: don't think, but look!—Look for example at board-games, with their multifarious relationships. Now pass to card-games; here you find many correspondences with the first group, but many common features drop out, and others appear.
> . . .
> And the result of this examination is: we see a complicated network of similarities overlapping and criss-crossing: sometimes overall similarities, sometimes similarities of detail.
>
> 67. I can think of no better expression to characterize these similarities than "family resemblances"; for the various resemblances between members of a family: build, features, colour of eyes, gait, temperament, etc. etc. overlap and criss-cross in the same way.— And I shall say: 'games' form a family. . . .
>
> 68. "All right: the concept of number is defined for you as the logical sum of these individual interrelated concepts: cardinal numbers, rational numbers, real numbers, etc.; and in the same way the concept of a game as the logical sum of a corresponding set

---

[2]There are further difficulties, e.g., prototype effects, which I pass over for the sake of brevity.

of sub-concepts."—It need not be so. For I *can* give the concept 'number' rigid limits in this way, that is, use the word "number" for a rigidly limited concept, but I can also use it so that the extension of the concept is *not* closed by a frontier. And this is how we do use the word "game".

69. How should we explain to someone what a game is? I imagine that we should describe *games* to him, and we might add: "This *and similar things* are called 'games'". And do we know any more about it ourselves? It is only other people whom we cannot tell exactly what a game is?—But this is not ignorance. We do not know the boundaries because none have been drawn. To repeat, we can draw a boundary—for a special purpose. Does it take that to make the concept usable? Not at all! (Except for that special purpose.) [334, pages 31e–33e]

The phrase *family resemblance* has stuck, as has the underlying idea. It is now thought that most of the categories we actually use have a family resemblance structure: the things in them (the particular shoes, or ships, etc.) do *not* have anything in common with one another; rather they have a loose, approximate, family resemblance to one another (and to a degree to things in other categories).

So far, so good. Now the question is "If categories work this way, then how do concepts work? How can we represent a category?" The definitional theory is out the window; we need a new theory.

The psychologist Eleanor Rosch (with colleagues) was the first to propose and develop an experimentally grounded theory of concepts for family resemblance categories.[3] In Rosch's *prototype theory* of concepts, a concept is represented by a prototypical member of a category. The concept prototype may describe a prominent member of the category (think: a robin is a prototypical bird), but this is not required. Prototypes may correspond to no actual member of the category, capturing instead a stereotypical or average member. A prototypical dog, for example, may not correspond very exactly to any breed or actual dog at all.

How do prototype concepts help us decide whether something is or is not in a given category? The basic idea is well expressed in this passage:

> The more prototypical a category member, the more attributes it has in common with other members of the category and the less attributes in common with contrasting categories. ...categories form to maximize the information-rich clusters of attributes in the environment ... [260, page 602]

Is X a dog? The idea is that we compare X to our prototype for dogs. If X is

---

[3]See [229] for a thorough review of the work. Her paper [260] is an excellent starting point.

similar enough we judge it to be a dog. If it is very dissimilar we judge it not to be a dog. If it is a borderline case, we have some difficulty in making the judgment; we may change our minds at different times; and we may disagree with others.

The prototype theory has achieved considerable empirical success and garnered good support. It is not without competitors, however. *Exemplar theory* holds that our concepts are represented by memories of example instances we have encountered of individuals in the category. Thus, one's concept of a dog would consist of memories of the things one has experienced and classified as dogs. When we encounter a new possible dog, we make our judgments based on similarities not to one or a few prototypes, but to an actual library of remembered dogs (things remembered as dogs).

A second competitor for prototype theory is called *knowledge theory*. Murphy gives the following summary:

> The knowledge approach [to concepts] argues that concepts are part of our general knowledge about the world. We do not learn concepts in isolation from everything else (as is the case in many psychology experiments); rather, we learn them as part of our overall understanding of the world around us. When we learn concepts about animals, this information is integrated with our general knowledge about biology, about behavior, and other relevant domains (perhaps cuisine, ecology, climate, and so on). This relation works both ways: Concepts are influenced by what we already know, but a new concept can also effect a change in our general knowledge. [229, page 60]

The theory of concepts is unsettled. What matters for present purposes is that the main contending theories are largely in agreement regarding their differences with the definitional (classical) theory. Generally it is the case that there are no necessary and sufficient conditions for defining concepts of interest. The members of a category may well be quite heterogeneous. Further, concept representations, and the judgments we can make using them, do not generally exclude the possibility of borderline cases. In such cases, it would be better to say that we decide whether to count something as belonging to the category, than to say that it is something we discover. Of course, to say that we decide is not to say that we cannot have good reasons for our decisions, or bad reasons for that matter.

We should expect, then, that lying may come in various forms, similar to one another in some ways and different in others. It is this terrain that we want to explore. Similarly, there are various forms of ought and we need to explore them, too. To that now.

17.3   Two Kinds of Ought

It is useful to distinguish two kinds of ought, two rather different uses of the word *ought* and its cognates. Consider:

1. You ought to eat healthy food, get a lot of exercise, and maintain a body mass index in the normal range.

2. You ought to tell the truth insofar as possible.

Sentence (1) is naturally interpreted as using *ought* in the *prudential* sense. Its meaning for "You ought" could be re-expressed as "It is prudent for you," "It is in your interest," "It is wise for you," and so on. Morality need have nothing to do with it. Sentence (2), in distinction, is not directly about one's interests. Instead, this is a *normative* use of *ought*. A norm is a rule of conduct that tells you what you should do on pain of violating the rule. (Think: the Ten Commandments.) As such, sentence (2) could be rephrased as "There is a governing norm (rule of conduct) for the present situation, and this norm requires that you tell the truth if you can." It may or may not be in your interest to tell the truth; the issue at hand is whether the relevant norm exists and what it tells you to do. Of course, whether you obey the norm is another matter entirely.

There exist many systems of norms affecting our lives. These include: laws, customs, systems of morality, "ways things are done" in a particular society, dress codes, and etiquette. We often think of *ought* in this sense as referring to morality. This is very often mistaken. There are lots of things you should not do that, if you do them anyway, are not immoral. A violation of etiquette is not in itself immoral; it is simply impolite. I like to use the neologisms *normal/normative, anormal, innormal* for the general case of normative oughts (akin to *moral, amoral, immoral*).

The distinction helps us see that the ought-never thesis is quite ambiguous. Is it asserting that lying in negotiations is imprudent or that it is norm-violating? And if the latter, what are the relevant norms?

As I read them, the standard texts on negotiations (such as [24, 98, 286]) all mainly emphasize the prudential interpretation. The implicit stance is something like "We all know that lying is morally wrong. The good news is that telling the truth in negotiations is the smart thing to do, so we can dispense with any problems of moral conflict." (I exaggerate.) Why is lying thought to be unwise? Simple: there is life after a negotiation and there are often good alternatives. Lies are often taken as repugnant when discovered and one's reputation may suffer in consequence. One can often achieve the benefits of a lie without risking heavy blowback upon discovery. We shall, in the sequel, be interested in exploring some of these alternatives.

What about the normative interpretation of the ought-never thesis? Richard Shell has a very useful short paper on the legal aspects of lying in negotiations [285] and there is other literature on lying and legal norms. Careful studies of non-legal norms regarding lying in negotiations are much needed.

We turn now to lying itself.

17.4   Lying

We begin by establishing an example prototype of lying, drawing it from the 1972 movie *The Godfather*. The movie opens with the wedding between Vito Corleone's daughter Connie and Carlo Rizzi. Vito is head—"Don"—of a Mafia family. Crime has made them prosperous. For nearly 3 hours of movie time, dramatic and violent events unfold, during which Carlo betrays the Corleone family and brutalizes Connie, even beating her when she is pregnant, the Corleone family is murderously attacked, Vito takes ill and eventually dies, and his son Michael reluctantly takes over. Michael arranges a counter-attack which is staged during the baptism of Connie and Carlo's new baby, with Michael as the godfather. During the ceremony, by plan, the Corleone enemies are decimated.

In the last part of the movie, Michael and his men arrange for the murder of Carlo. Shortly after, Connie confronts Michael for murdering her husband, the father of her infant. Kay, Michael's wife who has been kept in ignorance of Michael's dealings, is present.

> CONNIE
>
> Michael! You lousy bastard—you killed my husband! You waited until Papa died so nobody could stop you, and then you killed him. You blamed him for Sonny—you always did. Everybody did. But you never thought about me—you never gave a damn about me. Now what am I going to do?
>
> KAY
>
> Connie...
>
> CONNIE (to Kay, after Kay puts her arms around her)
>
> Why do you think he kept Carlo at the mall? All the time he knew he was gonna kill'im.
>
> (then, to Michael)
>
> And you stood Godfather to our baby – you lousy cold-hearted bastard. Want to know how many men he had killed with Carlo? Read the papers – read the papers!
>
> (then, after she picks up and slams down a newspaper)
>
> That's your husband! That's your husband!

[Connie goes toward Michael. Neri holds her back until Michael motions it's okay]

MICHAEL (taking Connie's arms as she cries)

Come on...

CONNIE (struggling out of Michael's arms)

No! No! No!

MICHAEL (to Neri)

Get her upstairs. Get her a doctor.

[Neri takes Connie out of the room. Michael sighs, then lights a cigarette]

MICHAEL (to Kay)

She's hysterical – hysterical.

KAY

Michael, is it true?

MICHAEL

Don't ask me about my business, Kay...

KAY

Is it true?

MICHAEL

Don't ask me about my business...

KAY

No.

MICHAEL (as he slams his hand on the desk)

Enough!

(then)

Alright. This one time [Michael points his finger] – this one time I'll let you ask me about my affairs...

KAY (whispering)

Is it true? – Is it?

MICHAEL (quietly, shaking his head)

No.

KAY (after a sigh of relief and Michael kisses and hugs her)

I guess we both need a drink, huh?

[Kay leaves the room to fix Michael a drink. At the same time, Rocco, Clemenza, and Neri enter the office. Clemenza shakes

Michael's hand. Kay turns her head to watch them. They embrace Michael, then kiss his hand.]

CLEMENZA (kissing Michael's hand)

Don Corleone...

[Rocco kisses Michael's hand as Neri shuts the door blocking Kay's view]

The screen blackens to indicate...

THE END

(From *The Godfather* `http://www.thegodfathertrilogy.com/gf1/transcript/gf1transcript.html`.)

Michael's soft "No" to Kay is as baldfaced a lie as one can ask for. Let us keep it in mind as we explore.

Lying by Sissela Bok [37] is a much-admired philosophical treatment of our subject. I do not know of a more authoritative source. Bok's core notion of lying emphasizes *intention to mislead* and is close to our standard account. Lying, Bok assumes, is basically wrong and generally to be lamented. Much of the book is concerned with laying out various kinds of lies and circumstances in which they occur, with an eye to assessing them morally. Throughout the book she incorporates historically important views of lying, often disagreeing with them. The scholarship is comprehensive, making the book all that more interesting. I will highlight a few sample points that serve present purposes. This only hints at the richness of the book.

Bok says she defines a lie as "any intentionally deceptive message which is *stated*" [37, page 13]. By *stated* she means to include spoken and written language, as well as gestures and other sorts of signals. She does not explicitly say very much about which intentionally deceptive practices she means to exclude from the category of lying. Keeping silent as a way of deception and providing a non-answer to a question would, I believe, usually save one from lying in her view. So, while her explicit definition would seem to be much broader than our standard account, much of what she does say fits well with it. And in the end she agrees that the precise definition is unimportant.

> I see nothing wrong with either a narrow or a wider definition of lying, so long as one retains the prerogative of morally evaluating the intentionally misleading statements, no matter whether they fall within the category of lying or outside it. [37, page 15]

Moral evaluation is her principal interest. Well, then, is it always wrong to lie? What if an innocent person is being pursued by a criminal—say a button man for the Corleones—and you are asked as to the whereabouts of the target? And you know. Should you tell the truth? We have here the moral issue as well as the prudential issue. Morally only, however, there have been traditions that say you have a duty to reveal the target to the criminal. What about so-called white lies? Lies told to children or the dying, lies told for the sake of comforting the afflicted? Lies to protect clients or customers? Lies to subjects in social science research? Lies for therapeutic purposes?

The list goes on and on. Bok is comprehensive but hardly exhaustive. Oddly enough, she has only a little to say about lies during negotiation or bargaining. Bok sees bargaining as a situation involving "mutual deceit." The following passage gives the essentials of her position.

> In a bazaar, for instance, false claims are a convention; to proclaim from the outset one's honest intention would be madness. If buyers and sellers bargain knowingly and voluntarily, one would be hard put to regard as misleading their exaggerations, false claims to have given their last bid, or words of feigned loss of interest. Both parties have then consented to the rules of the game. [37, page 131]

She does not quite wash her hands of moral considerations for lying in negotiations; she does come close.

Perhaps she is right when it comes to morality. Considering norms more expansively, however, reveals her account to be too limited. As we have seen [285] certain kinds of lying in negotiations is illegal. Customs and etiquette will draw additional lines between lies that fall above or below a norm.

With this as a necessarily quick overview, I want to close the section with a few comments on the lie Michael Corleone tells to his wife Kay. Michael lies, Kay seems to believe him, and the movie ends with Kay seeing that Michael is the new Don. Why would she ask in the first place? Does she expect him to tell her the truth if indeed he has just ordered a bloodbath including the murder of her brother-in-law? What is she thinking? That he will confess and with ministrations of love she will change him? What must Michael think? That Kay is hopelessly naïve and hence unreliable? Must he have her killed too? Or perhaps that she is manipulable, compromisable and so he can handle her? Is there any glimmer of right behavior to be had here? What would you do if you were Kay? Michael?

17.5   Spin

To *spin* is (roughly, as always), to describe events in a way that favors a particular party, such as oneself or one's employer. Someone who spins (such as a press agent) is said to be a *spin doctor* and someone who does it well is a *spin meister*. The language is current and in widespread use, yet seems to have originated only in the 1970s, although spinning has always been done. (See *The Concise Oxford Dictionary of Politics* http://www.oxfordreference.com/views/ENTRY.html?

`entry=t86.e1298&srn=2&ssid=513737036#FIRSTHIT.`) WordNet recognizes accurately the relevant sense of *spin*:

> Noun
>
> S: (n) spin (a distinctive interpretation (especially as used by politicians to sway public opinion)) "the campaign put a favorable spin on the story"
>
> Verb
>
> S: (v) spin (twist and turn so as to give an intended interpretation) "The President's spokesmen had to spin the story to make it less embarrassing"

Spinning need not require lying; the two are rather distinct. In negotiations it is accurate to say that often all parties will spin the situation—describe it in ways favorable to themselves. They may well do this without lying. (Think: glass half empty or half full?)

Lying is (we should agree) an abuse of something (truth perhaps?). We most often react to it with distaste. We are offended when someone lies to us. Spin, let me submit, usually provokes a different reaction. We rather expect spin in a negotiation. Of course the seller will put a rosy interpretation on that dump of a house; it's to be expected. Par for the course.

Is spin an abuse of anything? Would it ever be innormal to spin? Consider the case in which a friend asks for advice on a restaurant. Usually, you would be in violation of the norms of cooperative discourse (this is your friend you are talking to!) if you responded with heavy spin, one way or the other. Under these norms, your friend is expecting a balanced and fair account, not a one-sided hit. Spinning is in some situations innormal. It is an abuse of the pertinent presuppositions of discourse.

What about negotiations? Surely it is (often at least) part of their presuppositions that it is all right to spin. But how much? One can spin without lying by emphasizing the virtues of one's asset. What about admitting to its defects? Here too one can go far to deflect or pooh-pooh unfavorable aspects. What is too far? What about suggesting or insinuating that the norm of cooperative joint deliberation is in place (which forbids spinning), when the real situation is a negotiation (which permits spinning at least to some degree)?

These and many other fascinating questions I leave as exercises for the reader.

17.6 Misdirection

Lying and spin are just two forms of discourse abuse that occur with significant effect in negotiation. Other abuses have only vulgar names (see below) or have no standard names at all. In the latter category is what I'll call *passive misdirection*.

In one form of passive misdirection the perpetrator (the one doing the misdirecting) fails to disabuse (to correct or to set straight) the victim of a material falsehood or misunderstanding that—it is evident to the misdirector—the victim is subject to. We might call this a misdirection by silence. It might happen, for example, when the victim expresses a false assumption about the law favoring the victim's position and the perpetrator could, but fails to, correct the victim. Passive misdirection may also happen if the victim asks the perpetrator a material question and perpetrator gives a "no answer answer," say by changing the subject or in some other way wiggling out of directly lying. (Should we call this *evasive misdirection*?) In either case, the victim is left with the false impression that the perpetrator does not have knowledge that the victim's belief is false. In passive misdirection, the perpetrator knows that the victim believes something material that is false or flawed in some way, and the perpetrator does nothing to correct the victim's belief.

There is also active misdirection. Often in such cases the perpetrator will say something that is literally quite true, yet irrelevant or not pertinent to the issue to hand, thereby misleading the victim. Example: "Why didn't you come to the meeting yesterday?" "I had a terrible headache and had to lie down," when the headache came after the meeting ended. Another example: "How good is the mechanical condition of the car?" "I had my mechanic go over it thoroughly just last week" [which is why I'm hoping to unload it now].

Like spin, misdirection is an abuse of the norms of cooperative discourse, if these norms are in force. They may not be and negotiation is a case in point. Can we say how the norms of negotiation apply (or not) to misdirection? As Shell's article on when it is legal and not to lie in negotiations reminds us [285], there may well be governing norms that prohibit various forms of misdirection in negotiation. Misdirection seems to be much more problematic normatively than spin in negotiations.

17.7   Humbug

Humbug is an old-fashioned word, recognized, but not (I would guess) widely used in English today. (See http://en.wikipedia.org/wiki/Humbug (accessed 20110401) for incomplete but still useful comments. The article synonymizes *humbug* as hoax or jest.)

The *Oxford Dictionary of English* does better:

> 1. humbug noun
> [mass noun] deceptive or false talk or behaviour: his comments are sheer humbug. [count noun] a hypocrite.

And *WordNet* even better:

Noun

S: (n) baloney, boloney, bilgewater, bosh, drool, humbug, taradiddle, tarradiddle, tommyrot, tosh, twaddle (pretentious or silly talk or writing)

S: (n) humbug, snake oil (communication (written or spoken) intended to deceive)

S: (n) fraud, fraudulence, dupery, hoax, humbug, put-on (something intended to deceive; deliberate trickery intended to gain an advantage)

Verb

S: (v) humbug (trick or deceive)

Perhaps the most well-known use of the term is by Scrooge in Dickens's *A Christmas Carol* where he says of Christmas, "Bah! Humbug!", meaning Christmas is a fraud. My favorite use is also well known. At the end of *The Wizard of Oz* Dorothy and friends finally confront the Wizard. It does not go well. Here is the passage from the movie script:

MCU - The Wizard peering out from curtain - he ducks back out of sight and his voice booms out again -

OZ'S VOICE Oh - I - Pay no....

LS – Shooting past the Four at left to the Curtain in b.g. – Dorothy goes over to it and starts to pull it aside –

OZ'S VOICE ...attention to that man behind the curtain. Go - before I lose my temper! The Great and Powerful —....

MCS – Dorothy pulls back the curtain to reveal the Wizard at the controls – he reacts as he sees Dorothy – Dorothy questions him – the Wizard starts to speak into the microphone – then turns weakly back to Dorothy – CAMERA PULLS back slightly as the Lion, Scarecrow and Tin Man enter and stand behind Dorothy –

OZ'S VOICE ... – Oz – has spoken!

DOROTHY Who are you?

OZ'S VOICE Well, I – I – I am the Great and Powerful – Wizard of Oz.

DOROTHY You are?

WIZARD Uhhhh – yes...

DOROTHY I don't believe you!

> WIZARD No, I'm afraid it's true. There's no other Wizard except me.

> MCS – Dorothy and her three friends react – Camera shooting past the Wizard at left – the Scarecrow and Lion speak angrily –

> SCARECROW You humbug!

> LION Yeah!

> CS – Wizard – shooting past Dorothy, the Lion and Scarecrow – the Wizard speaks –

> WIZARD Yes-s-s – that...that's exactly so. I'm a humbug!

(`http://www.wendyswizardofoz.com/printablescript.htm`, 20110401.)

The philosopher Max Black has produced a widely cited essay, "The Prevalence of Humbug" [33], that begins to explore the concept beyond its dictionary definitions.

Humbug is, he thinks, an abuse of truth different from lying, lying in our basic sense ("intentional untruthful declaration" as Black quotes Kant). To account for, or articulate, the difference between lying and humbug, Black invokes a distinction between what someone says—this is the *message*—and what the speaker thinks about what the message says. This he calls the speaker's *stance* toward the message. Here is the money quote.

> We can usefully distinguish between the speaker's *message*, as I shall call it, and his or her *stance*. By the message I mean whatever is explicitly or implicitly said about the topic in question; while I reserve the term "stance" for the speaker's beliefs, attitudes, and evaluations, insofar as they are relevant to the verbal episode in question. [33, page 119]

Given the distinction, we can also distinguish evaluations of the message ("the *substance* of what is being said") and of the speaker's stance toward that substance. Black proposes that we use the word *humbug* for the latter, rather than the former.

> [In] rejecting the *substance* of what is being said... there need be no imputation about the sincerity of the speaker's stance. ... No matter: without impugning a speaker's stance, we can sometimes condemn what is being said as balderdash, claptrap, rubbish, cliché, hokum, drivel, buncombe, nonsense, gibberish, or tautology. With so rich a vocabulary for dismissing the substance of what is said, we could dispense with this use of "humbug." That useful work might well be reserved for criticism of a speaker's stance—to discredit the message's provenance rather than its content. [33, pages 119–120]

And what about the stance as the locus of humbug?

What then is the prima facie charge against a speaker accused of humbug? Well, some of the words that immediately suggest themselves are pretense, pretentiousness, affectation, insincerity, and deception. [33, pages 119–120]

He ends the essay with "The best [characterization of humbug] I can now provide is the following formula:"

HUMBUG: deceptive misrepresentation, short of lying, especially by pretentious word or deed, of somebody's own thoughts, feelings, or attitudes. [33, page 143]

What is the nature of Black's enterprise? He is not doing anything very empirical, he is not doing field linguistics, although real examples and real data matter. I suggest he and the other philosophers that follow below are engaged in *concept explication* (Rudolf Carnap's term). They identify an important concept (and category) and seek to sharpen it up, make it more precise, and in so doing to give us insight and understanding. Their method is to begin with some accepted uses and examples and then propose how we should understand the workings and meanings of the concept. This is what Black does with the word *humbug*.

The primary aim in concept explication is to say what the corresponding category is. Black is trying to answer the question *What is humbug?* Recalling our discussion of how concepts work, we should be wary of attempts at precise definition or characterization of the essence of anything. Humbug is almost surely a family resemblance sort of concept. While it is a good thing to get at its main characteristics (and perhaps Black has succeeded at this), there are other questions that are interesting and perhaps more important. One of them is *How does humbug work?* We might agree that humbug consists of deceptive misrepresentation, short of lying, but how does the perpetrator do that? It is easy enough to see how lying works. How does deceptive misrepresentation short of lying work?

Black gives us an intriguing hint. He speaks of "virtual lying" as being "in the close neighborhood of the kind of humbug that 'functionally or effectively' implants false belief" [33, page 134]. The perpetrator of humbug causes (or attempts to cause) the hearer to acquire false beliefs without at the same time actually stating them. The falsehoods are insinuated or suggested, not uttered directly. Further, in distinguishing between the speaker's message and the speaker's stance toward the message, Black locates humbug in the misrepresentation of the speaker's attitude toward the message.

The thought seems to be this or something like it. In humbug the message is not literally false; rather, it is designed to cause the hearer to acquire a false belief. Further, what makes the humbug work is the deceptive indication that the speaker endorses the message and its insinuated falsehood. The Wizard of Oz does not *say* he has supernatural powers for helping Dorothy and company. He *suggests* it by his appearance (he seems to be a supernatural power) and associated utterances. Lying? No. Humbug? Definitely.

Humbug surely abuses the norms of cooperative discourse. You should not be a humbug to a friend or intimate. Even so, is Black being censorious in saying this:

> I hope you will agree that while humbug has the short-term advantages of devious hypocrisy over naked felony, it is indeed an insidious and detestable evil. [33, pages 141–2]

Detestable evil?

And what of humbug in negotiation? Perhaps it is or is akin to spin beyond the norm? How would you, or should you, react to a discovery in negotiation of lying? Of spinning? Of humbug?

17.8   Bullshit

We are all familiar with bullshit by direct acquaintance and long experience. Even so, here are some authoritative reminders. From the *Oxford Dictionary of English*:

> bullshit (vulgar slang)
>
> noun [mass noun] stupid or untrue talk or writing; nonsense.
>
> verb (bullshits, bullshitting, bullshitted) [with obj.] talk nonsense to (someone) in an attempt to deceive them.
>
> - ORIGIN early 20th cent.: from bull3 + shit.

From *WordNet*

> Noun
>
> S: (n) bullshit, bull, Irish bull, horseshit, shit, crap, dogshit (obscene words for unacceptable behavior) "I put up with a lot of bullshit from that jerk"; "what he said was mostly bull"
>
> Verb
>
> S: (v) talk through one's hat, bullshit, bull, fake (speak insincerely or without regard for facts or truths) "The politician was not well prepared for the debate and faked it"

In a widely acclaimed essay, the philosopher Harry G. Frankfurt [110] has explored the concept of bullshit. Bullshit is, in Frankfort's explication, speech in disregard of the truth. That is, the bullshitter is trying to get the victim to believe (falsely) that the bullshitter is speaking the truth, while at the same time speaking without regard to truth. "In a standard case, he is trying to produce the impression that he knows what he is talking about when in reality he does not, and says whatever he thinks will aid that impression" [219]. Here is Frankfort's own summary [110, pages 53–5]:

What bullshit essentially misrepresents is neither the state of affairs to which it refers nor the beliefs of the speaker concerning that state of affairs. Those are what lies misrepresent, by virtue of being false. Since bullshit need not be false, it differs from lies in its misrepresentational intent. The bullshitter may not deceive us, or even intend to do so, either about the facts or about what he takes the facts to be. What he does necessarily attempt to deceive us about is his enterprise. His only indispensably distinctive characteristic is that in a certain way he misrepresents what he is up to.

This is the crux of the distinction between him and the liar. Both he and the liar represent themselves falsely as endeavoring to communicate the truth. The success of each depends upon deceiving us about that. But the fact about himself that the liar hides is that he is attempting to lead us away from a correct apprehension of reality; we are not to know that he wants us to believe something he supposes to be false. The fact about himself that the bullshitter hides, on the other hand, is that the truth-values of his statements are of no central interest to him; what we are not to understand is that his intention is neither to report the truth nor to conceal it. This does not mean that his speech is anarchically impulsive, but that the motive guiding and controlling it is unconcerned with how the things about which he speaks truly are.

So the bullshitter is one who speaks with little regard to whether what is said is true. Frankfurt, like Black on humbug, thinks bullshit is a bad thing, although he wonders aloud whether eliminating it would a good thing.

One context he approves of is the honored institution of the bull session. This is where we bullshit with abandon and it's ok because everyone knows it's harmless BS. Frankfurt notes that

> What tends to go on in a bull session is that the participants try out various thoughts and attitudes in order to see how it feels to hear themselves saying such things and in order to discover how others respond, without its being assumed that they are committed to what they say... [110, page 36]

Eric Ambler's novel *Dirty Story* (cited, as Frankfurt notes, by the *OED* [110, page 48]) has a father advise his son "Never tell a lie when you can bullshit your way through." This would seem to capture much that is important about the concept of bullshit. It is not lying; rather it can serve as a substitute for lying, allowing one to achieve aims that lying would achieve without actually having to lie. How exactly is this done? Frankfurt does not say a lot here except, as we have seen, to characterize bullshit as speech without regard for truth. And this is a bad thing indeed.

He [the bullshitter] does not reject the authority of the truth, as

the liar does, and oppose himself to it. He pays no attention to it
at all. By virtue of this, bullshit is a greater enemy of the truth
than lies are. [110, page 61]

Even so, Frankfurt recognizes that "our attitude toward bullshit is generally
more benign than our attitude toward lying" [110, page 50]. Why this should
be so puzzles him.

One explanation might be that much of our communicative intercourse
involves negotiation and the governing norms permit a certain amount of bullshitting. Surely, most of us spend more time negotiating than we do engaged
in cooperative inquiry and deliberation with others. To my ear (I leave the
decision to the reader) we use *bullshit* in negative as well as benign ways. The
benign sense is close in meaning to *small talk*. One might open a negotiation
or seek to smooth the conversation by engaging in a little bullshit, seeking
perhaps to project a certain image of oneself or to deflect close examination
or just to keep talking. On the other hand, when someone says or thinks
"Don't bullshit me!" this is a request to get serious and be forthcoming with
information. Here *bullshit* is used in a more negative sense. The suggestion or
presupposition of the speaker making the request is that the addressee has
been violating a norm of cooperation or a norm for being informative.

Finally, we can wonder whether the father's advice to "Never tell a lie when
you can bullshit your way through" is always wise. Are there negotiation
situations in which one's counter-parties would find it understandable that
you would lie, but would be disgusted with you if you engaged excessively in
bullshit? Is lying sometimes a more honest act than bullshitting?

17.9   Mindfucking

From the *Oxford Dictionary of English*:

1. mindfuck noun

esp. N Amer. coarse slang; transitive verb; manipulate or disturb
(someone) psychologically.

WordNet does not have an entry for *mindfucking* (20110401) but the
Urban Dictionary (http://www.urbandictionary.com/define.php?term=
mind%20fucking) does. (The following passage has been edited slightly for
purposes of presentation.)

mind fucking mind fuck mind fuck

1. mind fucking

buy mind fucking mugs, tshirts and magnets

The act of messing with someone's mind, usually to an extreme.

"The last time I mindfucked Peter, he thought he was an [sic] heroin addict for three days."

2. mind fucking

buy mind fucking mugs, tshirts and magnets

Convincing yourself that by over analyzing a situation you can gain control over it when, in reality, it is impossible to control.

"He was mind fucking himself trying to figure out what he could do to make her love him."

analyze convince control mind fuck think

(They list a third meaning, as does the *OED,* which need not concern us.) The *OED* cites this usage:

1975 H. Wentworth & S. B. Flexner Dict. Amer. Slang 723/2 Mind-fuck, to manipulate others to believe and act as one does, without any consideration for them.

Or "to believe and act as one prefers," which would be better as a definition.

In a recent, somewhat less well-known essay than Frankfurt's on *bullshit,* the philosopher Colin McGinn [219] explores the concept of "mindfucking." This is a relatively recent expression (at least in English) and seems to date from the 1960s. (I recall hearing such usage as "Tripping on LSD is a mind-fucking experience" meant as a good thing, and then thinking "Hey, I don't want my mind fucked with drugs.") The term has been appropriated for a number of purposes which do not concern us. There is, however, a core meaning of the word *mindfuck* and associated concept, "mindfuck," that is quite relevant to negotiation. In this core sense, which would appear to be its original sense, to mindfuck is to mess up someone's mind. The Urban Dictionary has this as its first sense, quoted above. From the *OED,* "mindfuck" as a noun:

A disturbing or revelatory experience, esp. one which is drug-induced or is caused by deliberate psychological manipulation. Also: deception.

Colin McGinn notes that there is a positive sense of mindfuck as well as a negative one. In the positive sense,

...the phrase is sometimes used to describe the positive sensation involved in having, or being presented with, some striking new idea, or having some sort of agreeably life-altering experience ...When a book or film or conversation is described as a mindfuck, this can be taken as a favorable evaluation...I may go to the cinema or to a lecture *hoping* for a mindfuck, but I cannot in this way (except masochistically) hope to be lied to or bullshitted to. [219, pages 5–6]

Radical new scientific theories, such as those from Copernicus, Darwin, and Einstein, as well as any Kuhnian paradigm change [188], fuck our minds in this positive sense. Indeed, I hope readers of this book will be positively mindfucked to some degree.

In the negative sense, with which McGinn is mainly concerned, a mindfuck is an unwelcome, even "nefarious," manipulation of the victim's mind, done by exploiting cognitive or emotional weaknesses in the victim. It is to be contrasted with rational persuasion. "[E]motional interference [with the victim] is the essential mechanism of the [mindfucker's] project" and as such "the mindfucker has a far more ambitious agenda" than the liar or the bullshitter [219, page 34]. The mindfucker manipulates the victim's emotions ("pushes their buttons").

> The mindfucker will typically play on the anxieties and insecurities of the victim in order to produce a set of false beliefs, which will then lead to the emotional disturbance that is sought as the end. [219, page 36]

This is not a good thing.

> The victim of the mindfuck is exploited, leaned on, invaded, imposed on, controlled and manipulated. Mindfucking is an inherently aggressive act. It is an act of psychological violence, more or less extreme. As such, it is clearly immoral. The intention behind it is morally objectionable: it is an intention to do harm. [219, pages 38–9]

As for an example, a prototype, of mindfucking in its negative sense, I can think of no better case than one McGinn highlights: Shakespeare's *Othello,* in which Iago mindfucks Othello into believing that his virtuous wife Desdemona is being unfaithful to him with Cassio. The entire play is an astute illustration of and meditation on mindfucking. As for examples from the real world, look to propaganda, advertising, the history of the mass media and PSYOPS, psychological operations pursued by armies and states.

No doubt the catalog is incomplete and will see new entries over time.

17.10   Discussion

Lumping humbug and bullshit together as being pretty much the same thing, we are left with lying, spinning, misdirection, bullshitting, and manipulative mindfucking as distinct related abuses of norms in bargaining and negotiation. A few points arising:

1. The normative aspects of deceptive practices (lying, spinning, misdirection, bullshitting, and manipulative mindfucking) vary by context. In

general, there will be more tolerance for them in bargaining and negotiation than in purportedly more cooperative situations.

2. It should be remembered that a common deceptive practice is to signal a cooperative context and behave appropriately for a bargaining context.

3. It is, perhaps, more than a triviality that as a noun *lie* is a count term ("How many lies did she utter yesterday?") while *bullshit* is a mass term ("How much of his talk was bullshit? A lot or a little?"). Our other terms are a bit murky in this regard, but *spin* and *mindfucking* seem to me closer to mass terms, while *misdirection* is closer to a count term. I think this has something to do with how we react to these various forms of deception. Lies that are crisp and definite are comparatively easy to handle if one has the information. A pack of lies is much harder and indeed I think we would tend to label a long string of lies as bullshit. There is something about spin, bullshit, and mindfucking that is exasperating because it is so hard to deal with. You can confront a liar with a fact. How do you confront a bullshit artist, a spin meister, or an accomplished mindfucker? Many will choose not to bargain at all with a bullshitting mindfucker.

4. On the other hand, it is often easier to detect spin, bullshit, and attempts at mindfucking than lying and misdirection. The bullshitter and compatriots have to say a lot and it all has to hold together. The liar is typically not discovered by examining what he or she says alone, but by finding facts in the world.

5. What happens when a prima facie form of deception—such as falsely saying that one cannot go lower in price—becomes openly permitted by the governing norms of negotiation? Is it even accurate to describe such actions as deceptive?

A bit of a summary now. Lying, of which there are many forms, is just one kind of deception in negotiations. Other kinds include spinning (marginally deceptive), misdirection, bullshitting, and mindfucking. Normatively, bargaining contexts permit higher levels of deception than other contexts, such as joint deliberation or friendly interaction with neighbors and relatives. Even so, there are limits, including legal limits.

Prudentially, lying is often thought to be the least advisable form of deception because it provokes a strong reaction if discovered and affords less cover than bullshit. I am not entirely convinced on this point. Because bullshit and mindfucking may be more readily detectable, the perpetrator risks follow on opprobrium, as does the liar.

So we arrive at a generalization of old advice. In the spirit of Ben Franklin's comment that "Honesty is the best policy," we can recommend this: It is wise to avoid insofar as possible deceptive practices in negotiation, including lying, spinning, misdirection, bullshitting, and mindfucking; better if you can to develop the skills you need to avoid them.

17.11   For Exploration

1. *Humbug* is by now something of an antiquated term. McGinn reports that he finds *bullshit* redolent of the 1950s (ancient history!). Are there newer terms than the ones we have discussed for kinds of deceptive, abusive practices in bargain and negotiation? If so, do they like *mindfuck* introduce new concepts as well?

2. Can you think of a situation or an example in which a lie is told for the sake of communicating a truth?

3. Various journalism organizations do fact checking on speeches by politicians and report deviations from truth. Discuss how one might go about constructing a bullshit index to measure the degree of bullshit in a given body of utterance.

4. What about a mindfuck index?

5. Granting that lying, bullshit and all are wrong, we must also admit that the victims have responsibility to do what they can to avoid being exploited. This is perhaps most strongly the case with mindfucking. *Othello* is a tragedy because its victim plays a key part in his own destruction. Read "The Ethics of Belief" (1877) by William K. Clifford and discuss what potential victims of deception in bargaining should do, both from a normative and from a prudential perspective.

6. How does horse shit differ from bullshit?

17.12   Concluding Notes

YouTube has a number of interviews with Harry Frankfurt. I recommend
`http://www.youtube.com/watch?v=W1RO93OSOSk`, part 1
and for part 2:
`http://www.youtube.com/watch?v=hp_c8-CfZtg`
The Onion does its customary exquisitely accurate sendup:
`http://www.youtube.com/watch?v=viVAAy_qkxO&feature=fvst`
George Carlin's famous "Religion is Bullshit" routine:
`http://www.youtube.com/watch?v=MeSSwKffj9o`
Not mentioned by McGinn, *Diary of a Seducer* by Søren Kierkegaard is a compelling and awful story of a mindfucker in action. More of a borderline case is presented in *Elmer Gantry* by Sinclair Lewis. Is it mindfucking or just bullshit or something else, perhaps the rampages of the hypocrite?

There is much in the literatures on rhetoric and persuasion that is relevant to detecting and assessing practices of deception. One can do much worse than starting with Albert O. Hirschman's *The Rhetoric of Reaction* [151].

Chapter 18

Evolutionary Models

18.1 Biological Evolution

The theory of biological evolution asserts that species change, driven by natural selection. Alfred Russel Wallace and Charles Darwin originated the theory independently. In 1858, Wallace sent Darwin a paper outlining his theory of evolution by natural selection. Darwin had been developing a very similar theory for years, keeping it private while working on planned comprehensive work to present it. That work was never completed. Stunned into a response, Darwin wrote a short paper on his views and the two papers, Darwin's and Wallace's, were published together that year [26]. Darwin immediately commenced to compose a short version of a planned larger work on evolution, one he could get into print quickly. *On the Origin of Species, or the Preservation of Favoured Races in the Struggle for Life* was the result, appearing in 1859 [68]. It is masterpiece enough for anyone.

Darwin customarily referred to his account of biological speciation as "the theory of descent with modification through natural selection" (e.g., [68, page 459, second paragraph of chapter XIV]). The celebrated concluding paragraph of the *Origin* states Darwin's views expansively enough for present purposes and clearly commits him to biological evolution.[1]

At the beginning of the summarizing chapter, Darwin summarizes the theory as follows:

> That many and grave objections may be advanced against the theory of descent with modification through natural selection, I do not deny. I have endeavoured to give to them their full force. Nothing at first can appear more difficult to believe than that the more complex organs and instincts should have been perfected, not by means superior to, though analogous with, human reason, but by the accumulation of innumerable slight variations, each good for the individual possessor. Nevertheless, this difficulty, though appearing to our imagination insuperably great, cannot be con-

---

[1]Oddly enough, although his and Wallace's accounts soon became known as theories of evolution, when I did a string search on 'evolution' in the online Project Gutenberg version, the word evolution did not appear.

sidered real if we admit the following propositions, namely,—that gradations in the perfection of any organ or instinct, which we may consider, either do now exist or could have existed, each good of its kind,—that all organs and instincts are, in ever so slight a degree, variable,—and, lastly, that there is a struggle for existence leading to the preservation of each profitable deviation of structure or instinct. The truth of these propositions cannot, I think, be disputed. [68, page 459, chapter XIV]

His celebrated concluding paragraph of the *Origin* gives a complementary summary.

It is interesting to contemplate an entangled bank, clothed with many plants of many kinds, with birds singing on the bushes, with various insects flitting about, and with worms crawling through the damp earth, and to reflect that these elaborately constructed forms, so different from each other, and dependent on each other in so complex a manner, have all been produced by laws acting around us. These laws, taken in the largest sense, being Growth with Reproduction; Inheritance which is almost implied by reproduction; Variability from the indirect and direct action of the external conditions of life, and from use and disuse; a Ratio of Increase so high as to lead to a Struggle for Life, and as a consequence to Natural Selection, entailing Divergence of Character and the Extinction of less-improved forms. Thus, from the war of nature, from famine and death, the most exalted object which we are capable of conceiving, namely, the production of the higher animals, directly follows. There is grandeur in this view of life, with its several powers, having been originally breathed into a few forms or into one; and that, whilst this planet has gone cycling on according to the fixed law of gravity, from so simple a beginning endless forms most beautiful and most wonderful have been, and are being, evolved. [68, pages 489–490]

Points arising:

1. Darwin's claim that speciation occurs by a process of "descent with modification through natural selection" is, or entails, something straightforwardly factual, if historical in nature. Just as there is a fact of the matter whether Oswald acted alone in killing Kennedy, although it may be difficult to ascertain, so it is a factual, contingent matter whether species have arisen by natural selection, that is, by the sort of process Darwin sketches in the quoted paragraph above [168]. That this is indeed what happened has achieved consensus in the scientific community.

2. The theory of evolution is described by Darwin in procedural terms, not mathematically with equations. There is a story that gets told in

order to explain how speciation comes about. In this story, there are organisms that can reproduce at a higher rate than can be sustained by the environment. This leads to a "struggle for existence." Some of the organisms leave more progeny for systematic reasons (not just by chance): they are better adapted to the current environment. Further, what it is that makes them better adapted is inherited to at least some degree by their offspring. The process continues and may slowly lead to very different forms of life [169].

Our point of departure is this second point. Inspired, as it is said, by biological evolution, we naturally ask whether the underlying process might be harnessed computationally to solve problems for us. This is hardly a new thought; we proceed with the benefit of an extensive body prior investigations. (See §18.8 for references.)

A bit of jargon is necessary and will be helpful. Procedures, computerized procedures normally, that model an evolutionary process are called *evolutionary algorithms* or EAs. In using them, we undertake *evolutionary computation* or EC in order to seek solutions for specified problems. There are several main kinds of evolutionary algorithms. We shall look into more than one of them. First, however, some common principles.

18.2   Representations

Told abstractly, as Darwin does, the theory of evolution story is clear enough. To implement it on a computer and apply it to a real problem or game—to do evolutionary computation—requires much more specificity. My aim at present is to provide a degree of specificity, by way of examples, sufficient for a basic understanding of how evolutionary algorithms work.

For starters, we will need three things: (1) a problem we wish to solve, (2) a way of representing a solution to the problem, and (3) a way of evaluating a solution and obtaining a numerical score, which we can use as a measure of the "fitness" of the solution. We will look at an example for obtaining these three elements. The reader should keep it in mind that there are many other, additional ways to do things in evolutionary computation.

First, then, our problem shall be a simple knapsack problem. Figure 18.1 gives its canonical schema. In any specific knapsack problem (instance) the parameters p_i, w_i ($i = 1, \ldots, n$), and c will have specific values given to them at the outset. That leaves the x_i values to be determined. The problem tells us in expression (18.3) that the x_is can individually take on only two possible values: 0 and 1.

We seek the best combination of items, x_is, to put into the knapsack. A

$$\max z = \sum_{i=0}^{n} p_i x_i \tag{18.1}$$

subject to the constraints

$$\sum_{i=0}^{n} w_i x_i \leq c \tag{18.2}$$

$$x_i \in \{0, 1\}, \quad i = 0, 1, 2, \ldots, n. \tag{18.3}$$

FIGURE 18.1: Canonical form for the simple knapsack (SK) model.

solution (aka: *chromosome*) abstractly presented is a vector of x_i values:

$$x = \langle x_1, x_2, \ldots, x_n \rangle \tag{18.4}$$

for a problem with n decision variables. Concretely, we use 1 to indicate that the item is to be put into the knapsack and 0 to indicate it is not. A specific solution (aka: *chromosome*), then, can look like this:

$$x = [0, 1, 1, 0, 1, 0, 0, 0, 0, 0] \tag{18.5}$$

for a knapsack problem with $n{=}10$ items. According to x, items 2, 3, and 5 are to be put into the knapsack, while the others are out. That is, in terms of the model, $n{=}10$, $x_1{=}0$, $x_2{=}1$, and so on. A note on terminology:

1. Something is a solution if it is a legal, possible setting for the *decision variables*, that is, the x_is, all of them from 1 to 10. Whether a given solution is a good or bad solution is something else entirely. We shall be asking our evolutionary algorithms to find good solutions for us.

2. A solution representation—a type such as (18.4) or a token such as (18.5)—is also called a *chromosome* in analogy to biological chromosomes.

3. An item in a chromosome, e.g., x_3, is said to be or be at a particular *locus* (plural: loci). Thus, if we count starting at 1, there is a 0 at locus 4 and a 1 at locus 5 for x in (18.5).

(This is all just jargon, nevertheless useful.) Setting $c{=}$ 38.5, p, and w—

$$p = [16.936, 31.87, 9.938, 67.334, 83.061, 74.642, 4.241, 40.666, 16.028, 66.306]$$

$$w = [39.628, 2.975, 30.141, 29.355, 6.824, 31.625, 3.415, 6.786, 7.732, 49.275]$$

—we complete our specification of the problem instance. This takes us to (3), evaluating solutions. Consider the following snippet of code in the Python programming language:

```
x = [0, 1, 1, 0, 1, 0, 0, 0, 0, 0]
c = 38.5
p = [16.936, 31.87, 9.938, 67.334, 83.061, 74.642, 4.241,
        40.666, 16.028, 66.306]
w = [39.628, 2.975, 30.141, 29.355, 6.824, 31.625, 3.415,
        6.786, 7.732, 49.275]

def KSFitness(ps, ws, rhs, soln):
    obj = 0
    lhs = 0
    for i in range(len(ps)):
        obj += ps[i] * soln[i]
        lhs += ws[i] * soln[i]
    print('The objective function value is ' + str(obj))
    print('The constraint left-hand side is ' + str(lhs))
    violation = max(lhs - rhs, 0)
    print('The amount of violation for the solution is ' +
        str(violation))
    fitness = obj - 50.0 * violation
    print('The net fitness value is ' + str(fitness))
    return fitness

KSFitness(p,w,c,x)
```

When we execute it, we evaluate the fitness of our solution for this problem. Here is the output produced by the Python interpreter:

```
>>>
The objective function value is 124.869
The constraint left-hand side is 39.94
The amount of violation for the solution is 1.44
The net fitness value is 52.869
>>>
```

This particular solution, x, violates the constraint. The left-hand side (lhs) of the constraint

$$\sum_{i=0}^{n} w_i x_i$$

comes to 39.94, which is greater than the right-hand side (rhs, $c = 38.5$) given in the problem. Because the problem requires that

$$\sum_{i=0}^{n} w_i x_i \leq c \tag{18.6}$$

but in fact for our solution

$$\sum_{i=0}^{n} w_i x_i = 39.94 \geq c = 38.5 \tag{18.7}$$

a constraint is violated. We say that the solution is *infeasible* when this happens. If the solution were feasible, then its fitness value would simply be the value of the objective function of the problem:

$$\sum_{i=0}^{n} p_i x_i = 124.869 \tag{18.8}$$

Since it is not, we choose to penalize the solution for violating the constraint. The penalty we have chosen is 50 times the amount of the violation, which comes to 72.0, yielding a net fitness of $124.869 - 72.0 = 52.869$.

---

1. Initialize the population.

2. Evaluate each member of the population to obtain its fitness value.

3. Repeat until a stopping condition obtains:

    (a) Select solutions as parents for the next generation.
    (b) Modify the selected solutions.
    (c) Evaluate the new candidate solutions.
    (d) Select solutions to constitute the next generation.

FIGURE 18.2: Prototypical schema for an evolutionary algorithm.

Figure 18.2 presents a typical evolutionary algorithm in schematic form. There are many variations in the details as well as many alternate schema. This one is, to emphasize, prototypical. Points arising:

1. Initialization, step (1), is usually done randomly.[2] Recalling our example solution to our example knapsack problem, a random solution could be created by randomly setting each of the values (1 or 0) in the solution. The initial population would simply consist of a number of such solutions, the number being set as a model parameter (here suppressed in the interests of simplicity).

---

[2]There are many ways to do something randomly. It comes down to which probability distribution or density one is drawing from. For present purposes this is a matter beyond the scope of the discussion. The reader should assume that some sensible distribution or density is being used.

2. Steps (2) and (3c) call for evaluating each member of the current population. This is done using the fitness evaluation function on each solution, much as we saw above with the KSFitness function in the Python code.

3. Selection of parents for the next generation, step (3a), may be done in many ways. A popular way, and a good one, is to use *tournament-2* selection. To get one parent by tournament-2 selection, the program randomly picks two solutions from the current population, compares them with regard to fitness, and chooses the solution with the better fitness value. To get n parents, the program repeats this process n times.

4. Selection of the current solutions to constitute the next generation population, step (3d), may also be done in any of a number of of ways. Often it does not need to be done at all since the size of the newly created population is right. If the number of individuals in the new population exceeds the desired population size for the next generation, then ranking by fitness, for example, is an appropriate method of selection.

   And often not. In one form, so-called (μ, λ) selection, $\lambda > \mu$ offspring are created at step (3a). Then at step (3d), the best μ are selected deterministically to constitute the next generation population. In an alternative, called $(\mu + \lambda)$ selection, the λ offspring created at step (3a) are pooled with the μ solutions from the current generation, the pool is ranked by fitness, and the best μ are used to constitute the next generation.

Thus, we see how evolution may be modeled computationally. Now to a few details and some specific kinds of evolutionary algorithms. As usual, we will discuss prototypical examples. The boundaries of each of the methods for evolutionary computation I discuss are fluid and themselves evolving.

18.3 Evolutionary Programming

Evolutionary programming (EP) covers a family of evolutionary algorithms (EAs). The boundaries between EP and other EAs are not distinct and indeed can be said to be dissolving. Nonetheless, prototypical cases of EP are distinctive and reward study.

Lawrence Fogel invented or discovered EP about 1960 [105] and was subsequently very active in the field. His sons, David and Gary, have also long been important participants in EC, in part through the firm Natural Selection, Inc. http://www.natural-selection.com/.

Solutions (chromosomes) in EP prototypically are vectors of real (i.e., floating point) numbers with the following form extending the simpler form of expression (18.4) on page 384.

$$x = \langle x_1, \ldots, x_n, \sigma_1, \ldots, \sigma_n \rangle \tag{18.9}$$

All items in x are floating point numbers. The x_is in (18.9) are the decision variables for the problem, just as in our earlier discussion with regard to expression (18.4). The σ_is are standard deviations. Each decision variable x_i has an associated standard deviation σ_i. Both the decision variables and their standard deviations are subject to genetic variation and evolutionary change.

Adverting to Figure 18.2 on page 386 again, let us see how EP works (in this prototypical incarnation).

1. Initialization, step (1), is done randomly. Each item in x will normally have a range of permitted values, for example non-negative and less than 1,000,000 for the x_is and between 0.05 and 0.3 for the σ_is. Solutions or chromosomes are generated by randomly drawing permitted values for their elements. In all, μ solutions are created; that is the population size.

2. Fitness evaluation of the individuals in the initial population occurs in step (2). Each of the μ solutions is evaluated with the fitness function.

3. Step (3a) is the selection of solutions as parents for the next generation. It works as follows. The current population has μ solutions (a setup parameter for the run). Each of the μ solutions duplicates itself. It's that simple.

4. Step (3b) has the purpose of introducing new genetic variation. In EAs generally, this is said to be done by applying *genetic operators* to the solutions. EP normally uses only one genetic operator: mutation. Here is how it is typically done. Given a solution or chromosome, $x = \langle x_1, \ldots, x_n, \sigma_1, \ldots, \sigma_n \rangle$, we transform it by mutation to produce a new chromosome, $x' = \langle x'_1, \ldots, x'_n, \sigma'_1, \ldots, \sigma'_n \rangle$, with

$$\sigma'_i = \sigma_i \cdot (1 + \alpha \cdot N(0,1)) \tag{18.10}$$

and

$$x'_i = x_i + \sigma'_i \cdot N(0,1) \tag{18.11}$$

$N(0,1)$ indicates the draw of random variable from a normal distribution with mean 0 and variance 1. α is a scaling parameter and is typically set to about 0.2 [87, page 92]. Put verbally, these equations say that we first create the σ_is (standard deviations for the decision variables) by perturbing them with draws from a normal distribution. Then we create each x_i by perturbing it with a new draw from a normal distribution having mean 0 and standard deviation σ_i.[3] Thus, each locus of the parent chromosome is subjected to mutational perturbation. Each of the μ

---

[3]Since we do not want a σ_i to be set less than or even very close to 0, an additional rule is usually imposed that sets σ_i to a specified small value, if the random draw would put it below that value, and similarly if an x_i would otherwise be set outside its permitted range.

solutions in the parent population is processed this way, creating μ new solutions for a total of $(\mu + \mu)$ solutions.

5. Fitness evaluation occurs in step (3c). Each of the $(\mu + \mu)$ solutions is evaluated with the fitness function (if it has not been evaluated already).

6. Selection occurs in step (3d). The standard approach is to use what is called *probabilistic* (or *stochastic*) $(\mu + \mu)$ *selection,* in which each solution is compared in a series of pairwise tournaments to (at least) $q \approx$ 10 randomly chosen solutions from the population. The solution's score might be set in any of a number of ways. A commonly used approach is to score a solution for a single tournament with say 2 for having a superior fitness to its counterpart, 1 for a tie and -1 for a loss. Then a solution's net score is its total across all pairwise tournaments in which it participated. This process is repeated for all $(\mu + \mu)$ solutions, that is the μ solutions in the parent population, and the μ mutated solutions created from them.

18.4 Genetic Algorithms

Genetic algorithms (GAs) were originated by John Holland in the 1960s and 1970s [154]. They are very widely used today and, like all members of the family of EAs, GAs in use exhibit an extensive range of design forms. Again, we will focus on a prototypical GA.

GAs may be described succinctly by indicating how they differ from EP. The primary difference lies in how introduction of genetic variation is handled (step (3b) in Figure 18.2). GAs usually employ two genetic operators: mutation (as in EP) and crossover (which does not normally appear in EP). I'll illustrate using our knapsack example. Solutions, recall (18.5), look like this:

$$x = [0, 1, 1, 0, 1, 0, 0, 0, 0, 0] \tag{18.12}$$

In implementing mutation on such solutions we might alter each locus with a certain small probability (≈ 0.03). If a mutation event occurred for a locus, say locus 10 in the example, a new value for the locus would be drawn randomly and the solution changed accordingly. So solution (18.12) might change to

$$x = [0, 1, 1, 0, 1, 0, 0, 0, 0, 1] \tag{18.13}$$

Crossover is the other important genetic operator in GAs. It may be effected in very many different ways. Perhaps the simplest is *single-point crossover,* in which two solutions are paired, a crossover point is randomly chosen and new daughter solutions are created by copying and exchanging genetic materials according to the crossover point. To illustrate, let our two solutions be:

$$x = [0, 1, 1, 0, 1, 0, 0, 0, 0, 0]$$

$$x = [0, 1, 1, 0, 1, 1, 1, 1, 1, 1]$$

Since the solutions are 10 loci long, there are 9 potential crossover points, between loci 1 and 2, 2 and 3, ..., and 9 and 10. If we cross these two solutions at point 7, the resulting daughter solutions are:

$$x = [0, 1, 1, 0, 1, 0, 0, 1, 1, 1]$$

$$x = [0, 1, 1, 0, 1, 1, 1, 0, 0, 0]$$

Not every solution undergoes crossover in every generation. The rate is normally set with a run parameter, ranging typically from probability 0.4 to 0.8.

As a secondary difference with EP, GAs tend to use other forms of selection. Tournament-2 selection is among those widely used. To choose a parent solution for the next generation two chromosomes from the current population are randomly drawn (with replacement) and compared with respect to fitness. The winner gets to be a parent. This is repeated μ times for a population of size μ.

18.5   Coevolution

'Coevolution' may be usefully defined as an evolutionary change in a trait of the individuals in one population in response to a trait of the individuals of a second population, followed by an evolutionary response by the second population to the change in the first. [160]

The author is referring to biological populations, and normally of distinct species coevolving. Many examples of mutualism and symbiosis found in nature may in fact be cases of coevolution, although whether or not they are in any one instance is of course an empirical matter.

Evolutionary computation (EC) has absorbed the idea and adapted it for evolutionary algorithms. Two or more populations are maintained, with separate evolutionary processes for each population. Crossover and selection, for example, only occur among individuals in a single interbreeding population (species). Fitness, however, is determined in combination with individuals of other species. The situation is inherently strategic, whether in biological settings or in computational settings. The fitness of an individual is now very much dependent upon the makeup of the individuals in other populations.

Coevolutionary setups, with strategic interpretations, have been used to find robust solutions to optimization problems. In general, something is robust if it performs well under varying conditions. Wagner's characterization is representative and applies to any system: "A biological system is robust if it continues to function in the face of perturbations" [326, page 1], although often we will want to add "well" after "function."

To illustrate, consider a scheduling problem called the (simple) *flowshop scheduling problem*. We are given m machines and n jobs. Each job has to be processed on each machine. The processing time for job j on each machine i is $p_{i,j}$. A solution for a flowshop scheduling problem is an ordering by which the n jobs are to be (sequentially) processed. Thus there are $n!$ possible schedules. To optimize, we seek a schedule with the smallest time required to process all of the jobs (called the *makespan*). See Figure 18.3 for how the makespan for any given schedule is calculated. The processing times, $p_{i,j}$, are, however, often not known with certainty. Instead we may have ranges on their values. If so, then an optimum solution under one assumed set of values for the $p_{i,j}$s may not be optimal or even very good when their values are actually realized. What to do? We seek a robust solution, one that will do well, in terms of makespan, regardless of how the $p_{i,j}$s come out.

Working with flowshop problems, Herrmann [149] assumed a minimax (worst-case-optimal) notion of robustness. Using what is now called a teacher-learner coevolution framework [96], Herrmann evolved two populations together, one of scenarios ($p_{i,j}$s, more generally, parameterizations of the optimization problem, the teachers in modern parlance) and one of solutions (learners). Heuristically, evolution of the teachers finds the most unfavorable scenarios and evolution of the learners finds the best solutions for the worst-case scenarios. In the end, the system evolved unfavorable scenarios and solutions for those that performed well under the circumstances.[4]

18.6 Examples of Applications to Games

18.6.1 Discovering Strategies for IIPD

It is natural to wonder, especially after Axelrod's second tournament (reported in [15]), whether a GA, or some other form of evolutionary computation, could learn or discover successful strategies in IIPD games. Not surprisingly, Axelrod wondered this and conducted a computational experiment, which is reported in [16, 17]. His representation scheme was as follows. Solutions considered the three previous rounds of play in the game (each round a PD). One round in the supergame has four possible outcomes (CC, CD, DC, and DD). So, for a series of three rounds, there are $4 \times 4 \times 4 = 64 = 2^6$

---

[4]See [174] for a non-coevolutionary GA treatment of flowshop scheduling, one that seeks robust schedules under risk, rather than under uncertainty.

1. Given:

   (a) m machines; n jobs;

   (b) $Decision_{n \times 1}$, a permutation vector of job IDs, $1 \ldots n$;

   (c) $Processing_{m \times n}$, array of processing times, $Processing(i, j) =$ processing time of job j on machine i;

   (d) $Start_{m \times n}$, array of starting times, $Start(i, j) =$ starting time of the jth job from $Decision$ $(Decision(j))$ on machine i, initialized to 0;

   (e) $Completion_{m \times n}$, array of completion times, $Completion(i, j) =$ completion time of the jth job from $Decision$ $(Decision(j))$ on machine i, initialized to 0;

   (f) $Availability_{m \times 1}$, array of next availability times for machines, $Availability(i) =$ next available starting time for machine i, initialized to 0.

2. For $j = 1$ to n:

   (a) For $i = 1$ to m:

      i. $Start(i, j) = Availability(i)$

      ii. $Completion(i, j) = Start(i, j) + Processing(i, Decision(j))$

      iii. $Availability(i) = Completion(i, j)$

3. $Makespan = Completion(m, n)$

FIGURE 18.3: Makespan calculation procedure, for standard simple flow-shop problems.

distinct paths/developments/sequences. For each one of these 64 sequences, the agent (with a three-game memory) has to have a policy of C or D. So we need 64 bits, each bit coded as 0 or 1, to specify a policy. Each of the 64 possibilities is assigned a locus on the solution chromosome. A 0 or 1 at that locus indicates whether the agent will cooperate or defect in response to the three-round pattern associated with that locus. Notice that each of the $64 = 2^6$ loci has a unique six-slot pattern associated with it: the six values for how the two players played three rounds ago, two rounds ago, and one round ago.

This leaves the question of the initial three moves. Axelrod's solution is to encode an assumption into the solution chromosome. The first six slots correspond to the plays from the (fictitious) three rounds prior to the first. An agent may be thought of as maintaining a record of play from the start of the game, with the three rounds of play before the start supplied by the agent's own strategy (solution chromosome). Let $s_i \in \{0,1\}$ be the play by oneself in round i and $c_i \in \{0,1\}$ be the play by one's counter-party in round i.

FIGURE 18.4: Schema for a player's play record at time 1.

Each player will maintain a history of play record as the game proceeds. See Figure 18.4. At the start of play, a player will fill in the first six slots of its history of play record by copying the contents of the first six loci of its solution chromosome. At each round of play, starting with round 1, the player will read off the six slots to the left from its position in its history of play record, then consult its solution chromosome (loci 7–70) to determine how to play in that round. At the end of the round, the player will record its play and the play by its counter-party, and then advance to the next play position in its history of play record.

The agent's chromosome thus represents one of the 64 possible outcomes from a three-move sequence, and an assumption of what happened prior to the start of play. We need 64+6=70 binary loci for this. That's the solution encoding. Note: this yields 2^{70} possibilities, a very large number ($\approx 10^{21}$). That's the size of the space the GA will search.

Axelrod used the GA to evolve populations of solutions played against the eight "indicative" (most successful) strategies from tournament 2. He used a population of 20 chromosomes. As for the remaining GA parameters, "Levels of crossover and mutation were chosen averaging one crossover and one-half mutation per chromosome per generation. Each game consisted of 151 moves, the average game length used in the tournament." And voilà!

> The results are quite remarkable: from a strictly random start, the genetic algorithm evolved populations whose median member was just successful as the best rule in the tournament, TIT FOR TAT.
>
> ...
>
> Although most of the runs evolve populations whose rules are very similar to TIT FOR TAT, in eleven of the forty runs, the median rule actually does substantially better than TIT FOR TAT. In these eleven runs, the populations evolved strategies that manage to exploit one of the eight representatives at the cost of achieving somewhat less cooperation with two others. But the net effect is a gain in effectiveness.

...

These very effective rules evolved by breaking the most important advice developed in the computer tournament, namely, to be
"nice," that is, never to be the first to defect. These highly effective rules always defect on the very first move, and sometimes n
the second move as well, and use the choices of the other player
to discriminate what should be done next. The highly effective
rules than had responses that allowed them to "apologize" and
get to mutual cooperation with most of the unexploitable representatives, and different responses that allowed them to exploit a
representative that was exploitable.

Although these rules are highly effective, it would not [be] accurate to say that they are better than TIT FOR TAT. Although they
are better in the particular environment consisting of fixed pro-
portions of the eight representatives of the second round of the
computer tournament, they are probably not very robust in other
environments. [17]

Axelrod also reports a second experiment, playing the GA-evolved rules
against each other. In a nutshell, he found that initially the non-cooperators
take over, reducing everyone's mean score, then gradually they are replaced
by cooperators and we get a generally increasing mean score.

The basic pattern on display here is one of representing on chromosomes
the individuals in a space of strategies, then using a GA (or other metaheuris-
tic) to search the represented space for effective strategies. It is a pattern well
suited for use throughout procedural game theory.

18.6.2   Bidding in Auctions

It is known experimentally that humans exhibit certain systematic biases
in auctions, generally tending to overbid. Andreoni and Miller [9] ran GA
experiments that essentially duplicated in GAs what has been found with
humans.

The GAs used a 20-bit encoding scheme: 10 bits each for two bidding
parameters, plugged into a theoretically correct bidding formula. The GAs
then searched for the optimal parameters. Each agent had its own private
value for the good at auction. There were five different auctions examined,
with both four- and eight-bidder groups (these determined different bidding
functions). Their experiments used a coevolution environment. They created
two populations of 40 bidders each. Each bidder was randomly initialized
with a private value. Bidders were drawn from each population, the winner
of the auction was determined and rewarded accordingly. This was repeated
for 1000 generations, with GA operators (mutation, recombination) applied
between generations. What happened? Behavior qualitatively the same as that
of human subjects.

18.6.3   Blondie24 and Friends

David B. Fogel and Kumar Chellapilla have created an impressive checkers-playing program by using EP to evolve strategies. The story is recounted in the very readable *Blondie24: Playing at the Edge of AI* by David Fogel [101]. See [59, 60, 61, 62] for supporting technical work. In a nutshell, Fogel and Chellapilla built a neural network of 1,741 arcs (connections between nodes) for the purpose of evaluating board positions [101, page 359]. The parameters of the network were thus the 1,741 weights on the arcs.

Fogel and Chellapilla used EP to evolve the weights. Starting from a randomly generated population of solutions, each providing values for all 1,741 weights. Each solution, thus, determined all of the weights for the otherwise fixed neural network. Fitnesses of the solutions were estimated by using the associated neural networks to play checkers against each other. Fogel and Chellapilla used a population of 15 solutions. Each member of the population produced a single offspring each generation. Tournaments were held among the resulting 30 solutions and the top 15 were selected to participate in the next generation. Following this process for several hundred generations resulted in a solution that performed at the grand master level on a popular checkers Web site.

Especially given the limited computational resources Fogel and Chellapilla had to work with (two late-1990s PCs), this is quite an impressive result. They went on to produce similarly good results in blackjack [102] and chess [103, 104].

18.7   For Exploration

1. Read "Computers Play the Beer Game: Can Artificial Agents Manage Supply Chains?" [179]. Describe the systems modeled (simple supply chains), the methodology used and the results achieved.

2. Design an extension to the work reported in [179]. Conduct experiments to discover effective strategies of play. Assess the performance of the resulting system.

3. Extend the techniques in [174] and [17] to find robust-under-risk strategies of play in various indefinitely iterated 2×2 games, such as Chicken, Stag Hunt, and especially various asymmetric games.

18.8   Concluding Notes

A word or two for the sake of disabusing those with certain commonly encountered misunderstandings about evolution and EAs. Evolution and EAs are powerful, impressive methods of discovering innovative, creative solutions to difficult problems. They are, for this reason alone, most interesting and worthy of investigation.

Evolutionary procedures are associated with at least two widely held misunderstandings. First, they are not universal solvents. Neither evolution nor EAs are guaranteed to be successful or even good for any finite level of effort. Often they are, but they can easily fail. Evolution meliorizes—it seeks incrementally improved solutions—rather than optimizes. Evolutionary procedures have no way to detect optimal solutions other than failure to find better solutions. They may well fail even to produce very good solutions.

Second, although evolutionary procedures work by rewarding comparatively better performance (fitness measures this), it is incorrect to assume that increasing the intensity of selection will hasten or improve evolved performance. The optimal level of selection intensity, and of reward for better performance, is problem-dependent. As a general rule, the more difficult the problem is the better it is to *relax* selection and allow relatively weaker performers to have chances at leaving descendants. Why? An intuitive argument may suffice. All learning involves a tradeoff between exploration and exploitation. What is best will change from problem to problem and normally cannot be known ahead of time (except perhaps by studying similar problems). To increase the intensity of selection, that is, to more toward a "winner take all" regime for contributions to the next generation, is to tilt heavily in the exploitation direction. This is appropriate and will work for certain kinds of simple problems. For other problems, the difficult problems, better solutions are to be had by being patient and engaging in more exploration. Turning down the pressure of selection is one way and a good way to do this. I invite the reader to conduct experiments and see these points in action.

* * *

The literature on evolutionary methods and games is very large and growing quickly. The new journal *IEEE Transactions on Computational Intelligence and AI in Games* http://www.ieee-cis.org/pubs/tciaig/ is a good source of examples and applications, as is the annual conference "IEEE Conference on Computational Intelligence and Games" http://www.ieee-cig.org/. *The European Journal of Operational Research* (http://www.elsevier.com/locate/ejor), *Evolutionary Computation* (http://www.mitpressjournals.org/loi/evco), and *IEEE Transactions on Evolutionary Computation* (http://ieeexplore.ieee.org/xpl/RecentIssue.jsp?reload=true&punumber=4235) are important journals in the field of EC and will publish articles on EC and strategic interaction. Much (most?) of the action in the field, however, is published at conferences. See GECCO, the Genetic and Evolutionary Computation Conference (e.g., `http://www.sigevo.org/gecco-2011/index.html`), Parallel Problem Solving from Nature (e.g., `http://home.agh.edu.pl/~ppsn/`), and the IEEE Congress on Evolutionary Computation (e.g., `http://www.cec2011.org/`).

Chapter 19

Backward Induction

19.1   Three Problems

19.1.1   The Surprise Examination

The scene is a classroom. The teacher announces that there will be a surprise examination given at one of 20 meetings of the class. The exam will be a surprise in the sense that on the morning of the day of the exam the students will not have enough information to know that the exam will occur that day with a probability at or above l_s, a given threshold (level or line for surprise), say $l_s \geq \frac{1}{3}$. That is, the students will be surprised if an exam is held on a given day *and* that morning the probability of having the exam is (so far as they know) less than $l_s = \frac{1}{3}$. If the exam is given but the probability is higher, the students are not surprised. There are, thus, two requirements: (1) an exam is given, and (2) on the day of the exam, from what the students know, the probability they will have an exam that day is less than l_s.

The students find the teacher's assertion puzzling. They reason as follows. "There cannot be a surprise exam on the 20th, the last day, since we would know that morning that the exam had not yet taken place and hence that it would have to be given that day. But if we know that the surprise exam cannot happen on the last day, by similar reasoning it cannot happen on the next-to-last day. If the morning of the nineteenth day arrives without our having had the exam, we would know that it has to be given then, since it can't happen on the 20th day. Therefore, the exam cannot be given on the 19th day either. By continuing this reasoning process we find that it is impossible for the teacher to give us a surprise exam during the 20 meetings of the class."

Some time later, the teacher opens the class session by handing out an exam. Naturally, the students are surprised. Surprise becomes distress when they see the first question on the exam:

1. Explain the fallacy in the reasoning that led you to believe it impossible for me to give you a surprise exam as announced or, if you now wish to maintain that I can give you a surprise exam, explain how I can do it.

How should *we* answer the examination question?

19.1.2   Backward Induction in Games

We may as well start with the ever-present Prisoner's Dilemma (PD) game, shown in Figures 19.1 and 19.2. Note, however, that this is for the sake of convenience. The issues here range well beyond the PD and apply quite broadly in games.

	Cooperate	Defect
Cooperate	3, 3	0, 5
Defect	5, 0	1, 1

FIGURE 19.1: Prisoner's Dilemma, a specific form.

	$\neg$Confess	Confess
$\neg$Confess	R_R, R_C	S_R, T_C
Confess	T_R, S_C	P_R, P_C

FIGURE 19.2: Prisoner's Dilemma, generic form; $T_x > R_x > P_x > S_x$; $R_X > 2T_x S_x$.

Now we are ready for the second question on the exam:

2. In a 100-shot Iterated Prisoner's Dilemma game, played between the teacher and an unknown, but fully competent human subject, the teacher announces that she will gain the reward from mutual cooperation at least 2 times, net. That is, if P is the penalty for mutual defection and R is the reward for mutual cooperation, the teacher is asserting that she will get at least $98 \cdot P + 2 \cdot R$ points from the 100 trials. Can this assertion plausibly be justified? Why or why not?

Let us agree for the moment that if we play PD once under the right conditions (anonymity, rationality, etc.), then it is rational for both players to defect. An argument from dominance does the trick.

But what about a supergame, in which the PD is the stage game? That is, what if we play PD iteratively, that is between the same players? There are two cases to distinguish:

1. We play indefinitely long. This might be infinitely long or the stopping rule may be probabilistic.

   In this case we have the Folk Theorems, which tell us that just about any possible strategy can belong to some equilibrium. A disturbing result.

2. We can play for a fixed and known (to the players) number of periods.

In this case, we have a proved result from game theory that says that there is only one (sub-game perfect) equilibrium and that is that everyone always defects. (See standard textbooks in game theory, such as [31, 114].)

The argument (in the proof) for universal defection is made on the basis of backward induction. If it were the last round of play, everyone would defect. The next-to-last round, everyone defects because they know that in the next round everyone will defect and so there can be no reward for cooperation. And so on for all the preceding rounds. So, game theory says that insofar as players are rational, they will always defect in finitely iterated PD. The details turn out to require rather careful, even fussy, reasoning and assumption of common knowledge among the agents [14]. Whether this is a credible assumption can be challenged. I shall not do so here. Suffice it to say that on the standard account from game theory, there is only one appropriate ('sub-game perfect') equilibrium for finitely iterated PD, and that is both agents always defect.

In short, reasoning by backward induction tells us that the teacher is wrong. She will not get at least $98 \cdot P + 2 \cdot R$ points from the 100 trials. Is this right? Does the fact that we are taking a surprise examination bear on the matter?

19.1.3   Dynamic Programming

There is a third and final question on the surprise examination.

3. Consider the network shown in Figure 19.3 (after [76, page 9]). The network consists of nodes, labeled 1 through 9. Directed arcs show permitted travel between nodes and the amount of time required to make the trip. For example, it is permitted to travel from node 4 to node 8 and the time required is 7 units. It is not permitted to travel directly between nodes 2 and 7, but you can go from 2 to 4 and then on to 7. It is not possible to reach node 3 starting from node 6. And so on. Find the shortest path from each node to node 9.

Reasoning by backward induction (called *dynamic programming* in this context), we find that the shortest route from node 8 to 9 is 10 units long, since there is only one route. Similarly, the shortest route from node 7 to 9 is 3 units long. The shortest route from 6 to 9 is the minimum of the lengths of the two possible routes, 6-9 for 15 instead of 6-8-9 for 17. The shortest route from 5 to 9 is 10, since that is the length of the only route. The shortest route from 4 to 9 is the minimum of the four possible routes: 4-8-9 for 17; 4-6-9 for 18 (notice that we already have 6-9 for 15, so we simply add 4-6 for 3 to it); 4-7-9 for 18; and 4-5-7-9 for 14. The shortest route from 2 to 9 via 5 is the time to go from 2 to 5 plus the cost from 5, which comes to 12 + 10 (from

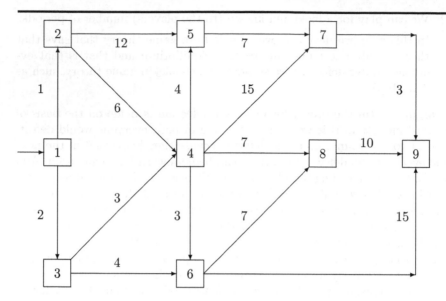

FIGURE 19.3: Network with travel times.

before). The shortest route from 2 to 9 via 4 is 6 plus the time from 4 to 9, which from before is 14. The shortest route from 3 to 9 is the shorter of 3 plus 4-9 = 3 + 14 = 17 and 4 plus 6-9 = 4 + 15 = 19. Finally the shortest route from 1 to 9 is the shorter of 1 plus 2-9 = 21 and 2 plus 3-9 = 19. Figure 19.4 displays the solution.

19.1.4   What to Make of All This?

We have three examples before us that involve the use of backward induction. In the Surprise Examination case, clearly the students have done something incorrectly. How did they go wrong? In the dynamic programming example, the reasoning seems correct and the results it finds seem correct. How is the reasoning here different than in the Surprise Examination case? This brings us to the case of backward induction in games. The argument for universal defection seems correct and certainly is accepted as correct, and still the situation produces unease. If the number of iterations were 1000, would it really be irrational to try cooperation just a little? 100,000? 1,000,000? What if we made the penalty for mutual defection minuscule, just ε larger than 0, the sucker's payoff? Say we made it I^{-1} where I is the number of iterations. What if we set the temptation to defect, T equal to $R+I^{-1}$, then increased R dramatically? For all such settings the backward induction argument remains in force, yet is there no point at which you would consider it irrational *not* to try some cooperation?

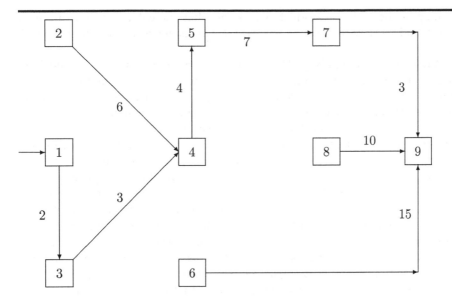

FIGURE 19.4: Shortest path solution by backward induction.

The similarity of this reasoning [for the Surprise Examination and the impossibility of giving one] to that of the argument for dominant defection [in Prisoner's Dilemma] throughout a series of known finite length is worth noting because of course the Surprise Examination is treated standardly in the philosophical literature as a *paradox,* thought to hide some fallacious piece of logical leg-erdemain. That the same form of reasoning is thought of as valid in the theoretical economics literature, though perhaps inapplicable in some practical sense, indicates that important work remains to be done in bridging the two bodies of work. [135, page 163]

What are we to make of all of this? Time to revisit the Surprise Examination problem.

19.2 Revisiting the Surprise Examination

To side with the students a bit, if only to sympathize with their position, let us consider an obvious play by the teacher. After making the announcement, the teacher secretly picks a day randomly, say from 1 to 19 (with an equal chance for each of these days). But what if either 19 or 18 comes up? On the morning of the 18th the students will know that the chance of an exam is at least $\frac{1}{3}$ and so they will not be surprised. Things get worse for the teacher if 19 comes up.

What about randomly picking a last possible date, one that could have been later and that leaves ample slack? The teacher might randomly pick a number from 2 to n ($2 < n < d_{\max}$), with $d_{\max} = 15$, say, and keep this secret from the students. Then the teacher might draw a second number d from 3 to n and give the exam on that day. The effect of this is to deny the students a starting point for reasoning backwards. This would seem to be a procedure sufficient for the teacher to give a surprise exam (providing the numbers work out). Even so, I want to explore a stronger, more fundamental position.

Suppose that instead of a span of 20 days (or whatever) the teacher announced that there would be a surprise exam on the very next meeting of the class. What would we make of that? Surely the teacher has said something false—that there will be a surprise exam at the next meeting. Perhaps it is worse than that; perhaps what the teacher said was necessarily false, a contradiction. Perhaps it is *impossible* that the teacher spoke truly.[1] If so, then the students should not be surprised no matter what the teacher does, since from a contradiction anything follows. Put more carefully, if the teacher has contradicted herself, then the students have received no information and no matter what the teacher does it will be surprising in the sense that the students do not have sufficient information to predict it.

Whether or not the teacher, in the one-day case, has spoken inconsistently, surely the teacher has spoken oddly. The teacher has made two assertions:

E: There will be an exam at the next class meeting, and

S: The exam will be a surprise, students will not have sufficient information on the morning of the next class meeting (before class) to know that the exam will occur that period.

Suppose now that the teacher rolls a pair of fair dice the morning of the next class period. The teacher's policy is that if a 7 a 2 or a 12 comes up (probability $= \frac{8}{36} = \frac{2}{9} < \frac{1}{3} = l_s$), the exam will be given; otherwise there will be no exam that day. The dice are thrown and in fact a 7 comes up, so the exam is given. The students are surprised and so is the teacher. Had the roll of the die produced anything but a 2, 7 or 12, no exam would have been given and the students would know that the teacher lied. The students did not have the information to know, with a probability above $\frac{1}{3}$, that the exam would be given that day.

The fundamental issue is not whether the teacher can give a surprise exam, but whether the teacher can speak truly (in saying that the exam will occur and that it will be a surprise) *and* give a surprise exam. If the teacher engages

---

[1] Barring annoying and ultimately irrelevant philosophy and science fiction eventualities, such as interventions on the students' memories, brains in vats, giving the exam without the students being aware of it, and so on.

1. $E \vee \neg E$

   There will be an exam tomorrow or not, a tautology.

2. $l_s \geq \frac{1}{3}$

   Arbitrary threshold; may be changed without loss of generality.

3. $(P(E) < l_s \wedge E) \to S$

   If the probability of an exam tomorrow is less than or equal to the stipulated threshold and the exam is held tomorrow, then surprise.

4. $V(a) \leftrightarrow (E \wedge S)$

   The teacher's assertion that there will be a surprise examination tomorrow is true if and only if there is an examination tomorrow and the students are surprised.

5. $P(E) < l_s$

   The probability of an exam tomorrow is less than l_s.

Thus $(E \wedge S \wedge V(a)) \vee (\neg E \wedge \neg V(a))$

   Either there is a surprise examination and the teacher speaks truly or there is no examination and the teacher speaks falsely.

FIGURE 19.5: Version 5 of the students' reasoning applied to the one-shot surprise exam problem.

in self-contradiction, or otherwise speaks falsely, then the students should not be surprised that an exam is given. At best, the students' reasoning shows that the teacher cannot speak truly and give a surprise exam.

How exactly did the students' reasoning go wrong (if it did go wrong)? We need to reconstruct the reasoning with some precision. Such reconstructions may of course miss the actual reasoning; our aim is to find the best possible reasoning. A change of idiom will help.

Figure 19.5 presents, formally, a logically valid argument, one that I think is most apt to the situation. (This is Version 5. §C among the appendices presents four other arguments that some might find tempting, and shows why they are wanting.)

The left disjunct of the conclusion (Figure 19.5) follows if there is indeed an exam the next day, while the right follows if there is no exam the next day. Version 5 is a valid argument with a consistent set of premises (that is, it is possible for each of the premises to be true). Notice that assumption (4) has been strengthened to a biconditional ("if and only if"). This is harmless and could have been done for the earlier versions. The strengthening amounts to

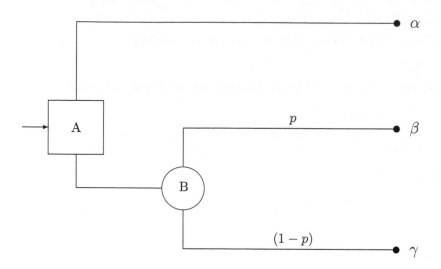

FIGURE 19.6: Decision tree for a one-shot surprise examination problem. $\alpha =$ Do not announce a surprise examination (and do not give one). $\beta =$ Announce a surprise examination and give one with probability p. $\gamma =$ Announce a surprise examination and not give one with probability $(1 - p)$.

accepting a rule that credits the teacher with speaking truthfully, $V(a)$, if and only if what she said—that $(E \wedge S)$—is in fact true.

The upshot is this. In the one-shot surprise exam problem, the teacher must either speak falsely (e.g., by making a self-contradictory statement) or speak truly but with a probability less than l_s. Only by putting herself at risk of falsehood is it possible for her to speak truly in this case. By taking a risk (of speaking falsely) the teacher expands her scope of action. This positions us to take an alternative, complementary look at the situation.

Figure 19.6 presents the situation for the teacher in the form of a decision tree. The teacher's choice is between α for sure, where α is not giving an exam and not announcing an exam, and a lottery between β with probability p and γ with probability $(1 - p)$. In either case, the teacher would announce that there will be a surprise examination. With probability p the exam is given; otherwise not. Presumably the teacher ranks the three outcomes in order of preference as $\beta \succ \alpha \succ \gamma$. In consequence for sufficiently high values of p, the teacher will prefer the lottery—at B—to α for sure. If there is such a value of p that is also less than l_s, then the teacher can arrange the requisite randomizing device and proceed to announce the surprise examination. It is unproblematic. Here's the picture:

$$0 \qquad\qquad l_r \quad l_s \qquad\qquad\qquad\qquad\qquad 1$$

If $p < l_s$ and the exam is held, the students are surprised. If $p \geq l_r$, where l_r is the minimum probability of speaking truly that the teacher will insist on, then the teacher is willing to risk speaking falsely. So, with $l_r \leq p < l_s$, the teacher can speak truly in saying there will be a surprise examination tomorrow, can do so with probability of speaking truly $= p$, and will prefer to choose down at A (taking the gamble at B), rather than up (taking α for sure), in Figure 19.6. Let $r = (1 - p)$ be the *risk* of speaking falsely.

A deeper lesson lurks. The soundness of version 5, Figure 19.5, relies on the teacher being willing to accept a risk of at least $r = (1 - l_s)$ of speaking falsely. Given this, many would choose not to utter the one-shot surprise exam assertion. Honesty, integrity, prudence, or whatever may well prevent a reasonable person from saying he or she knows to have a chance higher than $(1 - l_s)$ $(=\frac{2}{3}$ in our example) of being false. Better to keep silent.

What if the students know this? In the augmented one-shot case, we can construct a modified version of the correct reasoning for the original one-shot case. See Figure 19.7.

Concretely, if the threshold of surprise, l_s, is $\frac{1}{3}$ and the teacher's threshold of veracity, l_r, is $\frac{1}{2}$ (or anything greater than l_s), then the teacher cannot utter the one-shot surprise exam assertion without self-contradiction. We need $p < l_s$ for the teacher to prefer the risky choice (in order to guarantee surprise). More shots, however, attenuate the teacher's veracity scruples. All that is required is:

1. The total probability of having the exam is greater than or equal to l_r

2. The total probability of surprise is less than l_s.

This is trivially arranged for any $l_s : 0 < l_s < 1$, and for any $l_r : 0 < l_r < 1$, provided enough periods are available. Simply decide to give the exam with equal probability to every period. Concretely, let us say that there are $n = 20$ periods, $l_s = \frac{1}{3}$ and $l_r = \frac{4}{5}$. Each period has a $\frac{1}{20}$ chance of being picked for the time of the exam. The period is drawn and the exam is held.

Let us reason now by backward induction. If the exam is held on the 20th day, that morning the students will know that the exam will be held that day (assuming they are certain the exam will be held at all). There is only a 1 in 20 chance this will happen, however, and this is below l_r. If the exam occurs on day 19, the students have a 50% certainty that morning, which is above l_s, so they will not be surprised. The chance of this happening is 1 in 20, so counting just 19 and 20, we have a total chance of the teacher speaking falsely (there is an exam but not a surprise) of 1 in 10, which leaves another 10%. The teacher is willing to take a 20% chance of speaking falsely $(= 1 - l_r)$ and we've now eaten up half of that.

1. $E \vee \neg E$

   There will be an exam tomorrow or not, a tautology.

2. $l_s \geq \frac{1}{3}$

   Arbitrary threshold; may be changed without loss of generality.

3. $(P(E) \geq l_s \wedge E) \to \neg S$

   As before, if the probability of an exam tomorrow is greater than or equal to the stipulated threshold, and the exam is held, then there is no surprise.

4. $V(a) \leftrightarrow (E \wedge S)$

   Exam tomorrow and surprise, asserted by the teacher.

5. $P(E) \geq (1 - r)$

   The probability of an exam tomorrow is greater than or equal to $(1-r)$, from the teacher's policy on self-veracity.

6. $r < (1 - l_s)$.

   The teacher's willingness to risk falsehood is less than $(1 - l_s)$.

7. $(P(E) \geq (1 - r) \wedge r < (1 - l_s)) \to P(E) \geq l_s$

   Simple mathematical truth.

Thus $(\neg V(a) \wedge E \wedge \neg S) \vee (\neg E \wedge \neg V(a))$

   Contradiction, validly deduced.

FIGURE 19.7: Students' reasoning applied to the augmented one-shot surprise exam problem.

Continuing our backward induction, if the exam occurs on day 18, then the students have a chance of being surprised of $\frac{1}{3} = l_s$. This eats up another 5%. But that is it. If the exam occurs on day 17 or earlier, the students will be surprised. At day 17 there is a 1 in 4 chance the exam will be that day and $\frac{1}{4} < l_s$, so surprise. And more surprise if the exam occurs before day 17. In short, under this regime, there will be a surprise exam with probability 0.85. It is easy to see that if we increase n and keep all else constant, we can drive the probability of a surprise exam to as close to 1 as we want.

The n-period case generalizes the one-shot case. The teacher can speak truly provided the teacher is willing to undertake some risk, $(1 - l_r) = r > 0$, of speaking falsely and providing n is large enough (given r and l_s). The teacher cannot be certain of speaking truly, but in this respect the case is like most. Usually when we assert we take some chance of speaking falsely, even with the best of intentions. What is odd is to interpret a speaker otherwise. The only way the teacher could have spoken truthfully and given the surprise exam was to have spoken with some chance of speaking falsely. The students erred in interpreting the teacher.

19.3   Revisiting Definitely Iterated PD

What about Definitely Iterated PD? Are there paradoxes here as well, or not? There are. First, behavior is not predicted accurately. It is interesting, and significant, that in the first human experiment with Iterated Prisoner's Dilemma the human subjects were asked to record their thoughts as the game was being played [99].[2] Comments such as

- "Perverse!"

- "Oh ho! Guess I'll have to give him another chance."

- "In time he could learn, but not in ten moves so:"

- "What's he doing?!!"

- "I'm completely confused. Is he trying to convey information to me?" and

- "This is like toilet training a child—you have to be very patient."

appear throughout the 100 iterations of the game. Even so, the two subjects jointly cooperated in 60 of the 100 iterations. By the lights of classical game theory this was a remarkably rewarding triumph of irrational behavior. These results are consistent with subsequent empirical findings. By the available evidence, the teacher's claim in question 2 is well supported.

How are we to explain this? One possibility is that the agents believe their counter-players are irrational and so are responding rationally to that [186]. Perhaps. It certainly might happen. Whether it does is an empirical matter. I want to explore a somewhat different possibility, the possibility that the players are rational, that is to say warranted and acting sensibly in dealing with uncertainty ([186] examines situations under risk). Whether this happens is also an empirical matter. We shall have to await experiments to dispose of the question of what agents actually do.

Figure 19.6 (page 406) presents (the schema for) a decision tree for a decision under risk: node B indicates a chance event with probability p leading

---

[2]The relevant data are most conveniently reproduced in Poundstone's excellent and accessible treatment of the Prisoner's Dilemma game [247, pages 106–123].

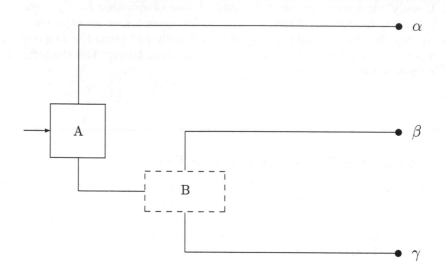

FIGURE 19.8: Schematic decision tree schema a decision under uncertainty. A is a decision node. B is an uncertainty node. Payoffs α, β, γ are in natural quantities.

to β and $(1-p)$ to γ. Figure 19.8 (page 410) is a decision tree for a decision under uncertainty: node B indicates an event that might lead to β or to γ, but no probability is available.

How can a decision maker rationally decide what to choose at A in Figure 19.8? Clearly, if the decision maker prefers β to α and prefers γ to α, then the decision is easy: choose down at A. Suppose, however, that α, β, and γ are given in natural quantities, for example money or energy, and that all things being equal, the decision maker prefers β to α ($\beta \succ \alpha$) and α to γ ($\alpha \succ \gamma$). Now we have something of a dilemma. Under risk, standard decision theory (Rational Choice Theory) would have us convert the payoffs (α, β, γ) to utilities and fold back the tree, taking the expected value at node B. That may be fine in the case of risk, Figure 19.6; it is inapplicable in the case of uncertainty, Figure 19.8.[3]

What remains is for decision makers to think hard about what they want to do, and then choose. There is no stricture from Rational Choice Theory preventing a decision maker from choosing down in Figure 19.8 even if all things being equal the decision maker's preferences on the (naturally measured) outcomes are $\beta \succ \alpha \succ \gamma$. The decision maker may simply decide that all things are not equal and that it is better to leave open the opportunity for

---

[3]I note that John Harsanyi and others would disagree. I do not think the disagreement is material for my present point.

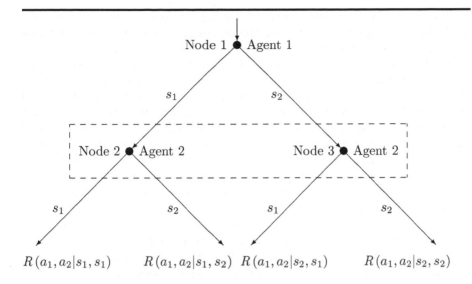

FIGURE 19.9: Extensive form game tree schema.

β. If γ obtains, the decision maker may well regret that without regretting the choice. We might say that such a decision maker is one who is guided by the L&L principle, as in " 'Tis better to have loved and lost than never to have loved at all."

We can now apply these ideas to a properly strategic situation. See Figure 19.9 (page 411), which is a schema for a 2×2 game but in extensive form. The uncertainty box now covers nodes 2 and 3. It determines what is called an *information set*. Agent 2, in this case, will be uncertain whether it is at node 2 or node 3. This is analogous to Figure 19.8 in which the decision maker was uncertain about the outcome of node B.

After agents 1 and 2 have made their decisions they receive rewards (a_1, a_2) contingent upon the strategies they played: $R(a_1, a_2|s_1, s_1)$ and so on. Let us suppose that we have a Prisoner's Dilemma game here with s_1 being the cooperative play and s_2 the defecting. The rewards then are $R(R, R|s_1, s_1)$, $R(S, T|s_1, s_2)$, $R(T, S|s_2, s_1)$, and $R(P, P|s_2, s_2)$ with all values, as always, given in money or some other natural quantity. Suppose that the values are in dollars and that $S = 0$, $P = \varepsilon$, and $T = R + \varepsilon$. Fix R at 3 and ε at 1. This creates a rather standard payoff schedule for a Prisoner's Dilemma and we would expect that people would choose to defect, s_2. Must they do so on pain of irrationality? No, not at all. While defecting is dominate as measured by dollar payoffs alone, there is nothing to say that an agent cannot prefer to take the act of cooperating or defecting into account in arriving at an overall decision about what to do.

For an agent that cares about defecting or cooperating (however described), any preference ordering on the outcomes is possible. Consider a new setting: $\varepsilon = 10^{-3}$ (that's in dollars) and $R = 10^3$ (also in dollars). It does not contradict rationality to invoke the L&L principle and decide to cooperation. I imagine that many people would. I suggest that often when people cooperate in laboratory experiments they are doing so for similar reasons. Further, I suggest that if we make R large enough and ε small enough (but > 0), almost no one would insist on defecting. Or switch the payoff currency. S is a cruel and painful death preceded by observation of cruel and painful deaths of all those you love and hold dear. ε is S, but with a very small modicum of mercy or comfort shown to someone. R is a long, rewarding, generally flourishing life for yourself and all those you love and hold dear. T is R plus another dollar of wealth. Still want to defect?

Now consider Figure 19.10, a decision tree with uncertainty on both branches from the starting decision node, A. Let us interpret the tree schematically. Choosing up at A is choosing to defect at every round in a Definitely Iterated Prisoner's Dilemma. B indicates uncertainty about the counter-party's play. Let α be the reward if the counter-party never cooperates either, and β be the reward if the counter-party does cooperate some (but we are uncertain how much). To choose down at A is to decide to try some cooperation. Let γ be the reward if the cooperation is reciprocated and δ if it is not (and the counter-party always defects).

Under this schematic interpretation for Figure 19.10, are there no circumstances—payoffs, number of rounds played—and no strategy—GRIM TRIGGER: cooperate until the counter-party defects, then defect ever after?—under which you would choose down at A and be satisfied with the L&L principle? To re-emphasize: nothing in Rational Choice Theory compels you not to on pain of irrationality.

19.4   Revisiting Dynamic Programming

Why is the DP problem seen as basically unproblematic, given what we find for the Surprise Examination problem and (taking my suggestion) for Definitely Iterated Prisoner's Dilemma?

There is a folk saying in the field of operations research to the effect that "We do not solve (or optimize) problems; we solve (or optimize) models." The reason this is a folk saying is that it is true and often forgotten. When it is forgotten, there is risk of having what I shall call a *wild-representation* problem. This is to say that we reason with representations (or models) of situations in the wild. There is the game in the wild and there is our representation of it with which we reason. No substantive conclusion, however correctly drawn from a representation, is necessarily correct in the wild, in the target

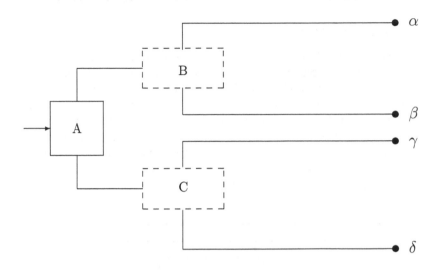

FIGURE 19.10: Schematic decision tree schema a decision under uncertainty. A is a decision node. B and C are uncertainty nodes. Payoffs α, β, γ are in natural quantities.

system. What we see in the Surprise Examination problem is that the students reasoned correctly on the wrong model.[4] Similarly, the argument I have made regarding the Definitely Iterated Prisoner's Dilemma is that there are circumstances in which a decision maker can be quite rational but for which the conclusion of the backward induction reasoning (to ALWAYS DEFECT) is incorrect because the model employed in the reasoning is incorrect. The modeler can declare that payoffs are in utilities and reason that the only rational strategy is ALWAYS DEFECT. The modeler, however, cannot force the decision maker to perceive the situation as the modeler would prefer.

In short, the dynamic programming backward induction is successful because the model happens to be accurate enough for the most part. The students got the model drastically wrong in the telling of the Surprise Examination. As for Definitely Iterated Prisoner's Dilemma and other situations that seem to afford a backward induction argument, there is no reason on grounds of rationality for the real decision maker to have preferences that take into account factors other than direct payoffs measured in utilities. And if the decision maker does this, it may well be rational not to act by the recommendation of backward induction on the inaccurate model.

---

[4]The folk term for this kind of mistake is "error of type III."

19.5   Rationality Redux

These points may reinforced by considering the matter from another perspective. Suppose we are playing five iterations of Prisoner's Dilemma with Axelrod's payoffs ($T=5$, $R=3$, $P=1$, $S=0$). The players have individually consideration sets of six strategies: ALWAYS DEFECT and GRIM-1–GRIM-5, where GRIM-n means to play GRIM TRIGGER (start by cooperating, cooperate until the counter-party defects, then defect always after that) until after the nth round of play, at which time you defect forever. So with five rounds of play, GRIM-5 = GRIM TRIGGER and GRIM-0 = ALWAYS DEFECT. Table 19.1 presents in strategic form the payoffs of these strategies under our present setup.

	5	4	3	2	1	0
5	15,15	12,17	10,15	8,13	6,11	4,9
4	17,12	13,13	10,15	8,13	6,11	4,9
3	15,10	15,10	11,11	8,13	6,11	4,9
2	13,8	13,8	13,8	9,9	6,11	4,9
1	11,6	11,6	11,6	11,6	7,7	4,9
0	9,4	9,4	9,4	9,4	9,4	5,5

TABLE 19.1: A cascading Prisoner's Dilemma in strategic form. Five rounds of Axelrod's stage game. Variations of GRIM TRIGGER.

Points arising:

1. We have what I will call a *Cascading Prisoner's Dilemma*. Every pair of adjacent strategies (GRIM-x,GRIM-$(x-1)$), $x = 1, 2, \ldots$, creates a Prisoner's Dilemma situation, favoring GRIM-$(x-1)$ over GRIM-x on the standard analysis. GRIM-$(x-1)$ is the best response to GRIM-x. Importantly, GRIM-0 is the best response only to two strategies: GRIM-1 and GRIM-0.

2. There is only one Nash equilibrium: (0,0), i.e., both players play GRIM-0.

3. The pattern extends straightforwardly if we increase the number of rounds of play.

4. A player that could move first and reveal its strategy would surely play GRIM-5; in the general case the player would play GRIM-n where n is the largest value of x available.

5. A player that played GRIM-0 against a GRIM-5 play suffers an opportunity cost of (15-9) for not matching with a play of GRIM-5 (and a higher opportunity cost of (17-9) for not playing GRIM-4). In general, the opportunity cost of playing GRIM-0 instead of matching a GRIM-x play is $(c_x - 9)$ where $c_0 = 5$ and $c_x = c_{x-1} + 2$, $x = 1, 2, \ldots$. Similarly, the opportunity cost of playing GRIM-0 against GRIM-x instead of playing GRIM-$(x-1)$ is $(2 + c_x - 9)$. And

$$\lim_{x \to \infty} (c_x - 9) = \lim_{x \to \infty} (2 + c_x - 9) = \infty \qquad (19.1)$$

6. At some point it becomes foolish for either player to play GRIM-0.

19.6 For Exploration

1. In light of our discussion of rationality, comment on these passages from David Hume's *Treatise of Human Nature*.

   > Reason is, and ought only to be the slave of the passions, and can never pretend to any other office than to serve and obey them.

   And

   > What may at first occur on this head, is, that as nothing can be contrary to truth or reason, except what has a reference to it, and as the judgments of our understanding only have this reference, it must follow, that passions can be contrary to reason only so far as they are accompany'd with some judgment or opinion. According to this principle, which is so obvious and natural, 'tis only in two senses, that any affection can be call'd unreasonable. First, When a passion, such as hope or fear, grief or joy, despair or security, is founded on the supposition or the existence of objects, which really do not exist. Secondly, When in exerting any passion in action, we chuse means insufficient for the design'd end, and deceive ourselves in our judgment of causes and effects. Where a passion is neither founded on false suppositions, nor chuses means insufficient for the end, the understanding can neither justify nor condemn it. 'Tis not contrary to reason to prefer the destruction of the whole world to the scratching of my finger. 'Tis not contrary to reason for me to chuse my total ruin, to prevent

the least uneasiness of an Indian or person wholly unknown to me. 'Tis as little contrary to reason to prefer even my own acknowledge'd lesser good to my greater, and have a more ardent affection for the former than the latter. A trivial good may, from certain circumstances, produce a desire superior to what arises from the greatest and most valuable enjoyment; nor is there any thing more extraordinary in this, than in mechanics to see one pound weight raise up a hundred by the advantage of its situation. In short, a passion must be accompany'd with some false judgment. in order to its being unreasonable; and even then 'tis not the passion, properly speaking, which is unreasonable, but the judgment.

—Part 3 Of the will and direct passions, Sect. 3 Of the influencing motives of the will

Note: Hume lived from 1711 to 1776. In the eighteenth century the word *passions* was used where we would today use *preferences*.

2. Construct and discuss other Cascading Prisoner's Dilemma games. Can you identify other ways to generate them? Are some more favorable to a degree of cooperation than others? If so, why?

3. Consider the Electronic Mail game introduced by Ariel Rubenstein [267]. Is the argument presented in this chapter relevant to an assessment of the Electronic Mail game? If so, why and how? If not, why not?

19.7   Concluding Notes

Good behavioral game theory work on backward induction is beginning to emerge. See, for example, "A Dynamic Level-k Model in Centipede Games" by Teck-Hua Ho and Xuanming Su [152]. They propose a model to predict violations of backward induction and do experiments to fit and test the model.

Chapter 20

Summing up

20.1 Taking Stock

Contexts of strategic interaction pervade and suffuse our lives, indeed the lives of all organisms. How are we are make sense of, come to grips with, understand these situations? Two encompassing themes have been with us throughout. The first is the focus on *problems of play* from the perspective of either the players themselves or of those who would manage contexts of strategic interaction. These would-be managing agents include regulators and society at large. The second theme is methodological. We have focused on *procedural and computational methods* for modeling strategic interaction. It is time to take stock and see where we are.

20.2 Problems of Play

Presented with a game in the wild, the response of *equilibrium* (i.e., classical) game theory is to model the situation—"represent the game"—formally and to seek a "solution" by finding the equilibrium of the game as represented (or if necessary its many equilibria). The predicted outcome is that some equilibrium or other will obtain; how it will come about is not part of the theory.[1] This important program of research is ongoing and has produced valuable insights into contexts of strategic interaction. Its results (when available) serve as important benchmarks for other strategic investigations. The predictions, as we have seen, however, have often failed in either of two ways. Players often fail to produce an equilibrium outcome or if they do, the classical theory is unable to predict which equilibrium will result. Nor does equilibrium theory

---

[1]Or rather, part of the standard theory. Game theorists have investigated and advocated various *equilibrium selection* or *equilibrium refinement* principles (see [142] and work on evolutionarily stable strategies, starting with [216, 217, 272]). This is interesting and important work. It has not, however, been fully accepted in mainstream game theory and does not in any event obviate the main points to hand. Connecting these models with results of learning by agents is, however, an important direction for future research.

explain how players will coordinate to select an equilibrium when there is more than one.

Behavioral game theory and experimental economics offer a different response to understanding strategic interaction. Game models are presented to subjects, who are then asked to play. Their behaviors are recorded and matched to predictions. Predictions from equilibrium game theory are especially salient and have often been disconfirmed in the behavioral laboratories. Positing behavioral models to explain the observed behavior is also undertaken. This important program of research is flourishing. Its results serve to confirm or disconfirm theories that predict outcomes in games. The results also provide data that theories of strategic interaction must ultimately explain.

Equilibrium game theory focuses on the question, What are the equilibria of a given model of a game? Behavioral game theory focuses on the question, Given a game model, how will real agents play when presented with it? These two questions are important and interesting. They do not, however, exhaust what is important and interesting to ask regarding strategic interaction.

I have focused in this book on other questions, two groups of related questions, which we may dub the *problems of play*. The first we can call *agent* (strategic) questions.[2] Given a context of strategic interaction, and a collection of strategies, what will happen? How can we evaluate the strategies an agent might use? By what principles can we rationally settle upon a specific strategy for use? How can we discover new strategies, ones we are not aware of? How, in general, can agents learn to play more effectively?

We can call the second group the *institutional* (strategic) questions. Given a society or system of interacting players—an institution such as a market governed by certain rules and constraints—what will happen? How can we manage its performance? Will it be stable or not? Will it be fair? Efficient? And so on. Answers to these two groups of questions might be said to constitute a body of *managerial game theory*, accounts of what we understand about how to play in, and about how to manage, a strategic context. They constitute accounts, in short, of the problems of play.[3] Points arising:

1. The term managerial game theory is a neologism. I mean it to be that body of knowledge addressing some part of the questions just outlined, the agent and the institutional strategic questions.

2. The suggestion behind the name managerial is that this is about how strategic contexts can be addressed, handled, dealt with, in short, managed.

3. There is no presumption here as to methods of study. In fact, all four methods we discussed in Chapter 1—analytic models ("a priori"), field studies ("in vivo"), laboratory experiments ("in vitro"), and procedures

---

[2]Agentive questions? Language impedes.

[3]And as such answers to these questions would have immediate implications for *mechanism design* and *social choice theory*.

and computations ("in silico")—are apt. We have focused primarily on procedures and computations, secondarily on field studies, and drawn upon all methods.

What have we found?

20.2.1   Agent Questions

It is convenient to divide agent questions broadly into two groups:

1. Strategy selection questions.

2. Learning and strategy discovery questions.

I will discuss them as such, recognizing that of course there is much overlap.

20.2.1.1   Strategy Selection

Given a strategic context and a consideration set of strategies, how can we find good strategies to choose for actual use? It is most appropriate to start with Axelrod's Iterated Prisoner's Dilemma (IPD) tournaments. The Folk Theorems teach us that in such indefinitely iterated cases there will be a very large number of equilibria. On equilibrium grounds, then, there is no obvious strategy or even very specific kind of strategy to be recommended. Before the tournaments were held, any recommendation for strategy choice would have to be on quite tenuous grounds.

The deep lesson of Axelrod's tournaments is that they open the door to a practical, workable way of finding warrant for selecting strategies, what I have called Pragmatic Strategic Rationality. Following a policy of "Play 'em and see how they do" we gain a method of evaluating strategies in a consideration set. All things being equal, the strategies to choose are those in the consideration set that do comparatively well.

The question then arises of how best to organize the tournaments. Axelrod's tournaments (to a good approximation) involved

- Only one game (Iterated Prisoner's Dilemma),

- Only one payoff regime ($P = 1$, $T = 5$, $S = 0$, $R = 3$),

- Only one context of play (certainty, no noise), and

- Only a simple, one-round tournament.

We have seen how each of these aspects can be generalized and applied more broadly to yield strategies that are robust to a range of conditions. Pragmatic warrant is increased to the degree that this happens. Strategies that emerge as robust to many different conditions are, all things being equal, better justified as choices.

Briefly, by way of framing what we have discussed, here are some of our findings pertaining to strategy selection.

1. In our discussions of Prisoner's Dilemma (Chapter 3) and IDS games (Chapter 14) we saw how multi-round tournaments can produce somewhat different results than single-round tournaments. A strategy that does well in earlier rounds and not so well later is (all things being equal) less attractive because we can expect other players to avoid the strategies it exploits at the beginning.

2. Also in Chapters 3 and 14 we discussed tournaments with hold-out sampling as a way of finding robust strategies. A strategy that does well in tournaments having different mixtures of strategies is plausibly robust.

3. Our investigations of the Cooperation Afforder game in Chapter 7 used a discrete replicator dynamics model (an alternative form of tournament, also considered by Axelrod) as well as analytic results to demonstrate that under certain conditions repeated play of Stag Hunt can lead reliably to a triumph of the cooperating (HUNT STAG) strategy. We then saw how this inevitable selection of a cooperating strategy might be leveraged into establishing cooperation, even mild altruism, on a widespread basis.

4. Chapter 15 on organizational ambidexterity illustrated how strategies derived as models from theory and observations can be played against each other to support decision making and to generate hypotheses for subsequent testing. Because the number of strategies was small, we relied on a simple tournament structure. The complexity of the model arose principally from the setup complexity of the modeled industry. Nevertheless clear lessons emerged. Similar models, when calibrated well (even if only subjectively), promise to provide material support for strategic decision making.

Together, these examples serve to illustrate the range and power of the turn to pragmatic rationality and, broadly speaking, tournaments for strategy evaluation and selection.

20.2.1.2   Learning and Strategy Discovery

Part III largely addressed models in which strategies of play were discovered based upon learning.

1. PROBE AND ADJUST proved successful in replicating outcomes in standard economic models of monopoly, Cournot competition, and Bertrand competition (Chapters 9, 10, and 11). It did this using a realistic and computationally tractable learning mechanism utterly distinct from the standard accounts. Remarkably, when the update policy of PROBE AND ADJUST is switched from Own Returns to MR-COR (Market Returns, Constrained by Own Returns) the players are able to achieve a high degree of tacit collusion, at the expense of the customers. Moreover,

MR-COR is robust in that a defecting player will have cause for regret. Very similar results obtained when we investigated more realistic supply curve bidding in Chapter 12.

2. In Chapter 13, on two-sided matching, we saw that both a genetic algorithm and an agent-based model were able to discover stable matches (equilibria) that were Pareto-superior in equity or social welfare (or both) to those found by the received deferred acceptance algorithm, a deterministic, non-learning procedure.

3. Chapter 18 on evolutionary models reviewed a number of cases. Perhaps most notable is Fogel's work on checkers and chess. See references therein for other examples, of which there are many. Evolutionary algorithms have proven effective in finding good strategies of play.

4. Chapter 6, from Part II, discussed play in iterated 2×2 games by agents using Q-learning (a form of reinforcement learning). The pattern that emerged was that the agents were able to achieve good overall returns and in doing so they were in large part negotiating positions on the Pareto frontier, rather then seeking equilibrium. Learning leans toward return, not equilibrium.

20.2.2 Institutional Questions

Institutional questions have to do with the performance of systems of strategically interacting players. Important (and overlapping) themes include:

1. Efficiency

2. Cooperation and collusion

3. Multiple objectives

20.2.2.1 Efficiency

It has been our practice to measure game payoffs in natural dimensions, such as money, time, numbers of descendants, and so on.[4] In doing so it becomes possible to distinguish two kinds of efficiency, as we can see with Chicken as our example.

If ε is small relative to 3, either positive or negative, then there are three Pareto efficient outcomes: (SWERVE, SWERVE), (SWERVE, DRIVE STRAIGHT), (DRIVE STRAIGHT, SWERVE). On the other hand if ε is positive, then (SWERVE, SWERVE) is not resource efficient (aka: Hicks efficient), because it fails to extract the last ε of value from the environment. Similarly, if ε is

---

[4]Affording thereby a ratio scale of measurement, in contrast to utility which is measured on an interval scale.

	Swerve	Drive Straight
Swerve	2 [P] 2	3 + ε [NP] 1
Drive Straight	1 [NP] 3 + ε	0 0

FIGURE 20.1: Chicken parameterized.

negative, then only (SWERVE, SWERVE) is resource efficient. Clearly, if an outcome is resource efficient, then it is also Pareto efficient. The converse need not be the case.

Representative findings:

1. In iterated 2×2 games agents using Q-learning arrived on average at highly resource efficient outcomes (Chapter 6). The agents collectively learned well to extract available resources from the environment.

2. The evolutionary fixation of cooperation in the Cooperation Afforder game (Chapter 7) maximizes extraction of available resources and is resource efficient.

Learning tends toward resource efficiency.

20.2.2.2 Cooperation and Collusion

Cooperation, which depending on circumstances may be a good or a bad thing, is very much an open issue. We have seen from the invasion inequalities how unconditional cooperators may conquer unconditional defectors in a spacial (gridscape) model. We have learned from common pool resource problems how the "tragedy of the commons" may prevail, leading to collapse of resources. We have seen how tacit collusion among PROBE AND ADJUST agents may lead to market failure. And we have seen, with the Cooperation Afforder game, how even imperfectly rewarded cooperation may triumph and be applied more broadly than its immediate scope of success.

20.2.2.3 Multiple Objectives

It is unusual for any system of real import to have only one dimension of interest. Two-sided matching may be taken as exemplary. Stability matters, of course, but so do equity and social welfare (or resource efficiency). Failure to recognize multiple objectives when they are there is an invitation to error. It is just such an error that lies behind the Surprise Examination paradox and the legitimate unease we have at times with backward induction (Chapter 19).

20.3   In a Nutshell

We have above all been interested in the problems of play, the problems that arise from engaging in strategic interaction. They, and our answers, may be most succinctly put as follows.

1. How can agents find a good strategies of play? (Agent questions.)

   Answer: Model the game, build a consideration set of strategies by research and/or learning, and use tournaments in one form or another to find robust strategies. Unless you have better information, go with a robust strategy. This is the essence of the idea of Pragmatic Strategic Rationality. Further, if strong behavioral evidence is available, see what the more successful players are doing.

2. How can we predict and control what happens in a context of strategic interaction? (Institutional questions.)

   Answer: See above. For prediction: If there is strong behavioral evidence, use it; absent that impute robust strategies to each of the players, play, and see what happens. For control: Alter the characteristics of the model that are subject to policy control, play, and see what happens; conduct a search for felicitous design parameters with robustly favorable results.

Classical game theory is not about the problems of play, as identified here. It is about equilibria in game representations when the players are epistemically gifted. Binmore's comments [31, pages 50–1] are representative (emphasis added):

> What is important here is that game theory does not pretend to tell you how to make judgments about the shortcomings [in terms of ideal rationality] of an opponent. **In making such judgments, you would be better advised to consult a psychologist than a game theorist.** Game theory is about what players will do when it is understood that both are rational in some [specific, very strong] sense. . . .

Besides consulting a psychologist, talking to a behavioral game theorist would be quite apt. So, I submit, would consulting with a computational modeler of strategic interaction.

Part V

Appendices

Appendix A

Game Concepts

A.1   Overview

My aim with this appendix is that it serve as a handy, rough-and-ready reference of "technical" concepts pertaining to the theory of games, particularly to game theory. My hope is that it will facilitate the flow of reading. The material here is rather like footnotes, and may be skipped or viewed at will. Also, I emphasize *rough-and-ready*. I have deliberately avoided fussiness and fine details in the characterizations given. I urge the reader to pursue matters further, as needed.

A.2   Concepts and Characterizations

A.2.1   Game

Synonymous with *context of strategic interaction* and *context of interdependent decision making*. Any situation in which two or more agents make choices and receive rewards that are at least in part dependent upon the choices of the other agents in the situation. Decision making (or choice, which I use as synonymous; neither implying consciousness or even a cognitive process) in games is said to be *strategic* or *non-parametric*. Non-strategic or parametric decision making occurs in situations that are not games. These are situations in which there is only one decision making agent.

A minor point of usage: insofar as convenient I prefer to use the term *game* for a model or representation of a context of strategic interaction (or "game in the wild"). It is, in any event, important to keep in mind the distinction between the game in the wild and the model of it. To forget is to invite *wild-representation* problems.

A.2.2   Cooperative and Noncooperative Games

These somewhat misleading terms signify an important distinction. In a *cooperative game* (also called a *negotiable game*) the players may communicate with each other and enter into mutually binding (enforced or enforceable) agreements. In a *noncooperative game,* depending on the rules, the players may or may not communicate with each other, but any communication is "cheap talk" in that it cannot be enforced. Unless otherwise indicated, the discussion in this book is in the context of noncooperative games. The main exception is the case of bargaining (aka: negotiation, Chapters 16 and 17), which is inherently a cooperative context.

A.2.3   Strategy, Policy

In its technical sense in game theory, a *strategy* is a complete set of instructions for directing play by a player in a game. WordNet's definition comes close: "an elaborate and systematic plan of action" (`http://wordnet.princeton.edu/`). There is a more colloquial sense, found in ordinary language, that is close in meaning to *plan of action*, which WordNet characterizes as "a plan for actively doing something."

As noted by WordNet, the term strategy has a second meaning, roughly "the branch of military science dealing with military command and the planning and conduct of a war." This meaning, of course, has been appropriated by other fields of endeavor, and so we commonly speak of business strategy, political strategy and so on.

I wish to reserve a technical use of *policy* as meaning (roughly) *temporary strategy*, or *instructions for directing play that if followed throughout the game would constitute a strategy.* For example, an agent playing a game multiple times is said to be playing a *supergame*, consisting of multiple *stage games*. An agent might change its rules of behavior over time, choosing one strategy in the stage game for a while, then switching to another. So, while it does not make sense to speak of an agent as changing its strategy during a game (given the technical sense of the term strategy), it is unproblematic to speak of an agent as changing its policy, or policy of play, during a game.

A.2.4   Games in Strategic Form

See Figure A.1 for the general, canonical *strategic form game representation* for two players each having two strategies. In a 2×2 game, there are two players, called Row and Column. The game is played once, and is said to be a *one-shot* game. Each player has two strategies from which it can choose. Row's strategies are labeled R_1 and R_2 and Column's C_1 and C_2 in the figure. The payoffs are given in the cells of the figure. If Row plays R_1 and Column plays C_1, then Row's payoff is r_1 and Column's is c_1, and similarly for the other cells. Unless stated otherwise, it is assumed that the game is one shot, that

	C_1	C_2
R_1	c_1 r_1	c_2 r_2
R_2	c_3 r_3	c_4 r_4

FIGURE A.1: Canonical strategic form matrix for the 2×2 game.

Row and Column do not communicate with each other, cannot form binding agreements,[1] and in fact are not known to each other. Play is anonymous. Also, we assume that the payoffs are specified as numbers and that uniformly more is better.[2]

The *strategic form* game representation on display in Figure A.1 is also called the *normal form* of the game, but this expression has fallen out of favor. See also §1.2, page 5.

The strategic form representation may be used for any finite number of players and strategies.

A.2.5 Games in Extensive Form

A game is represented in extensive form by a tree (network in which all nodes are connected by exactly one path). In this tree, nodes represent decision points for individual players, and the outgoing arcs represent their strategic decisions. See Figure 19.9 on page 411 for an example of a game in strategic form translated into extensive form.

Figure A.2 (after [261]) presents an example of a centipede game [261] in extensive form. Play begins at node 1 (N:1). Player 1 (A:1) has a choice of R ('right') or D ('down'). If player 1 chooses D the game ends, player 1 receives a payoff of 1, and player 2 receives a payoff of 0. If player 1 plays R, choice passes to player 2 at node 2. Player 2 may then choose R or D. If player 2 chooses D, the game ends, player 1 gets a payoff of 0, and player 2 gets a

---

[1]Our games, unless otherwise noted, are called *non-cooperative* games, in distinction to *cooperative* games. See §A.2.2, page 428.

[2]Some results in classical game theory assume only an ordering on the payoffs, other results assume the payoffs are given as utility values. This subtlety need not detain us at present. Unless otherwise noted, the payoffs for the games we consider will, we should assume, be in units of natural goods, such as money, apples, years out of jail, number of offspring, and so on. The natural goods are measured on a ratio scale and have a natural zero. See §A.2.8, page 431.

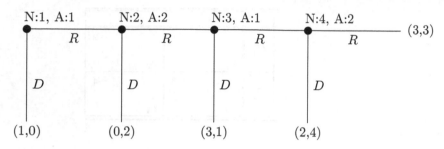

FIGURE A.2: A centipede game in extensive form. N: node number. A: agent number.

payoff of 2. If player 2 chooses R, choice passes back to player 1 at node 3, and the game continues in a similar fashion.

A.2.6   Games in Characteristic Function Form

The characteristic function form is used to represent (some) cooperative games §A.2.2. The representation of a game in characteristic function form has two elements: a list of players and a function on each subset of the players indicating the payoff value of that set as a coalition. This function is called the *characteristic function* of the game.

To illustrate, the games discussed in Chapter 16 are characteristic function games. In one case, we have three players, A, B, and C. The value of the AB coalition, $v(AB)$, is 18, $v(AC) = 24$, $v(BC) = 30$, and all other coalitions have 0 value. We can picture the game as in Figure A.3. Under various possible rules, the players negotiate for pairwise splits of the values of their coalitions.

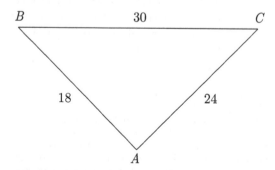

FIGURE A.3: Graph of a characteristic function game.

Once an agreement has been reached, the game is over and the agreement stands.

See [164] and [318] for fuller accounts of coalition formation games. The game in the figure has been explored extensively with agents represented with genetic programming. See [85, 86] as well as Chapter 16.

A.2.7 Risk and Uncertainty

A decision under risk is one in which the decision maker knows all aspects of the problem either with certainty or with a known probability distribution (or density). Either the parameters of the problem are given with certainty or with probability functions.

In a decision under uncertainty, at least some aspects of the problem are known neither with certainty nor up to a probability function. Decision making in games is normally viewed as decision making under uncertainty. Players do not have probability functions for how their counter-parties will choose strategies.

These terms are not used consistently in the literature. My use coheres with what is perhaps the dominant sense of these terms. In any event, it is mainstream.

A.2.8 Game Payoffs

When games are modeled, the payoffs commonly are given in one of three forms.

1. Ordinal or ranking scores. For example, the best of four outcomes might be given a 4, the second best a 3, the worst a 1, and so on.

2. Scores on a scale natural for the payoffs. For example, if the payoffs are monetary, then the scores might be given in euros or dollars. Payoffs on natural scales are often measurable on a ratio scale, having a natural 0 and being comparable on ratios. For example, if the payoffs are measured in euros, they have a ratio scale and it makes perfectly good sense to say that 200 euros is twice as much as 100 euros.

3. Cardinal (von Neumann-Morgenstern) utilities. Utilities are measured on interval scales. Temperature Fahrenheit or Celsius are familiar examples of quantities measured on an interval scale. Here 0 is arbitrary and the ratio of values is nonsense. It does not make sense to say that 200 degrees Fahrenheit or Celsius is twice as hot as 100 degrees.

In classical game theory, payoffs are usually given in utilities, sometimes as rank scores. In this book I shall normally assume that game payoffs are given on a natural scale.

A.2.9   Pareto Optimal

A *Pareto optimal* outcome of a game is one such that there is no other possible outcome in which every agent does as well as or better than it does in the Pareto outcome and no agent does worse. A strict Pareto optimal outcome is a Pareto optimal outcome in which at least one agent does strictly better than it would in the other outcomes.

Put more colloquially, starting with a Pareto optimal outcome, it is not possible to switch outcomes without making at least some player worse off.

A.2.10   Resource or Hicks Optimality

An outcome of a game is said to be *Hicks optimal* if there is no other outcome that results in greater total payoffs being realized, regardless of who gets them. Thus, a Hicks optimal outcome is always the point at which total payoffs across all players are maximized. A Hicks optimal outcome is always Pareto optimal.

I treat the term *resource optimal* as synonymous with Hicks optimal. Note that it is assumed that payoffs going to different players can sensibly be added to a common sum. This does make sense when the payoffs are in what I call natural units, for example, money, time, energy, shoes, ships, and sealing wax. See §A.2.8.

A.2.11   Equilibrium

Often decorated with the term Nash, an equilibrium of a game is an assignment (or *profile*) of strategies to players (one strategy to each player) such that no player has a better strategy, in the sense of yielding a better reward in the game, than the one it has, *given that the strategies of the other players are kept fixed.*

Put colloquially, in an equilibrium no agent has cause for regretting its choice of strategy.

Note:

1. An agent may have a "good" strategy and receive a "bad" outcome. The strategy taken by an agent in an equilibrium outcome may result, say by chance, in a very poor payoff. The agent may rue the payoff but not the strategy, if the strategy profile is in equilibrium.

2. It is often said of a strategy that it is an equilibrium strategy. This is loose talk at best. Individual strategies cannot be at equilibrium, only strategy profiles. There are two ways this kind of talk can be meaningful. First, if the strategy in question belongs to some equilibrium and no other strategy available to the agent belongs to any equilibrium. Second, if the strategy in question simply belongs to some equilibrium, in which case it is said to be *supported* by an equilibrium. These two meanings are

rather different. Watch out for authors who establish one and suggest the other.

3. A strategy profile may be in equilibrium, yet there may be alternate strategies available to two or more agents which, if chosen, would be jointly and individually better for them. The Prisoner's Dilemma game is a case in point.

There are other equilibrium concepts, special cases or *refinements* of Nash equilibrium. Prominent among them is ESS, evolutionarily stable strategy [217]. A strategy is an ESS if a population adopting it cannot be invaded by any other strategy appearing in small numbers. The usual reference context is a randomly playing population subject to an evolutionary dynamic.

A.2.12   Utility

Payoffs in standard game theory are normally assumed to be given as utility function values, i.e., as utilities. A utility function assigns a numerical value to a thing (object, state, event), with the intention that the assigned value represents the preference of a given decision maker in the specified context. Higher numbers indicate more preferred things. Utility functions can be either *ordinal* (outputting ranking scores) or *cardinal* (outputting real-valued scores).

Cardinal utilities apply to decisions under risk. They are measured on an interval scale (like temperature Fahrenheit and temperature Celsius) and in consequence cannot be validly used to compare across individuals. Bob's utility of 17 for the goods is simply not comparable to Carol's 29.

Utility has been axiomatized in various closely related ways. The core idea is to state a series of consistency or rationality axioms, and then demonstrate that *if* a decision maker is consistent and rational in this sense, *then* there will be a utility function for that decision on the outcomes in question, *such that* if A and B are two lotteries involving the outcomes, and the expected utility of A is larger than the expected utility of B, then the decision maker will prefer A to B.

Luce and Raiffa's book [203] remains an excellent introduction to utility theory. Keeney and Raiffa's book [167] is authoritative on utility models for decision making with multiple objectives. My own favorite treatment is in Richard Jeffrey's philosophically sophisticated book [161]. There are many, many other available presentations of utility theory.

A.2.13   Rational Choice Theory

Roughly, the claim that agents in a particular situation have well-formed utilities and act so as to actually maximize them.

Aka: the theory of rational self-interest (e.g., [269]).

A.2.14   Mutual and Common Knowledge

Players in a game are said to have *mutual knowledge* of the game if they all know the conditions or setup of the game (the payoffs, the strategies, the players, and so on) and if they all know that they all know this.

Players in a game are said to have *common knowledge* of the game if they all have mutual knowledge, and they all know that they all have mutual knowledge, and they all know that they all know that they all have mutual knowledge, and so on infinitely.

Common knowledge is a key assumption in classical game theory. It is an important component of rationality as conceived of in the theory. See David Lewis's *Convention* [196] for a seminal discussion of common knowledge.

A.2.15   The Folk Theorem

The Folk Theorem (or Theorems, since there are many versions) teaches us generally that if a (stage) game is infinitely or indefinitely iterated, then every outcome that could reasonably be negotiated for the stage game is an equilibrium of the iterated supergame. The upshot of this is that the number of equilibria in such games is extremely large, and in fact may be uncountable (i.e., the size of the real numbers).

There are many excellent presentations of the Folk Theorem. Binmore's introductory game theory text is a good source [31]. His presentation in [32] is very clear and penetrating. Fudenberg and Tirole [114] are also an excellent source.

A.2.16   Counting Strategies in Repeated Play

In a 2×2 game played once (see Figure A.1 on page 429), there are four possible outcomes: (R_1, C_1), (R_2, C_1), $R_1, C_2)$ and (R_2, C_2). We can read this off from the game representation in strategic form. If the agents play multiple rounds of the 2×2 game (called the *stage game*), then an agent has in pure strategies of the stage game at stage t:

$$2^{(2^2)^0} \times 2^{(2^2)^1} \times 2^{(2^2)^2} \times \ldots \times 2^{(2^2)^{(t-1)}} = 2^{\frac{((2^2)^t - 1)}{(2^2 - 1)}} \tag{A.1}$$

and more generally, if each agent has s strategies and $S = s^n$ where n is the number of agents, then

$$s^{S^0} \times s^{S^1} \times s^{S^2} \times \ldots \times s^{S^{(t-1)}} = s^{\frac{(s^t - 1)}{(S - 1)}} \tag{A.2}$$

In the simple case of a 2×2 stage game, with $s = 2$ and $n = 2$ (so $S = n^s = 4$), the number of pure strategies available to an agent with $t = 3$ is 2^{21}. When $t = 10$ this explodes to 2^{349525}. With only about 10^{80} or about $2^{20^{14}} = 2^{280}$ atomic particles in the universe, we see that this number is more than astronomically larger than astronomical.

* * *

Here briefly is the essential point. Say we have a 2×2 stage game. In each round of play there are $s^n = 2^2 = 4$ possible outcomes. Taking into account the previous round of play, an agent will have $s^{s^n} = 2^{2^2} = 2^4$ possible strategies. Why? Because a strategy specifies, for every possible outcome, what to play next. There are four possible outcomes and for each there are two possibilities (when $s = 2$). That gets us 2^4 strategies, and more generally s^o where o is the number of possible outcomes.

Now consider taking into account the previous two rounds of play. The number of possible outcomes is now $s^n \times s^n = s^{2n} = 2^2 \times 2^2 = 2^4 = 16$. And the number of possible strategies is

$$s^{(s^n)^2} = 2^{(2^2)^2} = 2^{16} \tag{A.3}$$

by the argument given immediately above. Generalizing this to $t - 1$ (instead of 2) previous periods (at time t) and setting $S = s^n$ as above, we get:

$$s^{(S)^{(t-1)}} \tag{A.4}$$

which is the key expression in the previous identities.

A.2.17 Repeated Play, Iterated Play

Even more so than in the case of risk and uncertainty (see §A.2.7) these terms are not used consistently in the literature.

I try to use them uniformly as follows. Repeated play: the players play multiple times, but with different, anonymous players. Each play is "one-shot." Iterated play: the players play multiple times, but with the same counter-players.

Terminology is not standard in the literature, so far as I can see. So my use is arbitrary, but I hope consistent. Beware of confusion when reading the literature.

When a game is played iteratively, it is said to be the *stage game* of the *supergame*, since the players play it again and again.

Appendix B

Useful Mathematical Results

B.1   Geometric Series

Here is a useful mathematical fact:

$$\sum_{i=0}^{\infty} w^i = (1 - w)^{-1}, \quad 0 < w < 1 \tag{B.1}$$

Obvious consequence of equation (B.1):

$$\sum_{i=0}^{\infty} c \cdot w^i = c \cdot (1 - w)^{-1}, \quad 0 < w < 1 \tag{B.2}$$

These are *geometric series*, with the general form:

$$\sum_{i=0}^{\infty} cw^i = c + cw + cw^2 + cw^3 + \cdots + cw^{n-1} + \cdots \tag{B.3}$$

(The ratio of succeeding terms is a constant, w.) Let:

$$s_n = c + cw + cw^2 + \cdots + cw^{n-1} \tag{B.4}$$

Multiplying both sides by w we get

$$s_n w = cw + cw^2 + cw^3 + \cdots + cw^n \tag{B.5}$$

Subtracting (B.5) from (B.4) and canceling (on the right) we get:

$$s_n - s_n w = c - cw^n \tag{B.6}$$

Or

$$s_n(1 - w) = c(1 - w^n) \tag{B.7}$$

Assuming $w \neq 1$ and rearranging:

$$s_n = \frac{c(1 - w^n)}{(1 - w)} \tag{B.8}$$

Assuming $|w| < 1$,

$$\lim_{n \to \infty} w^n = 0 \tag{B.9}$$

Then

$$\lim_{n \to \infty} s_n = \lim_{n \to \infty} \frac{c(1 - w^n)}{(1 - w)} = \frac{c}{(1 - w)}, \quad |w| < 1 \tag{B.10}$$

And so we have a proof of equation (B.2), and indeed something slightly stronger.

B.2   Present Value Factor

Useful summation:

$$\sum_{i=0}^{\infty} w^i = (1 - w)^{-1} \quad \text{for} \ \ 0 < w < 1 \tag{B.11}$$

(This is a geometric series; see §B.1.)
   PV factor =

$$\sum_{i=0}^{\infty} \frac{1}{(1 + r)^i} \tag{B.12}$$

where r = interest rate. Axelrod's discount rate [15], $w = (1 + r)^{-1}$
   Comment:

- r and w are inversely related.

- When the interest (hurdle) rate is small, the discount rate is large.

- When the interest rate is small, we are more interested in looking to the future.

B.3   Solving for Mixed Equilibria in 2×2 Games

Figure B.1 presents our canonical game matrix for the 2×2 game in strategic form. Row plays R_1 with probability x and R_2 with probability $(1 - x)$. Similarly, Column plays C_1 with probability y and C_2 with probability $(1 - y)$. The expected return for Row, G_R, is

$$G_R = xya_r + x(1 - y)b_r + (1 - x)yc_r + (1 - x)(1 - y)d_r \tag{B.13}$$

	C_1 y	C_2 $(1-y)$
R_1 x	a_c a_r	b_c b_r
R_2 $(1-x)$	c_c c_r	d_c d_r

FIGURE B.1: Canonical game matrix for the 2×2 game in strategic form.

The equilibrium values of x, y, $0 < x, y < 1$ if they exist, are found by taking the partial derivatives $\partial G_R/\partial x$ and $\partial G_C/\partial y$, setting the results to 0, and solving. We have

$$\frac{\partial G_R}{\partial x} = ya_r + (1-y)b_r - yc_r - d_r + yd_r = 0 \qquad (B.14)$$

Solving for y gives us:

$$y = \frac{d_r - b_r}{(a_r + d_r) - (b_r + c_r)} \qquad (B.15)$$

The analogous calculation for G_C yields

$$x = \frac{d_c - c_c}{(a_c + d_c) - (b_c + c_c)} \qquad (B.16)$$

If (and only if) the resulting values for x and y are legitimate probabilities, we have found a mixed equilibrium for the 2×2 game.

B.4   Boltzman Function

Here is the salient NetLogo code in the model, AmbidexterityStrategy-Explorer.nlogo. It occurs in the `ProbabilitySupplier0` reporter (function), which reports the probability that an agent will have supplier 0 as its focal supplier during the next epoch.

```
let p0 ([value0] of producer 0 / [price0] of producer 0)
let p1 ([value0] of producer 1 / [price0] of producer 1)

if (SwitchingPolicy = "Boltzman on value / price")
```

```
[let daProb (1 + exp(-1 * (p0 - p1) * 10 /
    BoltzmanFactor)) ^ -1
report daProb]
```

Here is what is going on. We need a function that will set the *probability* of a customer choosing, say producer 0, as its focal producer during the next epoch of the customer. This probability should depend upon (perhaps among other things) the unit prices offered by the producers and the product values offered by the producers. A simple ratio of value to price for each producer is an obvious candidate. That is what p0 and p1 are set to in the code above.

Now, given p0 and p1, we would like a function that assigns a probability to producer 0 of $\frac{1}{2}$ if they are equal, an increasing higher probability to producer 0 as (p0−p1) becomes increasingly positive, and an increasing higher probability to producer 1 as p1 becomes increasingly negative.

There are many sensible functions that will do this. Prominent among them and widely used is this function, called the *Boltzman function*:

$$P_A = \frac{1}{1 + e^{\frac{-\delta}{T}}} \qquad (B.17)$$

In our context, $\delta = (p0 - p1) \cdot 10$. (The 10 is just a scaling factor, for convenience.) Then the probability of accepting producer 0 is P_A. T is the temperature or BoltzmanFactor. We can plot the function. The resulting plot with $T = 3$ is given in Figure B.2.

If δ is kept constant and T decreased, the acceptance function becomes steeper in the middle. That is, a given advantage to one side will result in a higher probability of accepting the side with the advantage as T gets smaller. (We assume that $T > 0$. The temperature never goes to absolute 0, let alone below.) As T goes to positive infinity the curve becomes flatter in the middle and even large advantages on one side will leave substantial probability of accepting the disadvantaged option. Note that as $T \to \infty$, $e^{\frac{-\delta}{T}} \to 1$ and so $P_A \to \frac{1}{2}$.

B.5   Invasion Inequalities on the Gridscape

In Chapter 4 and elsewhere, we explore a two-dimensional imitate-the-best-neighbor strategy update regime for symmetric 2×2 games. Why just a two-dimensional gridscape? Why indeed.

Consider the Prisoner's Dilemma played on a one-dimensional (1-d) gridscape. As usual, 0s are Defectors and 1s are Cooperators. Suppose a field of 0s abuts a triple of cooperators:

$$\ldots 0000011100000 \ldots$$

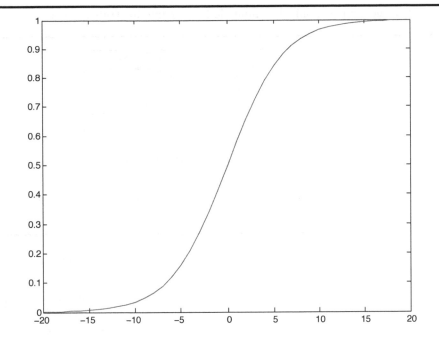

FIGURE B.2: Acceptance function with $T = 3$.

This triple is safe from invasion if

$$P + T < S + R \tag{B.18}$$

which is never true for the Prisoner's Dilemma. The triple is also safe if

$$P + T < 2R \tag{B.19}$$

which can happen, e.g., $T = 101, R = 100, P = 1, S = 0$. Notice, however, that two triples of 1s separated by a single 0 and surrounded by 0s are not safe from invasion. What about invasion of the 0s by the 1s? This will happen if

$$R + S > \max\{2P, T + P\} \tag{B.20}$$

In Prisoner's Dilemma, it is certainly possible that $R + S > 2P$, but the definition of the game disallows $R + S > T + P$. In 1-d, a clustered triple of 1s may or may not be safe from an invasion by a clustered triple of 0s, and the latter are always safe from invasion by the former.

In general, we are interested in what happens at the interface between 3-cubes of strategies. In one dimension the 3-cubes are three of one strategy in a row. In two dimensions, as we have seen, they are a 3×3 square of one strategy. In three dimensions we have a $3 \times 3 \times 3 = 3^3$ cube, and in general we have a 3^d hypercube.

Recall now our canonical symmetric 2×2 game in strategic form:

		S_1 x	S_2 $\bar{x}$
S_1		A	C
x	A	B	
S_2		B	D
$\bar{x}$	C	D	

FIGURE B.3: Canonical game matrix for the symmetric 2×2 game in strategic form.

In d dimensions, a hypercube of S_1s can invade an abutting hypercube of S_2s if

$$A[2 \cdot 3^{d-1} - 1] + B3^{d-1} > \max\{D[2 \cdot 3^{d-1} - 1] + C3^{d-1}, D[3^d - 1]\} \quad \text{(B.21)}$$

Similarly, a hypercube of S_2s can invade an abutting hypercube of S_1s if

$$D[2 \cdot 3^{d-1} - 1] + C3^{d-1} > \max\{A[2 \cdot 3^{d-1} - 1] + B3^{d-1}, A[3^d - 1]\} \quad \text{(B.22)}$$

Mapping these to Prisoners' Dilemma (0s for Defectors, 1s for Cooperators): $A \Rightarrow P, B \Rightarrow T, C \Rightarrow S$, and $D \Rightarrow R$. Then a cube of the Cooperators will invade an abutting cube of the Defectors if

$$R[2 \cdot 3^{d-1} - 1] + S3^{d-1} > \max\{P[2 \cdot 3^{d-1} - 1] + T3^{d-1}, P[3^d - 1]\} \quad \text{(B.23)}$$

Setting $S = 0$ and noting that $P[2 \cdot 3^{d-1} - 1] + T3^{d-1} > P[3^d - 1]$, this simplifies to

$$R[2 \cdot 3^{d-1} - 1] > P[2 \cdot 3^{d-1} - 1] + T3^{d-1} \quad \text{(B.24)}$$

or

$$R > P + \frac{T3^{d-1}}{[2 \cdot 3^{d-1} - 1]} \quad \text{(B.25)}$$

In 2-d:

$$5R > 5P + 3T \quad \text{(B.26)}$$

In 3-d:

$$17R > 17P + 9T \quad \text{(B.27)}$$

In the limit as $d \to \infty$:

$$2R > 2P + T \quad \text{(B.28)}$$

And unneglecting S gives us

$$2R > 2P + T - S \quad \text{(B.29)}$$

in the limit. So, the Cooperators *can* expand in 2-d and higher (depending on the actual values of T, R, P, and S), and things don't change much after 3-d.

Here are summary statistics from a run with $T = 150, R = 100, P = 5$ and $S = 0$. We begin with a single 3-cube (3×3 square) of Cooperators (1s) in a 20×20 gridscape of Defectors (0s).

Generation	0s	1s	Total Points
0	391	9	48560.0
1	379	21	68240.0
2	363	37	94000.0
3	339	61	132640.0
4	327	73	151600.0
5	319	81	164480.0
6	283	117	220640.0
7	243	157	282880.0
8	195	205	358000.0
9	159	241	413440.0
10	119	281	476400.0
11	115	285	485360.0
12	123	277	472480.0
13	107	293	495360.0
14	75	325	546160.0
15	95	305	517200.0
16	79	321	540800.0
17	79	321	541520.0
18	95	305	516480.0
19	63	337	565840.0
20	107	293	497520.0
21	43	357	594080.0
22	107	293	497520.0
23	43	357	594080.0
⋮	⋮	⋮	⋮

At generation 20 the system settles into a two-state oscillation, dominated by Cooperators.

Generation 0:
```
00000000000000000000
00000000000000000000
00000000000000000000
00011100000000000000
00011100000000000000
00011100000000000000
00000000000000000000
00000000000000000000
00000000000000000000
00000000000000000000
00000000000000000000
00000000000000000000
00000000000000000000
00000000000000000000
00000000000000000000
00000000000000000000
00000000000000000000
00000000000000000000
00000000000000000000
00000000000000000000
```

Generation 1:
```
00000000000000000000
00000000000000000000
00011100000000000000
00111110000000000000
00111110000000000000
00111110000000000000
00011100000000000000
00000000000000000000
00000000000000000000
00000000000000000000
00000000000000000000
00000000000000000000
00000000000000000000
00000000000000000000
00000000000000000000
00000000000000000000
00000000000000000000
00000000000000000000
00000000000000000000
00000000000000000000
```

Generation 2:
```
00000000000000000000
00011100000000000000
00111110000000000000
01111111000000000000
01111111000000000000
01111111000000000000
00111110000000000000
00011100000000000000
00000000000000000000
00000000000000000000
00000000000000000000
00000000000000000000
00000000000000000000
00000000000000000000
00000000000000000000
00000000000000000000
00000000000000000000
00000000000000000000
00000000000000000000
00000000000000000000
```

Generation 19:
```
11111111111111011111
11111111111110001111
11111111111111011111
11111111111110001111
11111111111110001111
11111111111110001111
11111111111111011111
11111111111110001111
11111111111111011111
11111111111111011111
11111111111111011111
11111111111111011111
10100010111110001111
00000000000000000000
10100010111110001111
11111111111111011111
11111111111111011111
11111111111111011111
11111111111111011111
```

Generation 20:
```
11111111111110001111
11111111111110001111
11111111111110001111
11111111111110001111
11111111111110001111
11111111111110001111
11111111111110001111
11111111111110001111
11111111111110001111
11111111111110001111
11111111111110001111
11111111111110001111
00000000000001010000
00000000000000000000
00000000000001010000
11111111111110001111
11111111111110001111
11111111111110001111
11111111111110001111
```

Generation 21:
```
11111111111111011111
11111111111111011111
11111111111111011111
11111111111111011111
11111111111111011111
11111111111111011111
11111111111111011111
11111111111111011111
11111111111111011111
11111111111111011111
11111111111111011111
11111111111111011111
11111111111110001111
00000000000000000000
11111111111110001111
11111111111111011111
11111111111111011111
11111111111111011111
11111111111111011111
```

The behavior is simplified on an infinite gradiscape. Assume:

1. An infinite gridscape in d dimensions.

2. Prisoner's Dilemma is being played.

3. Expansion of a Cooperating 3-cube obtains, as in Expression B.24, page 442.

4. The gridscape is filled with Defectors, except for a single 3-cube, V (in d dimensions), composed entirely of Cooperators.

   Remark: Results will hold if the 3-cube is extended to a larger hyper-rectangle (in d dimensions)

Then:

Proposition 1 *The 3-cube, V, cannot be invaded at any of its cells.*

Remark: Expression B.24 governs. Again:

$$R[2 \cdot 3^{d-1} - 1] > P[2 \cdot 3^{d-1} - 1] + T3^{d-1} \tag{B.30}$$

Invasion occurs on the surface of V. Every surface cell is adjacent to an interior cell. Every interior cell receives a return of $R[3^d - 1]$ each generation and $R[3^d - 1] > R[2 \cdot 3^{d-1} - 1]$. No cell on the surface of V is exposed to more than 3^{d-1} Defectors. Since $T > P$, V cannot be invaded.

Proposition 2 *Every expansion is permanent; conversion of a cell from Defect to Cooperate cannot be reversed.*

Remark: This is not true, as we have seen, in a finite gridscape, since a defector may abut more than one side of V. In support of the proposition, note that every converted cell belongs to some 3-cube in d dimensions and no 3-cube can be invaded.

Proposition 3 *Every cell adjacent to V will be converted to Cooperate within two generations.*

Remark: Corners and edges are even more favorable for expansion than other surface points, since they abut nearly internal cells; they have more than $[2 \cdot 3^{d-1} - 1]$ Cooperating neighbors.

Proposition 4 *V—a cluster of Cooperators playing Prisoner's Dilemma in a field of Defectors—will expand forever on the gridscape; every Defecting cell within a finite distance from V will be converted to Cooperate in a finite number of generations.*

There are other ways to generalize our gridscape models. So far, we have focused on the Moore neighborhood of depth 1. In two dimensions, this is the 8 immediately adjacent cells to a given cell. The depth 2 Moore neighborhood (in two dimensions) adds to this the 16 immediately surrounding cells. Whereas in the depth 1 Moore neighborhood we focused on 3-cubes in d dimensions, we attend to 5-cubes for the depth 2 Moore neighborhood. For the sake of simplicity (and without substantive consequence), let us assume that in each generation each cell plays itself. Thus, for the depth 1 neighborhood in d dimensions, each cell has 3^d neighbors (including itself). The revision of Expression B.21 (for depth 1 Moore models) is:

In d dimensions, a 3-hypercube of S_1s can invade an abutting 3-hypercube of S_2s if

$$A[2 \cdot 3^{d-1}] + B3^{d-1} > \max\{D[2 \cdot 3^{d-1}] + C3^{d-1}, D[3^d]\} \tag{B.31}$$

Setting $A \Rightarrow R, B \Rightarrow S, C \Rightarrow T$, and $D \Rightarrow P$ for Prisoner's Dilemma and rearranging gives us what was earlier the limiting formula:

$$2R + S > T + 2P \tag{B.32}$$

Under this regime even 3-cubes in one dimension may expand in Prisoner's Dilemma, depending on actual values of the rewards.

Generalizing to a wider Moore neighborhood, a 5-hypercube of S_1s can invade an abutting 5-hypercube of S_2s if

$$A[3 \cdot 5^{d-1}] + B[2 \cdot 5^{d-1}] > \max\{D[3 \cdot 5^{d-1}] + C[2 \cdot 5^{d-1}], D[5^d]\} \qquad \text{(B.33)}$$

With the usual translation to Prisoner's Dilemma this simplifies to

$$3R + 2S > 2T + 3P \qquad \text{(B.34)}$$

which is *less* favorable to expansion by the Cooperators. Generalizing further, let δ be the depth of the Moore neighborhood in use, then Expression B.33 goes to

$$A[(\delta+1)\cdot(2\delta+1)^{d-1}] + B[\delta\cdot(2\delta+1)^{d-1}] > \max \begin{cases} D[(\delta+1)\cdot(2\delta+1)^{d-1}] + \\ C[\delta\cdot(2\delta+1)^{d-1}] \\ D[(2\delta+1)^d] \end{cases}$$

$$\text{(B.35)}$$

With the usual translation to Prisoners' Dilemma this simplifies to

$$(\delta+1)R + \delta S > \delta T + (\delta+1)P \qquad \text{(B.36)}$$

and in the limit as $\delta \to \infty$

$$R + S > T + P \qquad \text{(B.37)}$$

As neighborhoods expand Cooperators have an increasingly difficult time expanding at the expense of Defectors.

There is more to say about the mathematics of play on the gridscape. Larger issues lurk, however. Although the two-dimensional gridscape is a good place to start, we should think more generally, in terms of social networks. The gridscape is such a network, with certain properties, such as regularity (e.g., everyone has the same number of neighbors). But societies of agents will often be able to find, design, or impose social networks with other forms, and these forms will matter. The creation, exercising, and destruction of social networks pervades social systems. It can be seen as among the primary drivers of human history. (The delightful and provocative essay by J. R. and William H. McNeill, *The Human Web: A Bird's-Eye View of World History*, argues just this [222].) It has also become part of common folklore, via the notion of "six degrees of separation" between typical individuals in our society. (Note that on the gridscape there are very many pairs of cells more than six cells apart.)

The gridscape is among the simplest of social networks. One should expect *greater* social effects to attend other less simple social network structures. That we have found as much of this as we have strikes me as remarkable.

Appendix C

Further Arguments on the Surprise Exam

Figure 19.5 on page 405 from Chapter 19 presents a valid, possibly sound, argument pertaining to the Surprise Examination problem. That argument is key to my analysis of the problem. Here I present four propaedeutic arguments, arguments one might be tempted to entertain but, for reasons given here, are incorrect.

* * *

Figure C.1 on page 448 presents version 1 of the reasoning we want to consider. Version 1 of the students' reasoning is valid, which is to say that if the premises (lines labeled 1–6) are true, then the conclusion must be true. Specifically, if an exam is given, E, then it follows that there is no surprise, $\neg S$, and that the teacher's utterance is not truthful, $\neg V(a)$. If an exam is not given, $\neg E$, then $\neg V(a)$ again follows. Either way, the teacher has spoken falsely. The students can legitimately be surprised, not that there was an exam given but that the teacher spoke falsely. Of course, by the argument, the students could conclude this well before learning of the exam. Knowing that the teacher's utterance was false, they perhaps should have withheld judgment as to whether there would actually be an exam.

But version 1 is unsound and premise (5) is the problem. It is simply false. Lots of things happen that have probabilities of happening other than 1. In our example, the exam itself could be a case in point. By construction the probability of the exam was $\frac{2}{9}$, although the students had no way of knowing this. Might the students protest that this misrepresents their assumption? Instead, they might offer $P(E|E) = 1$. To no avail, since we need $P(E) = 1$ (or $P(E) \geq l_s$). I, at least, do not see any very close relative of version 1 that is both sound and apt. We shall have to look further.

There are, I imagine, many who would agree in finding this formulation of the problem and the students' reasoning lacking. Surely the teacher is saying something stronger, viz., that there must and will be an exam tomorrow. That is necessarily E and S, or $\Box(E \wedge S)$ in symbols. Strengthening the case in this way we get version 2, Figure C.2 on page 449.

The new version of our assumption (5) is not so problematic ("If there must be an exam tomorrow, then the probability of an exam tomorrow is 1"). I am happy to accept it as true. Further, the argument remains valid. Now, however,

447

1. $E \vee \neg E$

   There will be an exam tomorrow or not, a tautology.

2. $l_s \geq \frac{1}{3}$

   Arbitrary threshold (line or level for surprise); may be changed so long as $0 < l_s < 1$.

3. $P(E) \geq l_s \to \neg S$

   If the probability of an exam tomorrow is greater than or equal to the stipulated threshold, then no surprise.

4. $V(a) \to (E \wedge S)$

   There will be an exam tomorrow and it will be a surprise, asserted by the teacher. a = the assertion by the teacher that there will be an exam on the next class (E) and that it will be a surprise (S). If the assertion is truthful or veridical, $V(a)$, then ($E \wedge S$).

5. $E \to P(E) = 1$

   If there is an exam tomorrow, then the probability of an exam tomorrow is 1.

6. $P(E) = 1 \to P(E) \geq l_s$

   A simple mathematical truth.

Thus $\neg V(a)$.

   The teacher's assertion that there will be a surprise examination is false.

FIGURE C.1: Version 1 of the students' reasoning applied to the one-shot Surprise Exam problem.

premise (4) is false. Or so we shall suppose. If the teacher were committed to the necessity of an exam tomorrow, then I would agree with the reasoning evident in version 2. But this trivializes the problem. Might nothing at all prevent an exam tomorrow? Related to the intuition that generates version 2 in Figure C.2 is the thought that the reasoning should be as in version 3 in Figure C.3 on page 450.

   Version 3 is valid, and although we might quibble with the truth of assumption (5), I do not want to challenge it for present purposes. There is real difficulty, however, with assumption (4). As in the case of version 2, if this is what the teacher said, or meant, then I agree that the teacher has spoken falsely. The teacher cannot say this and speak truly and give a surprise exam on the next day. Also as in the case of version 2, I want to dismiss this inter-

1. $E \vee \neg E$

   There will be an exam tomorrow or not, a tautology.

2. $l_s \geq \frac{2}{3}$

   Arbitrary threshold; may be changed without loss of generality.

3. $P(E) \geq l_s \rightarrow \neg S$

   If the probability of an exam tomorrow is greater than or equal to the stipulated threshold, then no surprise.

4. $V(a) \rightarrow \Box(E \wedge S)$

   There must be an exam tomorrow and it will be a surprise, asserted by the teacher.

5. $\Box E \rightarrow P(E) = 1$

   If there must be an exam tomorrow, then the probability of an exam tomorrow is 1.

6. $P(E) = 1 \rightarrow P(E) \geq l_s$

   A simple mathematical truth.

Thus $\neg V(a)$.

   The teacher's assertion that there will be a surprise examination is false.

FIGURE C.2: Version 2 of the students' reasoning applied to the one-shot Surprise Exam problem.

pretation from consideration. If this *is* what our teacher meant, then we will simply get another teacher, who will mean something else. The real question is whether when we find such a teacher she can speak truly and give the surprise exam.

Let us proceed, then, on the assumption that versions 2 and 3 fail for lack of soundness and the truth of premise (4).

Our next foray in explicating the students' reasoning, as applied to the one-shot surprise exam problem, treats the probability of the exam tomorrow as given but justified. See Figure C.4 on page 451.

Again, the argument is valid and the key premise is (5). Its justification is that there is no where else to put the probability mass. There will be an exam and there is only one day available for it, so all the probability has to be on that day. This latter inference is wrong. Things can happen, in particular there can be an exam, even if the probabilities are less than 1.

The lesson is that the teacher must either be untruthful or must take a

1. $E \lor \neg E$

   There will be an exam tomorrow or not, a tautology.

2. $l_s \geq \frac{1}{3}$

   Arbitrary threshold; may be changed without loss of generality.

3. $P(E) \geq l_s \rightarrow \neg S$

   If the probability of an exam tomorrow is greater than or equal to the stipulated threshold, then no surprise.

4. $V(a) \rightarrow (P(E) = 1 \land S)$

   There will be an exam tomorrow with probability $= 1$ and it will be a surprise, asserted by the teacher.

5. $P(E) = 1 \rightarrow E$

   If there is an exam tomorrow with probability $= 1$, then there will be an exam tomorrow.

6. $P(E) = 1 \rightarrow P(E) \geq l_s$

   A simple mathematical truth.

Thus $\neg V(a)$.

FIGURE C.3: Version 3 of the students' reasoning applied to the one-shot Surprise Exam problem.

chance of speaking falsely. Our next foray in explication, as applied to the one-shot surprise exam problem, gets it right. You will find it in Figure 19.5 on page 405.

1. $E \vee \neg E$

   There will be an exam tomorrow or not, a tautology.

2. $l_s \geq \frac{1}{3}$

   Arbitrary threshold; may be changed without loss of generality.

3. $P(E) \geq l_s \rightarrow \neg S$

   If the probability of an exam tomorrow is greater than or equal to the stipulated threshold, then no surprise.

4. $V(a) \rightarrow (E \wedge S)$

   Exam tomorrow and surprise, asserted by the teacher.

5. $P(E) = 1$

   The probability of an exam tomorrow is 1.

6. $P(E) = 1 \rightarrow P(E) \geq l_s$

   A simple mathematical truth.

Thus $\neg V(a)$.

FIGURE C.4: Version 4 of the students' reasoning applied to the one-shot Surprise Exam problem.

Appendix D

Resources on the Web

There are many. These are especially helpful.

1. Repository for this book:
   http://opim.wharton.upenn.edu/~sok/AGEbook/. NetLogo programs
   and Web pages are in the **nlogo/** subdirectory.

2. NetLogo home page: http://ccl.northwestern.edu/netlogo/.

3. "Glossary of Game Theory,"
   http://en.wikipedia.org/wiki/Glossary_of_game_theory.

4. "Stanford Encyclopedia of Philosophy," http://plato.stanford.
   edu/. Stable, but evolving, general repository of articles related to philosophy. A number of good articles pertaining to game theory, rationality, and so on.

5. Ross, Don, "Game Theory," *The Stanford Encyclopedia of Philosophy
   (Spring 2006 Edition)*, Edward N. Zalta (ed.), URL = http://plato.
   stanford.edu/archives/spr2006/entries/game-theory/.

6. Wikipedia article on game theory: http://en.wikipedia.org/wiki/
   Game_theory.

7. Game Theory.net: http://gametheory.net/.

8. Trading Agent Competition home page: http://www.sics.se/tac/.

9. Machine Learning in Games: http://satirist.org/learn-game/.
   Dated but useful.

10. IEEE Conference on Computational Intelligence and Games: http://
    www.ieee-cig.org/.

11. Open ABM http://www.openabm.org/.

12. GAMUT: Game-Theoretic Algorithms Evaluation Suite http://gamut.
    stanford.edu/.

13. Home page for "The Iterated Prisoner's Dilemma Competition: Cele-
    brating the 20th Anniversary": http://www.prisoners-dilemma.com/.
    Home page of Nick Jennings, whose group won the competition: http:
    //users.ecs.soton.ac.uk/nrj/.

Bibliography

[1] Jagdish Agrawal and Wagner A. Kamakura. The economic worth of celebrity endorsers: An event study analysis. *The Journal of Marketing*, 59(3):56–62, July 1995.

[2] George A. Akerlof and Rachel E. Kranton. Economics and identity. *The Quarterly Journal of Economics*, 115(3):715–53, August 2000. doi:10.1162/003355300554881.

[3] George A. Akerlof and Rachel E. Kranton. Identity and the economics of organizations. *Journal of Economic Perspectives*, 19(1):9–32, Winter 2005. doi:10.1257/0895330053147930.

[4] George A. Akerlof and Rachel E. Kranton. *Identity Economics: How Our Identities Shape Our Work, Wages, and Well-Being*. Princeton University Press, Princeton, NJ, 2010.

[5] Brian Aldershof and Olivia M. Carducci. Stable marriage and genetic algorithms: A fertile union. *Journal of Heuristics*, 5:29–46, 1999.

[6] F. Alkemade, H. La Poutr, and H. M. Amman. Robust evolutionary algorithm design for socio-economic simulation. *Computational Economics*, 28(4):355–370, November 2006.

[7] Graham T. Allison and Philip Zelikow. *Essence of Decision: Explaining the Cuban Missile Crisis*. Longman, New York, NY, second edition, 1999.

[8] D. G. Ancona, P. S. Goodman, B. S. Lawrence, and M. L. Tushman. Time: A new research lens. *Academy of Management Review*, 26:645–653, 2001.

[9] James Andreoni and John H. Miller. Auctions with artificial adaptive agents. *Games and Economic Behavior*, 10:39–64, 1995.

[10] J. Arifovic. Genetic algorithm learning and the cobweb model. *Journal of Economic Dynamics and Control*, 18(1):3–28, 1994.

[11] J. Arifovic and M. K. Maschek. Social vs. individual learning: What makes a difference? Working paper, Simon Fraser University, 2005.

[12] K. Atuahene-Gima. Resolving the capability-regidity paradox in new product innovation. *Journal of Marketing*, 69:61–83, 2005.

[13] R. J. Aumann and M. Maschler. The bargaining set for cooperative games. In M. Dresher, L. S. Shapley, and A. W. Tucker, editors, *Advances in Game Theory*. Princeton University Press, 1964.

[14] Robert J. Aumann. Backward induction and common knowledge of rationality. *Games and Economic Behavior*, 8:6–19, 1995.

[15] Robert Axelrod. *The Evolution of Cooperation*. Basic Books, Inc., New York, NY, 1984.

[16] Robert Axelrod. The evolution of strategies in iterated prisoner's dilemma. In Lawrence Davis, editor, *Genetic Algorithms and Simulated Annealing*, pages 32–41. Morgan Kaufman, Los Altos, CA, 1987.

[17] Robert Axelrod. The evolution of strategies in the iterated prisoner's dilemma. In Robert Axelrod, editor, *The Complexity of Cooperation: Agent-Based Models of Competition and Collaboration*, Princeton Studies in Complexity, pages 14–29. Princeton University Press, Princeton, NJ, 1997. Adapted from [16].

[18] Robert Axelrod and W.D. Hamilton. The evolution of cooperation. *Science*, 211:1390–1396, March 1981.

[19] Robert L. Axtell and Steven O. Kimbrough. The high cost of stability in two-sided matching: How much social welfare should be sacrificed in the pursuit of stability? In *Proceedings of the 2008 World Congress on Social Simulation (WCSS-08)*, 2008. http://mann.clermont.cemagref.fr/wcss/.

[20] Michael Bacharach. *Beyond Individual Choice: Teams and Frames in Game Theory*. Princeton University Press, Princeton, NJ, 2006. In Natalie Gold and Robert Sugden, editors.

[21] J. B. Barney. Firm resources and sustained competitive advantage. *Journal of Management*, 17:99–120, 1991.

[22] J. Barr and F. Saraceno. Cournot competition, organization and learning. *Journal of Economic Dynamics and Control*, 29(1):277–295, 2005.

[23] Raymond Battalio, Larry Samuelson, and John Van Huyck. Optimization incentives and coordination failure in laboratory stag hunt games. *Econometrica*, 69(3):749–764, May 2001.

[24] Max H. Bazerman and Margaret A. Neale. *Negotiating Rationally*. The Free Press, New York, NY, 1992.

[25] Neil J. Bearden. The evolution of inefficiency in a simulated stag hunt. http://www.unc.edu/~nbearden/Papers/staghunt.pdf. Accessed October 2003.

[26] B. G. Beddal. Wallace, Darwin, and the theory of natural selection. *Journal of the History of Biology*, 1(2):261–323, 1968. doi:10.1007/BF00351923.

[27] Jonathan Bendor, Dilip Mookherjee, and Debraj Ray. Reinforcement learning in repeated interaction games. *Advances in Theoretical Economics*, 1(1):article 3, 2001.

[28] M. J. Benner and M. L. Tushman. Exploitation, exploration, and process management: The productivity dilemma revisited. *Acad. Management Rev*, 28(2):238–256, 2003.

[29] Christina Bicchieri. *The Grammar of Society: The Nature and Dynamics of Social Norms*. Cambridge University Press, Cambridge, UK, 2005.

[30] Derek Bickerton. *Language and Species*. The University of Chicago Press, Chicago, IL, 1990.

[31] Ken Binmore. *Fun and Games: A Text on Game Theory*. D.H. Heath and Company, Lexington, MA, 1992.

[32] Ken Binmore. *Natural Justice*. Oxford University Press, Oxford, UK, 2005.

[33] Max Black. *The Prevalence of Humbug and Other Essays*, chapter The Prevalence of Humbug, pages 115–143. Cornell University Press, Ithaca, NY, 1983.

[34] Paul Bloom. *Descartes' Baby: How the Science of Child Development Explains What Makes Us Human*. Basic Books, New York, NY, 2004.

[35] Lawrence Bodin and Aaron Panken. High tech for a higher authority: The place of graduating rabbis from Hebrew Union College–Jewish Institute of Religion. *Interfaces*, 33(3):1–11, May–June 2003.

[36] Ian Bogost. *Persuasive Games: The Expressive Power of Videogames*. MIT Press, Cambridge, MA, 2007.

[37] Sissela Bok. *Lying: Moral Choice in Public and Private Life*. Vintage Books, New York, NY, 1978.

[38] T. Börgers and R. Sarin. Dynamic consistency and non-expected utility models of choice under uncertainty. *Journal of Economic Theory*, 77:1–14, 1997.

[39] T. Börgers and R. Sarin. Naïve reinforcement learning with endogenous aspirations. *International Economic Review*, 41:921–950, 2000.

[40] Robert Boyd and Peter J. Richerson. Punishment allows the evolution of cooperation (or anything else) in sizable groups. *Ethology and Sociobiology*, 13:171–195, 1992.

[41] Adam M. Brandenburger and Barry J. Nalebuff. *Co-opetition*. Currency Doubleday, New York, NY, 1996.

[42] Michael E. Bratman. Shared cooperative activity. *The Philosophical Review*, 101(2):327–341, April 1992.

[43] Thomas Brenner, editor. *Computational Techniques for Modelling Learning in Economics*. Kluwer Academic Publishers, Boston, MA, 1999.

[44] Thomas Brenner. *Modelling Learning in Economics*. Edward Elgar, Cheltenham, UK, 1999.

[45] Thomas Brenner. Agent learning representation: Advice on modelling economic learning. In Leigh Tesfatsion and Kenneth L. Judd, editors, *Handbook of Computational Economics, Volume 2, Agent-Based Computational Economics*, Handbooks in Economics, pages 895–948. North-Holland, Amsterdam, the Netherlands, 2006.

[46] Thomas Brenner and Ulrich Witt. Melioration learning in games with constant and frequency-dependent pay-offs. *Journal of Economic Behavior & Organization*, 50:429–448, 2003.

[47] Judith L. Bronstein. The scope for exploitation with mutualistic interactions. In Peter Hammerstein, editor, *Genetic and Cultural Evolution of Cooperation*, pages 185–202. MIT Press, Cambridge, MA, 2003.

[48] Stephanie L. Brown and R. Michael Brown. Selective investment theory: Recasting the functional significance of close relationships. *Psychological Inquiry*, 17(1):1–29, 2006.

[49] Charlotte Bruun, editor. *Advances in Artificial Economics*. Number 584 in Lecture Notes in Economics and Mathematical Systems. Springer, Berlin, 2006.

[50] D. W. Bunn and F. Oliveira. Evaluating individual market power in electricity markets via agent-based simulation. *Annals of Operations Research*, 121:57–78, 2003.

[51] R.A. Burgelman. Strategy as vector and the inertia of co-evolutionary lock-in. *Administrative Science Quarterly*, 47(2):325–357, 2002.

[52] Alberta Burgos. Learning to deal with risk: What does reinforcement learning tell us about risk attitudes? *Economics Bulletin*, 4(10):1–13, 2002.

[53] R. R. Bush and F. Mosteller. *Stochastic Models for Learning*. Wiley, New York, NY, 1955.

[54] Colin F. Camerer. *Behavioral Game Theory: Experiments in Strategic Interaction*. Russell Sage Foundation and Princeton University Press, New York, NY and Princeton, NJ, 2003.

[55] Colin F. Camerer and Teck-Hua Ho. Experience-weighted attraction learning in normal form games. *Econometrica*, 67(4):827–974, 1999.

[56] J. P. Carpenter. Evolutionary models of bargaining: Comparing agent-based computational and analytical approaches to understanding convention evolution. *Computational Economics*, 19(1):25–49, February 2002.

[57] Edward Castronova. *Exodus to the Virtual World: How Online Fun Is Changing Reality*. Palgrave MacMillan, New York, NY, 2007.

[58] Alex K. Chavez and Steven O. Kimbrough. A model of human behavior in coalition formation games. In *Proceedings of the Sixth Annual International Conference on Cognitive Modeling*, pages 70–5. Lawrence Erlbaum Associates, Mahwah, New Jersey, July 30–August 1, 2004.

[59] K. Chellapilla and David B. Fogel. Evolution, neural networks, games, and intelligence. *Proceedings of the IEEE*, 87(9):1471–1496, September 1999. Digital Object Identifier 10.1109/5.784222.

[60] K. Chellapilla and David B. Fogel. Evolving neural networks to play checkers without relying on expert knowledge. *IEEE Transactions on Neural Networks*, 10(6):1382–1391, November 1999. Digital Object Identifier 10.1109/72.809083.

[61] K. Chellapilla and David B. Fogel. Anaconda defeats Hoyle 6–0: a case study competing an evolved checkers program against commercially available software. *Evolutionary Computation, 2000. Proceedings of the 2000 Congress on*, 2:857–863, July 2000. Digital Object Identifier 10.1109/CEC.2000.870729.

[62] K. Chellapilla and David B. Fogel. Evolving an expert checkers playing program without using human expertise. *IEEE Transactions on Neural Networks*, 5(4):422–428, August 2001. Digital Object Identifier 10.1109/4235.942536.

[63] Christine Chou and Steven O. Kimbrough. On strategic choice and organizational ambidexterity: An agent-based modeling approach. In *CIEF 2008, Proceedings of the 7th International Conference on Computational Intelligence in Economics and Finance*. IEEE Computational Intelligence Society, 2008. http://www.aiecon.org/wehia/cief/cfp.htm.

[64] Caroline Claus and Craig Boutilier. The dynamics of reinforcement learning in cooperative multiagent systems. In *Proceedings of the Fifteenth National Conference on Artificial Intelligence*, pages 746–752. AAAI Press/MIT Press, Menlo Park, CA, 1998.

[65] Carlos A. Coello Coello, Gary B. Lamont, and David A. Van Veldhuizen. *Evolutionary Algorithms for Solving Multi-Objective Problems*. Springer Science+Business Media LLC, New York, NY, second edition, 2007.

[66] Andrew M. Colman, Briony D. Pulford, and Jo Rose. Collective rationality in interactive decisions: Evidence for team reasoning. *Acta Psychologica*, 128:387–397, 2008.

[67] Andrew M. Colman, Briony D. Pulford, and Jo Rose. Team reasoning and collective rationality: Piercing the veil of obviousness. *Acta Psychologica*, 128:419–412, 2008.

[68] Charles Darwin. *On the Origin of Species*. Harvard University Press, Cambridge, MA, 1964, facsimile edition of 1859 first edition. Available at Project Gutenberg, http://www.gutenberg.org/files/1228/1228-h/1228-h.htm.

[69] Dipankar Dasgupta, German Hernandez, Deon Garrett, Pavan Kalyan Vejandla, Aishwarya Kaushal, Ramjee Yerneni, and James Simien. A comparison of multiobjective evolutionary algorithms with informed initialization and Kuhn-Munkres algorithm for the sailor assignment problem. In *GECCO '08: Proceedings of the 2008 GECCO conference companion on Genetic and evolutionary computation*, pages 2129–2134, New York, NY, 2008. ACM.

[70] Robyn M. Dawes. Social dilemmas. *Annual Review of Psychology*, 31:169–193, 1980.

[71] Robyn M. Dawes and David M. Messick. Social dilemmas. *International Journal of Psychology*, 35(2):111–116, 2000.

[72] H. Dawid and J. Dermietzel. How robust is the equal split norm? On the de-stabilizing effect of responsive strategies. *Computational Economics*, 28:371–397, 2006.

[73] Richard Dawkins. *The Blind Watchmaker: Why the Evidence of Evolution Reveals a Universe without Design*. W. W. Norton & Company, New York, NY, 1985.

[74] Frans de Waal. *Primates and Philosophers: How Morality Evolved*. Princeton University Press, Princeton, NJ, 2006.

[75] Kalyanmoy Deb. *Multi-Objective Optimization using Evolutionary Algorithms*. John Wiley & Sons, LTD, Chichester, UK, 2001.

[76] Eric V. Denardo. *Dynamic Programming: Models and Applications.* Prentice-Hall, Inc., Englewood Cliffs, NJ, 1982.

[77] Jared Diamond. *Collapse: How Societies Choose to Fail or Succeed.* Viking, New York, NY, 2004.

[78] Avinash Dixit and Barry Nalebuff. *Thinking Strategically: The Competitive Edge in Business, Politics, and Everyday Life.* W.W. Norton & Company, New York, NY, 1991.

[79] Avinish K. Dixit and Barry J. Nalebuff. *The Art of Strategy: A Game Theorist's Guide to Success in Business and Life.* W.W. Norton & Company, New York, NY, 2008.

[80] John Duffy. Agent-based models and human subject experiments. In Leigh Tesfatsion and Kenneth L. Judd, editors, *Handbook of Computational Economics, Volume 2, Agent-Based Computational Economics*, Handbooks in Economics, pages 949–1012. North-Holland, Amsterdam, the Netherlands, 2006.

[81] Lee Alan Dugatkin. *Cooperation among Animals: An Evolutionary Perspective.* Oxford University Press, New York, NY, 1997.

[82] Robert B. Duncan. The ambidextrous organization: Designing dual structures for innovation. In Ralph H. Kilmann, Louis R. Pondy, and Dennis P. Slevin, editors, *The Management of Organization Design: Strategies and Implementation*, volume I, pages 167–188. North-Holland, New York, NY, 1976.

[83] Garett O. Dworman, Steven O. Kimbrough, and James D. Laing. Bargaining in a three-agent coalitions game: An application of genetic programming. In *Working Notes: AAAI-95 Fall Symposium Series, Genetic Programming*, pages 9–16, Boston, MA, November 10–12, 1995, 1995. AAAI.

[84] Garett O. Dworman, Steven O. Kimbrough, and James D. Laing. On automated discovery of models using genetic programming in game–theoretic contexts. In Jay F. Nunamaker, Jr. and Ralph H. Sprague, Jr., editors, *Proceedings of the Twenty-Eighth Annual Hawaii International Conference on System Sciences, Volume III: Information Systems: Decision Support and Knowledge-Based Systems*, pages 428–438, Los Alamitos, CA, 1995. IEEE Computer Society Press.

[85] Garett O. Dworman, Steven O. Kimbrough, and James D. Laing. On automated discovery of models using genetic programming: Bargaining in a three-agent coalitions game. *Journal of Management Information Systems*, 12(3):97–125, Winter 1995–96.

[86] Garett O. Dworman, Steven O. Kimbrough, and James D. Laing. Bargaining by artificial agents in two coalition games: A study in genetic programming for electronic commerce. In John R. Koza, David E. Goldberg, David B. Fogel, and Rick L. Riolo, editors, *Genetic Programming 1996: Proceedings of the First Annual Genetic Programming Conference, July 28–31, 1996, Stanford University*, pages 54–62. MIT Press, Cambridge, MA, 1996.

[87] A. E. Eiben and J. E. Smith. *Introduction to Evolutionary Computing*. Springer, Berlin, 2003.

[88] Robert C. Ellickson. *Order without Law: How Neighbors Settle Disputes*. Harvard University Press, Cambridge, MA, 1991.

[89] Jon Elster. *Ulysses and the Sirens: Studies in Rationality and Irrationality*. Cambridge University Press, Cambridge, UK, 1984.

[90] Jon Elster. *The Cement of Society: A Study of Social Order*. Studies in rationality and social change. Cambridge University Press, Cambridge, UK, 1989.

[91] Jon Elster. *Explaining Social Behavior: More Nuts and Bolts for the Social Sciences*. Cambridge University Press, Cambridge, UK, 2007.

[92] R. Entriken and S. Wan. Agent-based simulation of an automatic mitigation procedure. In *Proceedings of the 38th Hawaii International Conference on System Sciences*, 2005.

[93] Joshua M. Epstein and Robert Axtell. *Growing Artificial Societies: Social Science from the Bottom Up*. MIT Press, Cambridge, MA, 1996.

[94] Ido Erev and Alvin E. Roth. Predicting how people play games: Reinforcement learning in experimental games with unique, mixed strategy equilibria. *The American Economic Review*, 88(4):848–881, 1998.

[95] Ernst Fehr and Joseph Henrich. Is strong reciprocity a maladaptation? On the evolutionary foundations of human altruism. In Peter Hammerstein, editor, *Genetic and Cultural Evolution of Cooperation*, pages 55–82. MIT Press, Cambridge, MA, 2003.

[96] Sevan Gregory Ficici. Multiobjective optimization and coevolution. In Joshua Knowles, David Corne, and Kalyanmoy Deb, editors, *Multiobjective Problem Solving from Nature: From Concepts to Applications*, Natural Computing Series, pages 31–52. Springer, Berlin, 2008.

[97] Christopher D. Fiorillo, Philippe N. Tobler, and Wolfram Schultz. Discrete coding of reward probability and uncertainty by dopamine neurons. *Science*, 299:1898–1902, 2003.

[98] Roger Fisher, Willian Ury, and Bruce Patton. *Getting to Yes: Negotiating Agreement without Giving in*. Penguin Books, New York, NY, second edition, 1991.

[99] Merrill M. Flood. Some experimental games. Research Memorandum RM-789, RAND Corporation, Santa Monica, CA, 1952.

[100] S. Floyd and P. Lane. Strategizing throughout the organization: Managing role conflict in strategic renewal. *Academy of Management Review*, 25:154–177, 2000.

[101] David B. Fogel. *Blondie24: Playing at the Edge of AI*. Morgan Kaufmann, San Francisco, CA, 2002.

[102] David B. Fogel. Evolving strategies in blackjack. *Congress on Evolutionary Computation. CEC2004*, 2:1427–1434, June 2004. Digital Object Identifier 10.1109/CEC.2004.1331064.

[103] David B. Fogel, Timothy J. Hays, Sarah L. Hahn, and James Quon. A self-learning evolutionary chess program. *Proceedings of the IEEE*, 92(12):1947–1954, December 2004. Digital Object Identifier 10.1109/JPROC.2004.837633.

[104] David B. Fogel, Timothy J. Hays, Sarah L. Hahn, and James Quon. The Blondie25 chess program competes against Fritz 8.0 and a human chess master. *Computational Intelligence and Games, 2006 IEEE Symposium on*, pages 230–235, May 2006. Digital Object Identifier 10.1109/CIG.2006.311706.

[105] L. J. Fogel, A. J. Owens, and M. J. Walsh. *Artificial Intelligence through Simulated Evolution*. John Wiley & Sons, New York, NY, 1966.

[106] Robert H. Frank. *Passions within Reason: The Strategic Role of Emotions*. W. W. Norton & Co. Inc., New York, NY, 1988.

[107] Robert H. Frank. *The Winner-Take-All Society: Why the Few at the Top Get So Much More Than the Rest of Us*. Penguin, New York, NY, 1996.

[108] Robert H. Frank. *Luxury Fever: Why Money Fails to Satisfy in an Era of Excess*. The Free Press, New York, NY, 1999.

[109] Robert H. Frank. *What Price the Moral High Ground? Ethical Dilemmas in Competitive Environments*. Princeton University Press, Princeton, NJ, 2004.

[110] Harry G. Frankfurt. *On Bullshit*. Princeton University Press, Princeton, NJ, 2005.

[111] James W. Friedman. *Oligopoly Theory*. Cambridge University Press, Cambridge, UK, 1983.

[112] Jeffrey Friedman, editor. *The Rational Choice Controversy*. Yale University Press, New Haven, CT, 1996. Originally published as *Critical Review*, vol. 9, nos. 1–2, 1995.

[113] Drew Fudenberg and David K. Levine. *The Theory of Learning in Games*. MIT Press, Cambridge, MA, 1998.

[114] Drew Fudenberg and Jean Tirole. *Game Theory*. MIT Press, Cambridge, MA, 1991.

[115] Tomoko Fuku, Akira Namatame, and Taisei Kaizouji. Collective efficiency in two-sided matching. In Philippe Mathieu, Bruno Beaufils, and Olivier Brandouy, editors, *Artificial Economics: Agent-Based Methods in Finance, Game Theory and Their Applications*, pages 115–126. Springer-Verlag, Berlin, 2006.

[116] D. Gale and L. S. Shapley. College admissions and the stability of marriage. *The American Mathematical Monthly*, 69(1):9–15, January 1962.

[117] Charles R. Gallistel. *The Organization of Learning*. MIT Press, Cambridge, MA, 1990.

[118] GameTheory.net. Assurance game. http://www.gametheory.net/Dictionary/Games/AssuranceGame.html, Accessed 8 February 2005.

[119] GameTheory.net. Stag hunt. http://www.gametheory.net/Dictionary/Games/StagHunt.html, Accessed 8 February 2005.

[120] P. Ghemawat and J. Ricart i Costa. The organizational tension between static and dynamic efficiency. *Strategy Management Journal*, 15:91–112, 1993.

[121] C. B. Gibson and J. Birkinshaw. The antecedents, consequences, and mediating role of organizational ambidexterity. *Academy of Management Journal*, 47(2):209–226, 2004.

[122] Gerd Gigerenzer and Reinhard Selten. Rethinking rationality. In Gerd Gigerenzer and Reinhard Selten, editors, *Bounded Rationality: The Adaptive Toolbox*, pages 1–12. MIT Press, Cambridge, MA, 2001.

[123] Margaret Gilbert. *On Social Facts*. Princeton University Press, Princeton, NJ, 1992.

[124] Herbert Gintis. *Game Theory Evolving: A Problem-Centered Introduction to Modeling Strategic Interaction*. Princeton University Press, Princeton, NJ, 2000.

[125] Luc-Alain Giraldeau and Thomas Caraco. *Social Foraging Theory.* Princeton University Press, Princeton, NJ, 2000.

[126] Dhananjay K. Gode and Shyam Sunder. Allocative efficiency of markets with zero-intelligence traders: Market as a partial substitute for individual rationality. *Journal of Political Economy,* 101(1):119–137, 1993.

[127] Dhananjay K. Gode and Shyam Sunder. What makes markets allocationally efficient? *Quarterly Journal of Economics,* 112:603–630, 1997.

[128] Jacob K. Goeree and Charles A. Holt. Stochastic game theory: For playing games, not just for doing theory. *Proceedings of the National Academy of Sciences,* 96:10564–10567, September 1999.

[129] David E. Goldberg. *Genetic Algorithms in Search, Optimization & Machine Learning.* Addison-Wesley Publishing Company, Inc., Reading, MA, 1989.

[130] David Grann. Stealing time: What makes Rickey Henderson run? *The New Yorker,* pages 52–9, 12 September 2005.

[131] Donald P. Green and Ian Shapiro. *Pathologies of Rational Choice Theory: A Critique of Applications in Political Science.* Yale University Press, New Haven, CT, 1994.

[132] Richard L. Gregory. *Eye and Brain.* Princeton University Press, Princeton, NJ, fifth edition, 1997.

[133] Richard L. Gregory. *Seeing through Illusions.* Oxford University Press, Oxford, UK, 2009.

[134] Paul Grice. *Studies in the Way of Words,* chapter Logic and Conversation, pages 22–40. Harvard University Press, Cambridge, MA, 1989 (originally, 1967). ISBN: 0-674-85270-2.

[135] Patrick Grim, Gary Mar, and Paul St. Denis. *The Philosophical Computer: Exploratory Essays in Philosophical Computer Modeling.* MIT Press, Cambridge, MA, 1998.

[136] A. K. Gupta, K. G. Smith, and C. E. Shalley. The interplay between exploration and exploitation. *Academy of Management Journal,* 4:693–706, 2006.

[137] D. Gusfield and R. W. Irving. *The Stable Marriage Problem: Structure and Algorithms.* MIT Press, Cambridge, MA, 1989.

[138] Ian Hacking. *Representing and Intervening: Introductory Topics in the Philosophy of Natural Science.* Cambridge University Press, Cambridge, UK, 1983.

[139] Magnús M. Halldórsson, Kazuo Iwama, Shuichi Miyazaki, and Hiroki Yanagisawa. Improved approximation results for the stable marriage problem. *ACM Transactions on Algorithms*, 3(3):30, 2007.

[140] Peter Hammerstein, editor. *Genetic and Cultural Evolution of Cooperation*. MIT Press, Cambridge, MA, 2003.

[141] Garrett Hardin. The tragedy of the commons. *Science*, 162(3859):1243–1248, 13 December 1968.

[142] John C. Harsanyi and Reinhard Selten. *A General Theory of Equilibrium Selection in Games*. MIT Press, Cambridge, MA, 1988.

[143] Robert H. Hayes and William J. Abernathy. Managing our way to economic decline. *Harvard Business Review*, pages 138–149, July–August 2007.

[144] Z.-L. He and P.-K. Wong. Exploration vs. exploitation: An empirical test of the ambidexterity hypothesis. *Organization Science*, 15(4):481–494, 2004.

[145] G. Heal and H. Kunreuther. You only die once: Interdependent security in an uncertain world. In H. W. Richardson, P. Gordon, and J. E. Moore II, editors, *The Economic Impact of Terrorist Attacks*, pages 35–56. Edward Elgar, Cheltenham, UK, 2005.

[146] Phil Hellmuth, Jr. *Play Poker Like the Pros*. HarperCollins Publishers, Inc., New York, NY, 2003.

[147] Natalie Henrich and Joseph Henrich. *Why Humans Cooperate: A Cultural and Evolutionary Explanation*. Oxford University Press, Oxford, UK, 2007.

[148] R. B. Hergenhahn and M. H. Olson. *An Introduction to Theories of Learning*. Prentice-Hall, Upper Saddle River, NJ, fifth edition, 1997.

[149] Jeffrey W. Herrmann. A genetic algorithm for minimax optimization problems. In *Proceedings of the Congress on Evolutionary Computation*, volume 2, pages 1099–1103. IEEE Press, 1999.

[150] Richard J. Herrnstein. *The Matching Law: Papers in Psychology and Economics*. Harvard University Press, Cambridge, MA, 1997.

[151] Albert O. Hirschman. *The Rhetoric of Reaction: Perversity, Futility, Jeopardy*. Belknap Press, Cambridge, MA, 1991.

[152] Teck-Hua Ho and Xuanming Su. A dynamic level-k model in centipede games. Working paper, University of California at Berkeley, Haas School of Business, Berkeley, California, 15 March 2011.

[153] J. Hofbauer and K. Sigmund. *Evolutionary Games and Population Dynamics.* Cambridge University Press, Cambridge, UK, 1998.

[154] John H. Holland. *Adaptation in Natural and Artificial Systems.* University of Michigan Press, Ann Arbor, MI, 1975.

[155] C. H. Hommes, J. Sonnemans, J. Tuinstra, and H. van de Velden. Learning in cobweb experiments. Working paper TI 2003-020/1, University of Amsterdam, Tinbergen Institute, Amsterdam, the Netherlands, 2003.

[156] J. Hu and M. P. Wellman. Multiagent reinforcement learning: Theoretical framework and an algorithm. In *Fifteenth International Conference on Machine Learning*, pages 242–250, July 1998.

[157] Steffen Huck, Hans-Theo Normann, and Jörg Oechssler. Zero-knowledge cooperation in dilemma games. *Journal of Theoretical Biology*, 220:47–54, 2003.

[158] Steffen Huck, Hans-Theo Normann, and Jörg Oechssler. Two are few and four are many: number effects in experimental oligopolies. *Journal of Economic Behavior & Organization*, 53:435–446, 2004.

[159] Robert W. Irving and Paul Leather. The complexity of counting stable marriages. *SIAM Journal on Computing*, 15(3):655–667, August 1986.

[160] Daniel H. Janzen. When is it coevolution? *Evolution*, 34(3):611–612, 1980.

[161] Richard C. Jeffrey. *The Logic of Decision.* University of Chicago Press, Chicago, Il, second edition, 1983.

[162] John H. Kagel and Alvin E. Roth, editors. *The Handbook of Experimental Economics.* Princeton University Press, Princeton, NJ, 1995.

[163] James P. Kahan and Amnon Rapoport. Test of the bargaining set and kernel models in three-person games. In Anatol Rapoport, editor, *Game Theory as a Theory of Conflict Resolution*, pages 119–160. D. Reidel, Dordrecht, the Netherlands, 1974.

[164] James P. Kahan and Amnon Rapoport. *Theories of Coalition Formation.* Lawrence Earlbaum Associates, Hillsdale, NJ, 1984.

[165] Daniel Kahneman and Amos Tversky. Prospect theory: An analysis of decision under risk. *Econometrica*, XLVII:263–291, 1979.

[166] R. Katila and G. Ahuja. Something old, something new: a longitudinal study of search behavior and new product introduction. *Academy of Management Journal*, 45:1183–1194, 2002.

[167] Ralph L. Keeney and Howard Raiffa. *Decisions with multiple objectives: Preferences and value tradeoffs.* Cambridge University Press, Cambridge, UK, 1993.

[168] Steven O. Kimbrough. The concepts of fitness and selection in evolutionary biology. *Journal of Social and Biological Structures,* 3:149–170, 1980.

[169] Steven O. Kimbrough. Computational modeling and explanation: Opportunities for the information and management sciences. In Hemant K. Bhargava and Nong Ye, editors, *Computational Modeling and Problem Solving in the Networked World: Interfaces in Computing and Optimization,* Operations Research/Computer Science Interfaces Series, pages 31–57. Kluwer, Boston, MA, 2003.

[170] Steven O. Kimbrough. A note on exploring rationality in games. Working paper, University of Pennsylvania, Philadelphia, PA, March 2004. Presented at SEP (Society for Exact Philosophy), spring 2004. http://opim-sun.wharton.upenn.edu/~sok/comprats/2005/exploring-rationality-note-sep2004.pdf.

[171] Steven O. Kimbrough. Notes on MLPS: A model for learning in policy space for agents in repeated games. working paper, University of Pennsylvania, Department of Operations and Information Management, December 2004. http://opim-sun.wharton.upenn.edu/~sok/sokpapers/2005/markov-policy.pdf.

[172] Steven O. Kimbrough. Foraging for trust: Exploring rationality and the stag hunt game. In P. Hermann, Valérie Issarny, and Simon Shiu, editors, *Trust Management: Third International Conference, iTrust 2005, Paris, France, May 23-26, 2005. Proceedings,* volume 3477 / 2005 of *LNCS: Lecture Notes in Computer Science,* pages 1–16. Springer-Verlag, Berlin, 23–26 May 2005. ISSN: 0302-9743. ISBN: 3-540-26042-0.

[173] Steven O. Kimbrough and Ann Kuo. On heuristics for two-sided matching: Revisiting the stable marriage problem as a multiobjective problem. In *Proceedings of the Genetic and Evolutionary Computation Conference (GECCO-2010),* New York, NY, 2010. Association for Computing Machinery.

[174] Steven O. Kimbrough, Ann Kuo, and Hoong Chuin LAU. Finding robust-under-risk solutions for flowshop scheduling. In *MIC 2011: The IX Metaheuristics International Conference,* Udine, Italy, 25–28 July 2011.

[175] Steven O. Kimbrough and Ming Lu. Simple reinforcement learning agents: Pareto beats Nash in an algorithmic game theory study. *Information Systems and e-Business Management,* 3(1):1–19, March 2005. http://dx.doi.org/10.1007/s10257-003-0024-0.

[176] Steven O. Kimbrough, Ming Lu, and Ann Kuo. A note on strategic learning in policy space. In Steven O. Kimbrough and D. J. Wu, editors, *Formal Modelling in Electronic Commerce: Representation, Inference, and Strategic Interaction*, pages 463–475. Springer, Berlin, 2005.

[177] Steven O. Kimbrough, Ming Lu, and Frederic Murphy. Learning and tacit collusion by artificial agents in Cournot duopoly games. In Steven O. Kimbrough and D. J. Wu, editors, *Formal Modelling in Electronic Commerce*, pages 477–492. Springer, Berlin, 2005.

[178] Steven O. Kimbrough and Frederic H. Murphy. Learning to collude tacitly on production levels by oligopolistic agents. *Computational Economics*, 33(1):47–78, February 2009. http://dx.doi.org/10.1007/s10614-008-9150-6 and http://opim.wharton.upenn.edu/~sok/sokpapers/2009/oligopoly-panda-r2.pdf.

[179] Steven O. Kimbrough, D. J. Wu, and Fang Zhong. Computers play the beer game: Can artificial agents manage supply chains? *Decision Support Systems*, 33(3):323–333, 2002.

[180] A. M. Knott. Exploration and exploitations as complements. In N. Bontis and C. W. Choo, editors, *The Strategic Management of Intellectual Capital and Organization Knowledge: A Collection of Readings*, pages 339–358. Oxford University Press, New York, NY, 2002.

[181] Donald E. Knuth. *Marriages Stables*. Les Presses de l'Université de Montreal, Montreal, Canada, 1976.

[182] Donald E. Knuth. *Stable Marriage and Its Relation to Other Combinatorial Problems: An Introduction to the Mathematical Analysis of Algorithms*, volume 10 of *CRM Proceedings & Lecture Notes*. American Mathematical Society, Providence, RI, 1997. Originally published as [181].

[183] Hilary Kornblith, editor. *Epistemology: Internalism and Externalism*. Blackwell Publishers, Malden, MA, 2001.

[184] David M. Kreps. *A Course in Microeconomic Theory*. Princeton University Press, Princeton, NJ, 1990.

[185] David M. Kreps. *Game Theory and Economic Modeling*. Clarendon Press, Oxford, England, 1990.

[186] David M. Kreps, P. Milgrom, J. Roberts, and Robert Wilson. Rational cooperation in the finitely repeated Prisoner's Dilemma. *Journal of Economic Theory*, 27:245–52, 1982.

[187] Robert E. Kuenne. *Price and Nonprice Rivalry in Oligopoly: The Integrated Battleground*. St. Martin's Press, New York, NY, 1998.

[188] Thomas S. Kuhn. *The Structure of Scientific Revolutions.* University of Chicago Press, Chicago, IL, 1962.

[189] Howard Kunreuther and Geoffrey Heal. Interdependent security. *The Journal of Risk and Uncertainty,* 26(2/3):231–249, 2003.

[190] Howard Kunreuther, Gabriel Silvasi, Eric T. Bradlow, and Dylan Small. Bayesian analysis of deterministic and stochastic prisoner's dilemma games. *Judgment and Decision Making,* 4(5):363–384, August 2009.

[191] Christian Lebiere, R. Gray, Dario Salvucci, and R. West. Choice and learning under uncertainty: A case study in baseball batting. In *Proceedings of the 25th Annual Conference of the Cognitive Science Society,* 2003.

[192] Olof Leimar and Richard C. Connor. By-product benefits, reciprocity, and pseudoreciprocity in mutualism. In Peter Hammerstein, editor, *Genetic and Cultural Evolution of Cooperation,* pages 203–222. MIT Press, Cambridge, MA, 2003.

[193] D. A. Levinthal and J. G. March. The myopia of learning. *Strategic Management J,* 14(Special Issue):95–112, 1993.

[194] Daniel A. Levinthal. Adaptation on rugged landscapes. *Management Science,* 43(7):934–950, 1997.

[195] B. Levitt, J. G. March, W. R. Scott, and J. F. Short. Organizational learning. *Annual Review of Sociology,* 14:319–340, 1988.

[196] David Lewis. *Convention: A Philosophical Study.* Basil Blackwell, Oxford, UK, 1969/1986.

[197] Michael Lewis. *Liar's Poker: Rising through the Wreckage on Wall Street.* W.W. Norton & Company, New York, NY, 1989.

[198] Michael Lewis. *Moneyball: The Art of Winning an Unfair Game.* W.W. Norton & Company, New York, NY, 2003.

[199] Michael Lewis. The ballad of Big Mike. *The New York Times,* 24 September 2006. http://select.nytimes.com/gst/abstract.html?res=F60B14F83B550C778EDDA00894DE404482.

[200] B. H. Liddell Hart. *Strategy.* New American Library, New York, NY, 1988.

[201] Leonard M. Lodish and Carl F. Mela. If brands are built over years, why are they managed over quarters? *Harvard Business Review,* pages 104–112, July–August 2007.

[202] M. H. Lubatkin, Z. Simsek, Y. Ling, and J. F. Veiga. Ambidexterity and performance in small- to medium-sized firms: The pivotal role of top management team behavioral integration. *Journal of Management*, 32:646–672, 2006.

[203] R. Duncan Luce and Howard Raiffa. *Games and Decisions*. John Wiley, New York, NY, 1957. Reprinted by Dover Books, 1989.

[204] Ben Macintyre. *Operation Mincemeat: How a Dead Man and a Bizarre Plan Fooled the Nazis and Assured an Allied Victory*. Crown, New York, NY, 2010.

[205] Ian C. MacMillan, Alexander B. van Putten, and Rita Gunther Mc-Grath. Global gamesmanship. *Harvard Business Review*, May 2003. Reprint R0305D.

[206] Margaret Olwen MacMillan. *Paris 1919: Six Months That Changed the World*. Random House, New York, NY, 2002.

[207] Michael W. Macy and Andreas Flache. Learning dynamics in social dilemmas. *Proceedings of the National Academy of Science (PNAS)*, 99(suppl. 3):7229–7236, May 14, 2002. www.pnas.org/cgi/doi/10.1073/pnas.092080099.

[208] Janet Mann, Richard C. Connor, Peter L. Tyack, and Hal Whitehead, editors. *Cetacean Societies: Field Studies of Dolphins and Whales*. University of Chicago Press, Chicago, IL, 2000. ISBN-10: 0226503410 ISBN-13: 978-0226503417.

[209] James G. March. Exploration and exploitation in organizational learning. *Organization Science*, 2:71–87, 1991.

[210] James G. March. Exploration and exploitation in organizational learning. *Organization Science*, 2:71–87, 1991.

[211] Lynn Margulis and Dorion Sagan. *Acquiring Genomes: A Theory of the Origins of Species*. Basic Books, New York, NY, 2002.

[212] R. E. Marks and D. F. Midgley. Using evolutionary computing to explore social phenomena: Modeling the interactions between consumers, retailers and brands. Working paper, Australian Graduate School of Management, 2006.

[213] Robert Marks. Market design using agent-based models. In Leigh Tesfatsion and Kenneth L. Judd, editors, *Handbook of Computational Economics, volume 2, Agent-Based Computational Economics*, Handbooks in Economics, pages 1339–1380. North-Holland, Amsterdam, the Netherlands, 2006.

[214] Dennis M. Marlock. *How to Become a Professional Con Artist.* Paladin Press, Boulder, CO, 2001.

[215] Ernest R. May. *Strange Victory: Hitler's Conquest of France.* Hill and Wang, New York, NY, 2000.

[216] J. Maynard Smith and G. R. Price. The logic of animal conflict. *Nature, London,* 246:15–18, 1973.

[217] John Maynard Smith. *Evolution and the Theory of Games.* Cambridge Univesity Press, New York, NY, 1982.

[218] James E. Mazur. *Learning and Behavior.* Prentice-Hall, Upper Saddle River, NJ, fifth edition, 2002.

[219] Colin McGinn. *Mindfucking: A Critique of Mental Manipulation.* Acumen, Stocksfield, UK, 2008.

[220] Jane McGonigal. *Reality Is Broken: Why Games Make Us Better and How They Can Change the World.* Penguin Press, New York, NY, 2011.

[221] James McManus. *Positively Fifth Street: Murderers, Cheetahs, and Binion's World Series of Poker.* Farrar, Straus, Giroux, New York, NY, USA, 2003.

[222] J. R. McNeill and William H. McNeill. *The Human Web: A Bird's-Eye View of World History.* W. W. Norton & Company, New York, NY, 2003.

[223] William H. McNeill. *Keeping Together in Time: Dance and Drill in Human History.* Harvard University Press, Cambridge, MA, 1995.

[224] Judith Mehta, Chris Starmer, and Robert Sugden. The nature of salience: An experimental investigation of pure coordination games. *American Economic Review,* 84:658–73, 1994.

[225] Zbigniew Michalewicz. *Genetic Algorithms + Data Structures = Evolution Programs.* Springer, Berlin, third edition, 1996.

[226] D. F. Midgley, R. E. Marks, and L. G. Cooper. Breeding competitive strategies. *Management Science,* 43(3):257–275, March 1997.

[227] Dilip Mookherjee and Barry Sopher. Learning and decision costs in experimental constant sum games. *Games and Economic Behavior,* 19(1):97–132, 1997.

[228] Rajatish Mukherjee and Sandip Sen. Towards a Pareto-optimal solution in general-sum games. In *Proceedings of the Second International Joint Conference on Autonomous Agents and Multiagent Systems,* pages 153–160, 2003. `citeseer.nj.nec.com/591017.html`.

[229] Gregory L. Murphy. *The Big Book of Concepts.* MIT Press, Cambridge, MA, 2002.

[230] Thomas Nagel. *The Possibility of Altruism.* Princeton University Press, Princeton, NJ, 1979.

[231] Thomas T. Nagle and John E. Hogan. *The Strategy and Tactics of Pricing: A Guide to Growing More Profitably.* Pearson/Prentice Hall, Upper Saddle River, NJ, fourth edition, 2006.

[232] M. Nakamura, K. Onaga, S. Kyan, and M. Silva. Genetic algorithm for sex-fair stable marriage problem. In *Circuits and Systems, 1995. ISCAS '95., 1995 IEEE International Symposium on,* volume 1, pages 509–512, May 1995.

[233] Nancy J. Nersessian. The cognitive basis of model-based reasoning. In Peter Carruthers, Stephen Stich, and Michael Siegal, editors, *The Cognitive Basis of Science,* pages 133–153. Cambridge University Press, Cambridge, UK, 2003.

[234] Walter Nicholson. *Microeconomic Theory.* The Dryden Press, Hinsdale, IL, second edition, 1978.

[235] Noam Nisan, Tim Roughgarden, Eva Tardos, and Vijay V. Vazirani, editors. *Algorithmic Game Theory.* Cambridge University Press, Cambridge, UK, 2007.

[236] Donald A. Norman. *The Design of Everyday Things.* Basic Books, New York, NY, 2002. ISBN-10: 0465067107. ISBN-13: 978-0465067107.

[237] Michael J. North and Charles M. Macal. *Managing Business Complexity: Discovering Strategic Solutions with Agent-Based Modeling and Simulation.* Oxford University Press, Oxford, UK, 2007.

[238] Martin A. Nowak. *Evolutionary Dynamics: Exploring the Equations of Life.* Belknap Press, Cambridge, MA, 2006.

[239] Martin A. Nowak and Karl Sigmund. Evolution of indirect reciprocity. *Nature,* 437:1291–1298, 27 October 2005. doi:10.1038/nature04131.

[240] Elinor Ostrom. *Governing the Commons: The Evolution of Institutions for Collective Action.* Cambridge University Press, Cambridge, UK, 1990.

[241] Elinor Ostrom. *Understanding Institutional Diversity.* Princeton University Press, Princeton, NJ, 2005.

[242] Elinor Ostrom, Roy Gardner, and James Walker. *Rules, Games, & Common-Pool Resources.* University of Michigan Press, Ann Arbor, MI, 1994.

[243] David Owen. Turning tricks: The rise and fall of contract bridge. *The New Yorker*, pages 90–93, 17 September 2007.

[244] Michael Pollan. *The Omnivore's Dilemma: A Natural History of Four Meals*. Penguin Books, New York, NY, 2006.

[245] Michael Pollan. Unhappy meals. *The New York Times*, 28 January 2007.

[246] Michael E. Porter. *Competitive Strategy: Techniques for Analyzing Industries and Competitors*. The Free Press, New York, NY, 1980/1998.

[247] William Poundstone. *Prisoner's Dilemma: John von Neumann, Game Theory and the Puzzle of the Bomb*. Anchor Books, Doubleday, New York, NY, 1992.

[248] William Poundstone. *Fortune's Formula: The Untold Story of the Scientific Betting System That Beat the Casinos and Wall Street*. Hill & Wang, New York, NY, 2005.

[249] A. Pyka and G. Fagiolo. Agent based modeling: A methodology for neo-Shumpeterian economics. Working paper 272, University of Augsburg, Germany, 2005.

[250] R Development Core Team. *R: A Language and Environment for Statistical Computing*. R Foundation for Statistical Computing, Vienna, Austria, 2007. ISBN 3-900051-07-0.

[251] Sebastian Raisch and Julian Birkinshaw. Organizational ambidexterity: Antecedents, outcomes, and moderators. *Journal of Management*, 34(3):375–409, 2008.

[252] Jagmohan Raju and Z. John Zhang. *Smart Pricing: How Google, Priceline, and Leading Businesses Use Pricing Innovation for Profitability*. Pearson Prentice Hall, Upper Saddle River, NJ, 2010.

[253] Anatol Rapoport, Melvin J. Guyer, and David G. Gordon. *The 2×2 Game*. The University of Michigan Press, Ann Arbor, MI, 1976.

[254] Peter J. Richerson and Robert Boyd. *Not by Genes Alone: How Culture Transformed Human Evolution*. The University of Chicago Press, Chicago, IL, 2005.

[255] Peter J. Richerson, Robert T. Boyd, and Joseph Henrich. Cultural evolution of human cooperation. In Peter Hammerstein, editor, *Genetic and Cultural Evolution of Cooperation*, pages 357–388. MIT Press, Cambridge, MA, 2003.

[256] Thomas Riechmann. Cournot or Walras? Agent based learning, rationality, and long run results in oligopoly games. Discussion paper 261, University of Hannover, Faculty of Economics, Hannover, Germany, August 2002.

[257] Frank E. Ritter and Dieter P. Wallach. Models of two-person games in ACT-R and Soar. In *Proceedings of the 2nd European Conference on Cognitive Modelling*, 1998.

[258] Jan W. Rivkin. Imitation of complex strategies. *Management Science*, 46(6):824–844, June 2000.

[259] Jan W. Rivkin and Nicolaj Siggelkow. Balancing search and stability: Interdependencies among elements of organizational design. *Management Science*, 49(3):290–311, March 2003.

[260] Eleanor Rosch and Carolyn B. Mervis. Family resemblances: Studies in the internal structure of categories. *Cognitive Psychology*, 7:573–605, 1975.

[261] R. Rosenthal. Games of perfect information, predatory pricing, and the chain store. *Journal of Economic Theory*, 25(1):92–100, 1981.

[262] Alvin E. Roth. Deferred acceptance algorithms: History, theory, practice, and open questions. *International Journal of Game Theory*, 36:537–569, March 2008.

[263] Alvin E. Roth and Ido Erev. Learning in extensive-form games: Experimental data and simple dynamic models in the intermediate term. *Games and Economic Behavior*, 8:164–212, 1995.

[264] Alvin E. Roth and Marilda A. Oliveira Sotomayor. *Two-Sided Matching: A Study in Game-Theoretic Modeling and Analysis*. Cambridge University Press, Cambridge, UK, 1990.

[265] Joan Roughgarden. *The Genial Gene: Deconstructing Darwinian Selfishness*. The University of California Press, Berkeley, CA, 2009.

[266] Jean Jacques Rousseau. A discourse upon the origin and the foundation of the inequality among mankind. http://www.gutenberg.org/etext/11136, 17 February 2004. Originally published in French in 1755.

[267] Ariel Rubenstein. The electronic mail game: A game with almost common knowledge. *American Economic Review*, 79(3):385–391, June 1989.

[268] B. Sallans, A. Pfister, A. Karatzoglou, and G. Dorffner. Simulation and validation of an integrated markets model. *Journal of Artificial Societies and Social Simulation*, 6(4):http://jasss.soc.surrey.ac.uk/6/4/2.html, 2003.

[269] David Sally. Conversation and cooperation in social dilemmas: A meta-analysis of experiments from 1958 to 1992. *Rationality and Society*, 7(1):58–92, January 1995.

[270] Andrea Saltelli, Marco Ratto, and Terry Andres. *Global Sensitivity Analysis: The Primer*. John Wiley & Sons, Chichester, UK, 2008.

[271] Andrea Saltelli, Stefano Tarantola, Francesca Campolongo, and Marco Ratto. *Sensitivity Analysis in Practice: A Guide to Assessing Scientific Models*. John Wiley & Sons, Chichester, UK, 2004.

[272] Larry Samuelson. *Evolutionary Games and Equilibrium Selection*. MIT Press, Cambridge, MA, 1997.

[273] T. Sandholm and R. Crites. Multiagent reinforcement learning in iterated prisoner's dilemma. *Biosystems*, 37:147–166, 1995. Special Issue on the Prisoner's Dilemma.

[274] Robert M. Sapolsky. *Monkeyluv: And Other Essays on Our Lives as Animals*. Scribner, New York, NY, 2005.

[275] Robert M. Sapolsky. A natural history of peace. *Foreign Affairs*, 85(1):http://www.foreignaffairs.org/20060101faessay85110/robert--m--sapolsky/a--natural--history--of--peace.html, January–February 2006.

[276] R. Sarin and F. Vahid. Predicting how people play games. *Games and Economic Behavior*, 34:104–122, 2001.

[277] Thomas C. Schelling. *Arms and Influence*. Yale University Press, New Haven, CT, 1967.

[278] Thomas C. Schelling. *Micromotives and Macrobehavior*. W. W. Norton & Company, New York, NY, 1978.

[279] Thomas C. Schelling. *The Strategy of Conflict*. Harvard Univ Pr, Cambridge, MA, 1980 (originally published 1960).

[280] Thomas C. Schelling. *Choice and Consequence*. Harvard University Press, Cambridge, MA, 1985.

[281] John R. Searle. *The Construction of Social Reality*. The Free Press, New York, NY, 1997.

[282] Reinhard Selten, Michael Mitzkewitz, and Gerald R. Uhlich. Duopoly strategies programmed by experienced players. *Econometrica*, 65(3):517–555, May 1997.

[283] Amartya K. Sen. Rational fools: A critique of the behavioural foundations of economic theory. *Philosophy and Public Affairs*, 6:317–344, 1977.

[284] Richard Sennett. *The Conscience of the Eye: The Design and Social Life of Cities*. W. W. Norton, New York, NY, 1992.

[285] G. Richard Shell. When is it legal to lie in negotiations? *Sloan Management Review*, 32:93–101, Spring 1991.

[286] G. Richard Shell. *Bargaining for Advantage: Negotiation Strategies for Reasonable People*. Penguin, New York, NY, 2006.

[287] Martin Shubik. *Game Theory in the Social Sciences*. MIT Press, Cambridge, MA, 1982.

[288] N. Siggelkow and D. A. Levinthal. Temporarily divide to conquer: Centralized, decentralized, and reintegrated organizational approaches to exploration and adaptation. *Organization Science*, 14:650–669, 2003.

[289] N. Siggelkow and J. Rivkin. When exploration backfires: Unintended consequences of organizational search. *Academy of Management Journal*, 49:779–796, 2006.

[290] Joan B. Silk. Cooperation without counting. In Peter Hammerstein, editor, *Genetic and Cultural Evolution of Cooperation*, pages 37–54. MIT Press, Cambridge, MA, 2003.

[291] Herbert Simon. From substantive to procedural rationality. In Spiro J. Latsis, editor, *Method and Appraisal in Economics*. Cambridge University Press, New York, NY, 1976.

[292] Herbert A. Simon. *Models of Bounded Rationality*, volume 2. MIT Press, Cambridge, MA, 1982.

[293] Herbert A. Simon. Rationality in psychology and economics. *The Journal of Business*, 59(4):S209–S224, 1986.

[294] Brian Skyrms. *Evolution of the Social Contract*. Cambridge University Press, Cambridge, UK, 1996.

[295] Brian Skyrms. The stag hunt. World Wide Web, 2001. Proceedings and Addresses of the American Philosophical Association. http://www.lps.uci.edu/home/fac-staff/faculty/skyrms/StagHunt.pdf. Accessed December 2009.

[296] Brian Skyrms. The stag hunt. *Proceedings and Addresses of the American Philosophical Association*, 75(2):31–41, 2001.

[297] Brian Skyrms. *The Stag Hunt and the Evolution of Social Structure*. Cambridge University Press, Cambridge, UK, 2004.

[298] Brian Skyrms. *Signals: Evolution, Learning & Information*. Oxford University Press, Oxford, UK, 2010.

[299] Elliott Sober and David Sloan Wilson. *Unto Others: The Evolution and Psychology of Unselfish Behavior.* Harvard University Press, Cambridge, MA, 1998.

[300] D. W. Stephens, C. M. McLinn, and J. R. Stevens. Discounting and reciprocity in an iterated prisoner's dilemma. *Science*, 298:2216–2218, 2002.

[301] David W. Stephens and Jorn R. Krebs. *Foraging Theory.* Princeton University Press, Princeton, NJ, 1986.

[302] Steven Stoft. *Power System Economics: Designing Markets for Electricity.* John Wiley & Sons, New York, NY, 2002.

[303] Robert Sugden. Nash equilibrium, team reasoning and cognitive hierarchy theory. *Acta Psychologica*, 128:402–404, 2008.

[304] Richard S. Sutton and Andrew G. Barto. *Reinforcement Learning: An Introduction.* MIT Press, Cambridge, MA, 1998.

[305] Hideyuki Takagi. Interactive evolutionary computation: Fusion of the capabilities of EC optimization and human evaluation. *Proceedings of the IEEE*, 89(9):1275–1296, September 2001.

[306] Nicholas S. P. Tay and Robert F. Lusch. Agent-based modeling of ambidextrous organizations: Virtualizing competitive strategy. *IEEE Intelligent Systems*, pages 50–57, September/October 2007. www.computer. org/intelligent.

[307] P. Taylor and L. Jonker. Evolutionarily stable strategies and game dynamics. *Mathematical Biosciences*, 40:145–156, 1978.

[308] Leigh Tesfatsion. Guest editorial: Agent-based modeling of evolutionary economic systems. *IEEE Transactions on Evolutionary Computation*, 5(5):437–441, October 2001.

[309] Leigh Tesfatsion. Agent-based computational economics: A constructive approach. In Leigh Tesfatsion and Kenneth L. Judd, editors, *Handbook of Computational Economics, Volume 2, Agent-Based Computational Economics*, Handbooks in Economics, pages 831–880. North-Holland, Amsterdam, the Netherlands, 2006.

[310] Leigh Tesfatsion and Kenneth L. Judd, editors. *Handbook of Computational Economics, Volume 2, Agent-Based Computational Economics.* Handbooks in Economics. North-Holland, Amsterdam, the Netherlands, 2006.

[311] Leigh Testatsion and Kenneth L. Judd, editors. *Handbook of Computational Economics, Volume 2, Agent-Based Computational Economics.* Handbooks in Economics. North-Holland, Amsterdam, the Netherlands, 2006.

[312] Michael Tomasello. *Why We Cooperate*. MIT Press, Cambridge, MA, 2009.

[313] Raimo Tuomela. *The Philosophy of Sociality: The Shared Point of View*. Oxford University Press, New York, NY, 2007.

[314] M. L. Tushman and C. O'Reilly. Ambidextrous organizations: Managing evolutionary and revolutionary change. *California Management Review*, 38:8–30, 1996.

[315] M. L. Tushman and W. K. Smith. Organizational technology. In J. Baum, editor, *Companion to Organization*, pages 386–414. Blackwell, Malden, MA, 2002.

[316] Michael L. Tushman and Charles A. O'Reilly III. *Winning through Innovation: A Practical Guide to Leading Organization Change and Renewal*. Harvard Business School Press, Boston, MA, 1997.

[317] Amos Tversky and Daniel Kahneman. The framing of decision and the psychology of choice. *Science*, 211:453–458, 1981.

[318] Gerald R. Uhlich. *Descriptive Theories of Bargaining*. Springer-Verlag, Berlin, 1990.

[319] John B. Van Huyck, Raymond C. Battalio, and Richard O. Beil. Tacit coordination games, strategic uncertainty, and coordination failure. *The American Economic Review*, 80(1):234–248, March 1990.

[320] John B. Van Huyck, Raymond C. Battalio, and Richard O. Beil. Strategic uncertainty, equilibrium selection, and coordination failure in average opinion games. *The Quarterly Journal of Economics*, 106(3):885–910, August 1991.

[321] Hal R. Varian. *Microeconomic Analysis*. W. W. Norton & Company, New York, NY, third edition, 1992.

[322] Hal R. Varian. *Intermediate Microeconomics: A Modern Approach*. W. W. Norton & Company, New York, NY, sixth edition, 2003.

[323] Ngo Anh Vien and Tae Choong Chung. Multiobjective fitness functions for stable marriage problem using genetic algrithm. In *SICE-ICASE, 2006. International Joint Conference*, pages 5500–5503, Oct. 2006.

[324] Ngo Anh Vien, Nguyen Hoang Viet, Hyun Kim, SeungGwan Lee, and TaeChoong Chung. Ant colony based algorithm for stable marriage problem. In Khaled Elleithy, editor, *Advances and Innovations in Systems, Computing Sciences and Software Engineering*, pages 457–461. Springer, Dordrecht, the Netherlands, 2007. DOI: 10.1007/978-1-4020-6264-3.

[325] Nicholas J. Vriend. An illustration of the essential difference between individual learning and social learning and its consequences for computational analysis. *Journal of Economic Dynamics and Control*, 24:1–19, 2000.

[326] Andreas Wagner. *Robustness and Evolvability in Living Systems*. Princeton Studies in Complexity. Princeton University Press, Princeton, NJ, 2005.

[327] L. Waltman and U. Kaymak. Q-learning agents in a Cournot oligopoly model. Working paper, Erasmus University, Faculty of Economics, Rotterdam, the Netherlands, 2005.

[328] C. Watkins. *Learning from Delayed Rewards*. PhD thesis, King's College, Oxford, Oxford, UK, 1989.

[329] C. J. C. H. Watkins and P. Dayan. Q-learning. *Machine Learning*, 8:279–292, 1992.

[330] Jörgen W. Weibull. *Evolutionary Game Theory*. MIT Press, Cambridge, MA, 1995.

[331] B. Wernerfelt and C.A. Montgomery. Tobin's QR and the importance of focus in firm performance. *American Economic Review*, 78:246–250, 1988.

[332] Edward O. Wilson. *Sociobiology: The New Synthesis, Twenty-fifth Anniversary Edition*. Belknap Press of Harvard University Press, Cambridge, MA, 2000.

[333] Wayne L. Winston. *Operations Research Applications and Algorithms*. Brooks/Cole, Belmont, CA, fourth edition, 2004.

[334] Ludwig Wittgenstein. *Philosophical Investigations*. Basil Blackwell & Mott, Ltd., The Macmillan Company, New York, NY, third edition, 1958.

[335] Erte Xiao and Howard Kunreuther. Punishment and cooperation in stochastic social dilemmas. Working paper, Wharton School, University of Pennsylvania, 14 July 2010.

[336] Brent Zenobia, Charles Weber, and Tugrul Daim. Artificial markets: A review and assessment of a new venue for innovation research. *Technovation*, 29:338–350, 2009.

[337] Mitchell Zuckoff. The perfect mark: How a Massachusetts psychotherapist fell for a Nigerian e-mail scam. *The New Yorker*, 15 May 2006. http://www.newyorker.com/fact/content/articles/060515fa_fact.

Index

Milton Keynes UK
Ingram Content Group UK Ltd.
UKHW021914071024
449327UK00022B/1666